MULTIVARIATE
BAYESIAN
STATISTICS

Models for Source Separation and Signal Unmixing

MULTIVARIATE
BAYESIAN
STATISTICS

Models for Source Separation and Signal Unmixing

Daniel B. Rowe

CRC Press
Taylor & Francis Group
Boca Raton London New York

CRC Press is an imprint of the
Taylor & Francis Group, an **informa** business

A CHAPMAN & HALL BOOK

Chapman & Hall/CRC Press
Taylor & Francis Group
6000 Broken Sound Parkway NW, Suite 300
Boca Raton, FL 33487-2742

© 2003 by Taylor & Francis Group, LLC
CRC Press is an imprint of Taylor & Francis Group, an Informa business

First issued in paperback 2019

No claim to original U.S. Government works

ISBN-13: 978-0-367-45466-1 (pbk)
ISBN-13: 978-1-58488-318-0 (hbk)

Visit the Taylor & Francis Web site at
http://www.taylorandfrancis.com

and the CRC Press Web site at
http://www.crcpress.com

Library of Congress Card Number 2002031598

Library of Congress Cataloging-in-Publication Data

Rowe, Daniel B.
    Multivariate Bayesian statistics : models for source separation and signal unmixing / Daniel B. Rowe.
      p. cm.
    Includes bibliographical references and index.
    ISBN 1-58488-318-9 (alk. paper)
    1. Bayesian statistical decision theory. 2. Multivariate analysis. I. Title.

QR749.H64G78 2002
519.5'42—dc21   2002031598

To Gretchen and Isabel

Preface

This text addresses the Source Separation problem from a Bayesian statistical approach. There are two possible approaches to solve the Source Separation problem. The first is to impose constraints on the model and likelihood such as independence of sources, while the second is to incorporate available knowledge regarding the model parameters. One of these two approaches has to be followed because the Source Separation model is what is called an overparameterized model. That is, there are more parameters than can be uniquely estimated. In this text, the second approach which does not impose potentially unreasonable model and likelihood constraints is followed. The Bayesian statistical approach not only allows the sources and mixing coefficients to be estimated, but also inferences to be drawn on them.

There are many problems from diverse disciplines such as acoustics, EEG, FMRI, genetics, MEG, portfolio allocation, radar, and surveillance, just to name a few which can be cast into the Source Separation problem. Any problem where a signal is believed to be made up of a combination of elementary signals is a Source Separation problem. The real-world source separation problem is more difficult than it appears at first glance.

The plan of the book is as follows. First, an introductory chapter describes the Source Separation problem by motivating it with a description of a cocktail party. In the description of the Source Separation problem, it is assumed that the mixing process is instantaneous and constant over time.

Second, statistical material that is needed for the Bayesian Source Separation model is introduced. This material includes statistical distributions, introductory Bayesian Statistics, specification of prior distributions, hyperparameter assessment, Bayesian estimation methods, and Multivariate Regression.

Third, the Bayesian Regression and Bayesian Factor Analysis models are introduced to lead us to the Bayesian Source Separation model and then to the Bayesian Source Separation model with unobservable and observable sources.

In all models except for the Bayesian Factor Analysis model which still retains a priori uncorrelated factors from its Psychometric origins, the unobserved source components are allowed to be correlated or dependent instead of constrained to be independent. Models and likelihoods are described with these specifications and then Bayesian statistical solutions are detailed in which available prior knowledge regarding the parameters is quantified and incorporated into the inferences.

Fourth, in the aforementioned models, it is specified that the observed mixed vectors and also the unobserved source vectors are independent over time (but correlated within each vector). Models and likelihoods in which the mixing process is allowed to be delayed and change over time are introduced. In addition, the observed mixed vectors along with the unobserved source vectors are allowed to be correlated over time (also correlated within each vector). Available prior knowledge regarding the parameters is quantified and incorporated into the inferences and then Bayesian statistical solutions are described.

When quantifying available prior knowledge, both Conjugate and generalized Conjugate prior distributions are used. There may exist instances when the covariance structure for the Conjugate prior distributions may not be rich enough to quantify the prior information and thus generalized Conjugate distributions should be used.

Formulas or algorithms for both marginal mean and joint modal or maximum a posteriori estimates are derived. In the Regression model, large sample approximations are made in the derivation of the marginal distributions, and hence, the marginal estimates when generalized Conjugate prior distributions are specified. In this instance, a Gibbs sampling algorithm is also derived to compute exact sampling based marginal mean estimates.

More formally, the outline of the book is as follows.

Chapter 1 introduces the Source Separation model by motivating it with the "cocktail party" problem. The cocktail party is an easily understood example of a Source Separation problem.

Part I is a succinct but necessary coverage of fundamental statistical knowledge and skills for the Bayesian Source Separation model.

Chapter 2 contains needed background information on statistical distributions. Distributions are used in Statistics to model random variation and uncertainty so that it can be understood and minimized. Several common distributions are described which are also used in this text.

Chapter 3 gives a brief introduction to Bayesian Statistics. Bayesian Statistics is an approach in which inferences are made not only from information contained in a set of data but also with available prior knowledge either from a previous similar data set or from an expert in the form of a prior distribution.

Chapter 4 highlights the selection of different common types of prior distributions used in Bayesian Statistics. Knowledge regarding values of the parameters from our available prior information is quantified through prior distributions.

Chapter 5 elaborates on the assessment of hyperpameters of the prior distributions used in Bayesian Statistics to quantify our available knowledge. Upon assessing the hyperparameters of the prior distribution, the entire prior distribution is completely determined.

Chapter 6 describes two estimation methods commonly used for Bayesian Statistics and in this text, namely, Gibbs sampling for marginal posterior mean estimates and the iterated conditional modes algorithm for joint maximum a posteriori estimates. After quantifying available knowledge about parameter values in the form of prior distributions, this knowledge is combined with the information contained in a set of data through its likelihood. A joint posterior distribution is obtained with the use of Bayes' rule. This joint distribution is evaluated to determine estimates of the model parameters.

Chapter 7 builds up from the Scalar Normal model to the Multivariate (Non-Bayesian) Regression model. The buildup includes Simple and Multiple Regression. The Regression model is preliminary knowledge which is necessary to successfully understand the material of the text.

Part II considers the instantaneous constant mixing model where both the observed vectors and unobserved sources are independent over time but allowed to be dependent within each vector. The source components are correlated or dependent.

Chapter 8 considers the sources to be known or observable for a description of (Multivariate) Bayesian Regression. The Bayesian Regression model will assist us in the progression toward the Bayesian Source Separation model.

Chapter 9 considers the sources to be unknown or unobservable and details the Bayesian Factor Analysis model while pointing out its model differences with Bayesian Regression.

Chapter 10 details the specifics of the Bayesian Source Separation model and highlights its subtle but important differences from Bayesian Regression and Factor Analysis.

Chapter 11 discusses the case when some sources are observed while others are unobserved. This is a model which is a combination of Bayesian Regression with observable sources and Bayesian Source Separation with unobservable sources. Both the Bayesian Regression and Bayesian Source Separation models can be found by setting either the number of observable or unobservable sources to be zero.

Chapter 12 consists of a case study example applying Bayesian Source Separation to functional magnetic resonance imaging (FMRI).

Part III details more general models in which sources are allowed to be delayed and mixing coefficients to change over time. This corresponds to the speakers at the party being a physical distance from the microphones, thus their conversation is not mixed instantaneously, and to speakers at a party moving around the room, thus their mixing coefficient increases and decreases as they move closer or further away from the microphones. Also, observation vectors as well as source vectors are allowed to be correlated over time. If a person were talking (not talking) at a given time increment, then in the next time increment this person is most likely talking (not talking).

Chapter 13 generalizes the model to delayed sources and dynamic mixing as well as Regression coefficients. Occasionally the speakers are a physical distance from the microphones and thus their conversations do not instantaneously enter into the mixing process. Although this Chapter is presented prior to the Chapter 14 on correlated vectors which is due to mathematical coherence, a reader may wish to read Chapter 14 before this one.

Chapter 14 expands the model to allow the observed mixed conversation vectors in addition to observed and unobserved source vectors to be correlated over time. There may be instances where the observation vectors and also the source vectors are not independent over time.

Chapter 15 brings the text to an end with some concluding remarks on the material of the text.

Appendix A presents methods for determining activation in FMRI.

Appendix B outlines methods for assessing hyperparameters in the Bayesian Source Separation FMRI case study.

This text covers the basics of the necessary Multivariate Statistics and linear algebra before delving into the substantive material. For each model, I give two distinct ways to estimate the parameters.

It is my hope that I have provided enough information for the reader to learn the fundamental statistical material. It is assumed that the reader has a good knowledge of linear algebra, multivariable calculus, and calculus-based Statistics. Those with sufficient breath and depth in the fundamental material may skip directly to Chapter 8 and use the fundamental chapters as reference material.

It is my belief that the most important topic in statistics is statistical distribution theory. Everything can be derived from statistical distribution theory. This text has many different uses for many different audiences. A short course on Classical Multivariate Statistics can be put together by considering Chapter 2 and Chapter 7. A larger course on Multivariate Bayesian Statistics can be assembled by considering Part I and Chapter 8 of Part II. A one year long course on Multivariate Bayesian Statistics can be made by considering the whole text.

Daniel B. Rowe
Biophysics Research Institute
Medical College of Wisconsin
Milwaukee, Wisconsin

Contents

xvi

List of Figures

List of Tables

1

Introduction

1.1 The Cocktail Party

The Source Separation model will be easier to explain if the mechanics of a "cocktail party" are first described. The cocktail party is an easily understood example of where the Source Separation model can be applied. There are many other applications where the Source Separation model is appropriate. At a cocktail party there are partygoers or speakers holding conversations while at the same time there are microphones recording or observing the speakers also called underlying sources. The cocktail party is illustrated in adapted Figure 1.1. Partygoers, speakers, and sources will be used interchangeably as will be recorded and observed. The cocktail party will be returned to often when describing new concepts or material.

At the cocktail party there are typically several small groups of speakers holding conversations. In each group, typically only one person is speaking at a time. Consider the closest two speakers in Figure 1.1. In this group, person one (left) speaks, then person two (right) speaks, then person one again, and so on. The speakers are obviously negatively correlated. In the Bayesian Source Separation model of this text, the speakers are allowed to be correlated and not constrained to be independent.

At a cocktail party, there are p microphones that record or observe m partygoers or speakers at n time increments. This notation is consistent with traditional Multivariate Statistics. The observed conversations consist of mixtures of true unobservable conversations. A given microphone is not placed to a given speakers' mouth and is not shielded from the other speakers. The microphones do not observe the speakers' conversation in isolation. The recorded conversations are mixed. The problem is to unmix or recover the original conversations from the recorded mixed conversations.

Consider the following example. There is a party with $m = 4$ speakers and $p = 3$ microphones as seen in Figure 1.2. At time increment i, where $i = 1, \ldots, n$, the conversation emitted from speaker 1 is s_{i1}, speaker 2 is s_{i2}, speaker 3 is s_{i3}, and speaker 4 is s_{i4}. Then, the recorded conversation at microphone 1 is x_{i1}, at microphone 2 is x_{i2}, and at microphone 3 is x_{i3}.

There is an unknown function f as illustrated in Figure 1.3 called the mixing function which takes the emitted source signals and mixes them to produce

FIGURE 1.1
The cocktail party.

the observed mixed signals.

1.2 The Source Separation Model

The p microphones are recording mixtures of the m speakers at each of the n time increments. What is emitted from the m speakers at time i are m distinct values collected as the rows of vector s_i and presented as

$$s_i = \begin{pmatrix} s_{i1} \\ \vdots \\ s_{im} \end{pmatrix} \qquad (1.2.1)$$

and what is recorded at time i by the p microphones are p distinct values collected as the rows of vector x_i and presented as

FIGURE 1.2
The unknown mixing process.

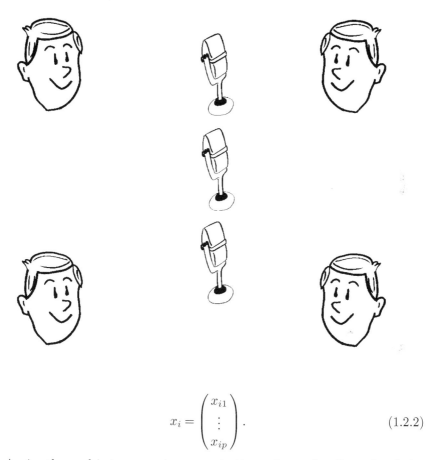

$$x_i = \begin{pmatrix} x_{i1} \\ \vdots \\ x_{ip} \end{pmatrix}. \tag{1.2.2}$$

Again, the goal is to separate or unmix these observed p-dimensional signal vectors into m-dimensional true underlying and unobserved source signal vectors.

The process that mixes the speakers' conversations is instantaneous, constant, and independent over time. The number of speakers is known. Relaxation of these assumptions is discussed in Part III. The separation of sources model for all time i is

$$\begin{array}{cccc} (x_i|s_i) = & f(s_i) & + & \epsilon_i, \\ (p \times 1) & (p \times 1) & & (p \times 1) \end{array} \tag{1.2.3}$$

where $f(s_i)$ is a function which is depicted in Figure 1.3 that mixes the source signals and ϵ_i is the random error. Using a Taylor series expansion [11], the function f, with appropriate smoothness conditions can be expanded about the vector c, written as

FIGURE 1.3
Sample mixing process.

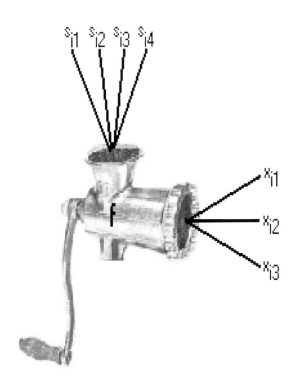

$$f(s_i) = f(c) + f'(c)(s_i - c) + \dots \, , \tag{1.2.4}$$

and by considering the first two terms (as in the familiar Regression model) becomes

$$\begin{aligned}
f(s_i) &= f(c) + f'(c)(s_i - c) \\
&= [f(c) - f'(c)c] + f'(c)s_i \\
&= \mu + \Lambda s_i, \tag{1.2.5}
\end{aligned}$$

where $f'(c)$ and Λ are $p \times m$ matrices. This is called the linear synthesis model. As implied in Figure 1.4, the source signal emitted from each of the speakers' mouths gets multiplied by a mixing coefficient which determines the strength of its contribution but there is also an overall mean background noise level at each microphone and random error entering into the mixing process which is recorded. More formally, the adopted model is

$$
\begin{array}{ccccccc}
(x_i|\mu,\Lambda,s_i) = & \mu & + & \Lambda & s_i & + & \epsilon_i, \\
(p\times 1) & (p\times 1) & & (p\times m) & (m\times 1) & & (p\times 1)
\end{array}
\tag{1.2.6}
$$

FIGURE 1.4
The mixing process.

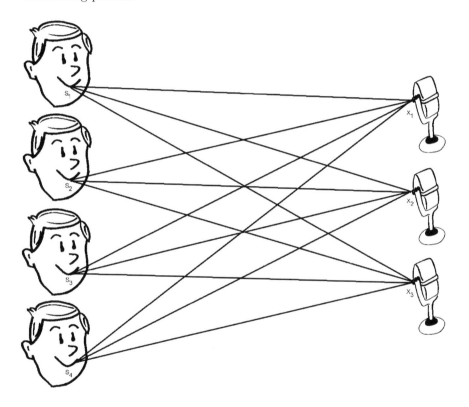

where x_i is as previously described,

$$
\mu = \begin{pmatrix} \mu_1 \\ \vdots \\ \mu_p \end{pmatrix}
\tag{1.2.7}
$$

is a p-dimensional unobserved population mean vector,

$$\Lambda = \begin{pmatrix} \lambda_1' \\ \vdots \\ \lambda_p' \end{pmatrix} \qquad (1.2.8)$$

is a $p \times m$ matrix of unobserved mixing coefficients, s_i is the ith m-dimensional unobservable source vector as previously described, and

$$\epsilon_i = \begin{pmatrix} \epsilon_{i1} \\ \vdots \\ \epsilon_{ip} \end{pmatrix} \qquad (1.2.9)$$

is the p-dimensional vector of errors or noise terms of the ith observed signal vector.

The observed mixed signal x_{ij} is the jth element of the observed mixed signal vector i, which may be thought of as the recorded mixed conversation signal at time increment i, $i = 1, \ldots, n$ for microphone j, $j = 1, \ldots, p$. The observed signal x_{ij} is a mixture of the sources or true unobserved speakers conversations s_i with error, at time increment i, $i = 1, \ldots, n$. The unobserved source signal s_{ik} is the kth element of the unobserved source vector s_i, which may be thought of as the unobserved source signal conversation of speaker k, $k = 1, \ldots, m$ at time increment i, $i = 1, \ldots, n$.

The model describes the mixing process by writing the observed signal x_{ij} as the sum of an overall (background) mean part μ_j plus a linear combination of the unobserved source signal components s_{ik} and the observation error ϵ_{ij}. Element j of x_{ij} may be found by simple matrix multiplication and addition as the jth element of μ, μ_j; plus the element-wise multiplication and addition of the jth row of Λ, λ_j' and s_i; and the addition of the jth element of ϵ_i, ϵ_{ij}. This is written in vector notation as

$$\begin{aligned} (x_{ij} | \mu_j, \lambda_j, s_i) &= \mu_j + \sum_{k=1}^{m} \lambda_{jk}\, s_{ik} + \epsilon_{ij} \\ &= \mu_j + \lambda_j'\, s_i + \epsilon_{ij}. \end{aligned} \qquad (1.2.10)$$

Simply put, the observed conversation for a given microphone consists of an overall background mean at that microphone plus contributions from each of the speakers and random error. The contribution from the speakers to a recorded conversation depends on the coefficient for the speaker to the microphone. The problem at hand is to unmix the unobserved sources and obtain information regarding the mixing process by determining the remaining parameters in the model.

Exercises

1. Describe in words x_{ij}, λ_{jk}, s_{ik}, μ_j, and ϵ_{ij} including subscripts.

2. Assuming that the mixing process described in Equation 1.2.6 is known to have values as listed in Table 1.1, where s_i is the vector valued speakers signal at time i, compute the observed vector value x_i.

TABLE 1.1
Sample data.

μ	Λ	s_i	ϵ_i
1	5 5 3 1	2	3
2	3 5 5 3	4	4
3	1 3 5 5	6	5
		8	

3. Assume that a quadratic term is also kept in the Taylor series expansion. Write down the model for x_{ij} with a quadratic term.

Part I

Fundamentals

2

Statistical Distributions

In this Chapter of the text, various statistical distributions used in Bayesian Statistics are described. Although scalar and vector distributions can be found as special cases of their corresponding matrix distribution, they are treated individually for clarity.

2.1 Scalar Distributions

2.1.1 Binomial

A Bernoulli trial or experiment is an experiment in which there are two possible outcomes, one labeled "success" with probability ϱ, the other labeled "failure" with probability $1 - \varrho$. The Binomial distribution [1, 22] gives the probability of x successes out of n independent Bernoulli trials, where the probability of success in each trial is ϱ. A Bernoulli trial is an experiment such as flipping a coin where there are only two possible outcomes.

A random variable that follows a Binomial distribution is denoted

$$x|\varrho \sim Bin(n, \varrho) \tag{2.1.1}$$

where (n, ϱ) parameterize the distribution which is given by

$$p(x|\varrho) = \frac{n!}{(n-x)!x!} \varrho^x (1-\varrho)^{n-x} \tag{2.1.2}$$

with

$$x \in \{x : 0, 1, \ldots, n\}, \quad \varrho \in (0, 1). \tag{2.1.3}$$

Properties

The mean, mode, and variance of the Binomial distribution are

$$E(x|\varrho) = n\varrho \tag{2.1.4}$$

$$Mode(x|\varrho) = x_0 \quad \text{(as defined below)} \tag{2.1.5}$$

$$var(x|\varrho) = n\varrho(1-\varrho) \tag{2.1.6}$$

which can be found by summation and differencing.

Since the Binomial distribution is a discrete distribution, differentiation in order to determine the most probable value is not appropriate. However, it is well known that the Binomial distribution increases monotonically and then decreases monotonically [47]. The most probable value is when $x = x_0$, where x_0 is an integer such that

$$(n+1)\varrho - 1 < x_0 \leq (n+1)\varrho \tag{2.1.7}$$

which can be found by differencing instead of taking the derivative.

2.1.2 Beta

The Beta distribution [1, 22] gives the probability of a random variable ϱ having a particular value between zero and one. The Beta distribution is often used as the prior distribution (see the Conjugate procedure in Chapter 4) for the probability of success ϱ in a Binomial experiment.

A random variable that follows a Beta distribution is denoted

$$\varrho|\alpha, \beta \sim B(\alpha, \beta), \tag{2.1.8}$$

where (α, β) parameterize the distribution which is given by

$$p(\varrho|\alpha, \beta) = \frac{\Gamma(\alpha+\beta)}{\Gamma(\alpha)\Gamma(\beta)} \varrho^{\alpha-1}(1-\varrho)^{\beta-1}, \tag{2.1.9}$$

where

$$\varrho \in (0,1), \quad \alpha \in (0,+\infty), \quad \beta \in (0,+\infty) \tag{2.1.10}$$

and $\Gamma(\cdot)$ is the gamma function which is given by

$$\Gamma(\alpha) = \int_0^{+\infty} t^{\alpha-1}e^{-t} \, dt \tag{2.1.11}$$

for $\alpha \in \mathbb{R}^+$, where $\mathbb{R}$ denotes the set of real numbers, $\mathbb{R}^+$ the set of positive real numbers,

$$\Gamma(\alpha) = \alpha\Gamma(\alpha-1) \tag{2.1.12}$$

and

$$\Gamma(\alpha) = (\alpha-1)! \tag{2.1.13}$$

for $\alpha \in \mathbb{N}$, where $\mathbb{N}$ denotes the set of natural numbers. Another property of the gamma function is that

$$\Gamma\left(\frac{1}{2}\right) = \pi^{\frac{1}{2}}. \tag{2.1.14}$$

Properties

The mean, mode, and variance of the Beta distribution are

$$E(\varrho|\alpha,\beta) = \frac{\alpha}{\alpha+\beta}, \tag{2.1.15}$$

$$Mode(\varrho|\alpha,\beta) = \frac{\alpha-1}{\alpha+\beta-2}, \tag{2.1.16}$$

$$var(\varrho|\alpha,\beta) = \frac{\alpha\beta}{(\alpha+\beta)^2(\alpha+\beta+1)}, \tag{2.1.17}$$

which can be found by integration and differentiation. The mode is defined for $\alpha+\beta>2$. Note that the Uniform distribution is a special case of the Beta distribution with $\alpha=\beta=1$.

Generalized Beta

A random variable $\rho = (b-a)\varrho + a$ (where $\varrho \sim B(\alpha,\beta)$) is said to have a generalized Beta (type I) distribution [17, 41] given by

$$p(\rho|\alpha,\beta) = \frac{\Gamma(\alpha+\beta)}{\Gamma(\alpha)\Gamma(\beta)} \left(\frac{\rho-a}{b-a}\right)^{\alpha-1} \left(1-\frac{\rho-a}{b-a}\right)^{\beta-1}, \tag{2.1.18}$$

with

$$\rho \in (a,b), \quad \alpha \in (0,+\infty), \quad \beta \in (0,+\infty). \tag{2.1.19}$$

Note that the range of ρ is in the interval (a,b), and if $(a,b)=(-1,1)$, then this is a Beta distribution over the interval $(-1,1)$, which has been used as the prior distribution for a correlation. However, as shown in Chapter 4, a Beta prior distribution is not the Conjugate prior distribution for the correlation parameter ρ. The normalizing constant is often omitted in practice in Bayesian Statistics.

Properties

The mean, mode, and variance of the generalized Beta (type I) distribution are

$$E(\rho|\alpha,\beta) = (b-a)\frac{\alpha}{\alpha+\beta} + a, \tag{2.1.20}$$

$$Mode(\rho|\alpha,\beta) = \frac{(\alpha-1)b-(\beta-1)a}{\alpha+\beta-2}, \tag{2.1.21}$$

$$var(\rho|\alpha,\beta) = \frac{(b-a)^2\alpha\beta}{(\alpha+\beta)^2(\alpha+\beta+1)}, \tag{2.1.22}$$

which can be found by integration and differentiation. The mode is defined for $\alpha+\beta>2$.

2.1.3   Normal

The Normal or Gaussian distribution [1, 17, 22, 41] is used to describe continuous real valued random variables.

A random variable that follows a Normal distribution is denoted

$$x|\mu,\sigma^2 \sim N(\mu,\sigma^2), \qquad (2.1.23)$$

where (μ,σ^2) parameterize the distribution which is given by

$$p(x|\mu,\sigma^2) = (2\pi\sigma^2)^{-\frac{1}{2}} e^{-\frac{(x-\mu)^2}{2\sigma^2}} \qquad (2.1.24)$$

with

$$x \in (-\infty,+\infty), \quad \mu \in (-\infty,+\infty), \quad \sigma \in (0,+\infty). \qquad (2.1.25)$$

Properties

The mean, mode, and variance of the Normal distribution are

$$E(x|\mu,\sigma^2) = \mu, \qquad (2.1.26)$$
$$Mode(x|\mu,\sigma^2) = \mu, \qquad (2.1.27)$$
$$var(x|\mu,\sigma^2) = \sigma^2, \qquad (2.1.28)$$

which can be found by integration and differentiation.

The Normal distribution, also called the bell curve, is that distribution which any other distribution with finite first and second moments tends to be on average according to the central limit theorem [1].

2.1.4   Gamma and Scalar Wishart

A Gamma variate [1, 22] is found as a variate which is the sum of the squares of ν_0 centered independent Normal variates with common mean μ and variance υ^2, $g = (x_1 - \mu)^2 + \cdots + (x_{\nu_0} - \mu)^2$. The variance of the Normal random variates along with the number of random variates characterize the distribution which is presented in a more familiar Multivariate parameterization. A random variable that follows a Gamma distribution is denoted

$$g|\alpha,\beta \sim G(\alpha,\beta), \qquad (2.1.29)$$

where (α,β) parameterize the distribution which is given by

$$p(g|\alpha,\beta) = \frac{g^{\alpha-1} e^{-g/\beta}}{\Gamma(\alpha)\beta^\alpha}, \qquad (2.1.30)$$

with $\Gamma(\cdot)$ being the gamma function and

$$g \in \mathbb{R}^+, \quad \alpha \in \mathbb{R}^+, \quad \beta \in \mathbb{R}^+, \tag{2.1.31}$$

where $\mathbb{R}$ denotes the set of real numbers and $\mathbb{R}^+$ the set of positive real numbers.

Properties

The mean, mode, and variance of the Gamma distribution are

$$E(g|\alpha,\beta) = \alpha\beta, \tag{2.1.32}$$
$$Mode(g|\alpha,\beta) = (\alpha-1)\beta, \tag{2.1.33}$$
$$var(g|\alpha,\beta) = \alpha\beta^2, \tag{2.1.34}$$

which can be found by integration and differentiation. The mode is defined for $\alpha > 1$.

A more familiar parameterization used in Multivariate Statistics [17, 41] which is the Scalar Wishart distribution is when

$$\alpha = \frac{\nu_0}{2}, \quad \beta = 2\upsilon^2. \tag{2.1.35}$$

The Wishart distribution is the Multivariate (Matrix variate) generalization of the Gamma distribution. A random variate g that follows the one-dimensional or Scalar Wishart distribution is denoted by

$$g|\upsilon^2, \nu_0 \sim W(\upsilon^2, 1, \nu_0), \tag{2.1.36}$$

where (υ^2, ν_0) parameterize the distribution which is given by

$$p(g|\upsilon^2, \nu_0) = \frac{(\upsilon^2)^{-\frac{\nu_0}{2}} g^{\frac{\nu_0-2}{2}} e^{-\frac{g}{2\upsilon^2}}}{\Gamma\left(\frac{\nu_0}{2}\right) 2^{\frac{\nu_0}{2}}}, \tag{2.1.37}$$

where

$$g \in \mathbb{R}^+, \quad \upsilon^2 \in \mathbb{R}^+, \quad \nu_0 \in \mathbb{R}^+. \tag{2.1.38}$$

Although the Gamma and Scalar Wishart distributions were derived from ν_0 (an integer valued positive number) Normal variates, there is no restriction that ν_0 in these distributions be integer valued.

Properties

The mean, mode, and variance of the Scalar Wishart distribution are

$$E(g|\upsilon^2, \nu_0) = \nu_0 \upsilon^2, \tag{2.1.39}$$
$$Mode(g|\upsilon^2, \nu_0) = (\nu_0 - 2)\upsilon^2, \tag{2.1.40}$$
$$var(g|\upsilon^2, \nu_0) = 2\nu_0 \upsilon^4, \tag{2.1.41}$$

which can be found by integration and differentiation. The mode is defined for $\nu_0 > 2$.

This parameterization will be followed in this text. Note that the familiar Chi-squared distribution with ν_0 degrees of freedom results when

$$\alpha = \frac{\nu_0}{2}, \quad \beta = 2. \tag{2.1.42}$$

2.1.5 Inverted Gamma and Scalar Inverted Wishart

An Inverted Gamma variate [1, 22] is found as a variate which is the reciprocal of a Gamma variate, $\sigma^2 = g^{-1}$. A random variable σ^2 that follows an Inverted Gamma distribution is denoted

$$\sigma^2 | \alpha, \beta \sim IG(\alpha, \beta), \tag{2.1.43}$$

where (α, β) parameterize the distribution which is given by

$$p(\sigma^2 | \alpha, \beta) = \frac{(\sigma^2)^{-(\alpha+1)} e^{-\frac{1}{\beta \sigma^2}}}{\Gamma(\alpha) \beta^\alpha}, \tag{2.1.44}$$

with $\Gamma(\cdot)$ being the gamma function,

$$\sigma^2 \in \mathbb{R}^+, \quad \alpha \in \mathbb{R}^+, \quad \beta \in \mathbb{R}^+. \tag{2.1.45}$$

Properties

The mean, mode, and variance of the Inverted Gamma distribution are

$$E(\sigma^2 | \alpha, \beta) = \frac{1}{(\alpha-1)\beta}, \tag{2.1.46}$$

$$Mode(\sigma^2 | \alpha, \beta) = \frac{1}{(\alpha+1)\beta}, \tag{2.1.47}$$

$$var(\sigma^2 | \alpha, \beta) = \frac{1}{(\alpha-1)^2(\alpha-2)\beta^2}, \tag{2.1.48}$$

which can be found by integration and differentiation. The mean is defined for $\alpha > 1$ and the variance for $\alpha > 2$.

A more familiar parameterization used in Multivariate Statistics [17, 41] which is the scalar version of the Inverted Wishart distribution is

$$\alpha = \frac{\nu-2}{2} \left(= \frac{\nu_0}{2}\right), \quad \beta = \frac{2}{q} \left(= 2v^2\right). \tag{2.1.49}$$

A random variable which follows a Scalar Inverted Wishart distribution is denoted by

$$\sigma^2 | q, \nu_0 \sim IW(q, 1, \nu), \tag{2.1.50}$$

where (q, ν) parameterize the distribution which is given by

$$p(\sigma^2 | q, \nu) = \frac{(\sigma^2)^{-\frac{\nu}{2}} q^{\frac{\nu-2}{2}} e^{-\frac{q}{2\sigma^2}}}{\Gamma\left(\frac{\nu-2}{2}\right) 2^{\frac{\nu-2}{2}}}, \tag{2.1.51}$$

where

$$\sigma^2 \in \mathbb{R}^+, \quad q \in \mathbb{R}^+, \quad \nu \in \mathbb{R}^+. \tag{2.1.52}$$

Note: In the transformation of variable from g to σ^2,

$$q = \upsilon^{-2}, \quad \nu_0 = \nu - 2, \tag{2.1.53}$$

and the Jacobian of the transformation is

$$J(g \rightarrow \sigma^2) = \sigma^{-4}. \tag{2.1.54}$$

Although the Scalar Inverted Wishart distribution was derived from $\nu - 2$ (an integer valued positive number) Normal variates, there is no restriction that ν in the Scalar Inverted Wishart distribution be integer valued.

Properties
    The mean, mode, and variance of the Scalar Inverted Wishart distribution are

$$E(\sigma^2 | q, \nu) = \frac{q}{\nu - 2 - 2}, \tag{2.1.55}$$

$$Mode(\sigma^2 | q, \nu) = \frac{q}{\nu}, \tag{2.1.56}$$

$$var(\sigma^2 | q, \nu) = \frac{2q^2}{(\nu - 2 - 2)^2 (\nu - 2 - 4)}, \tag{2.1.57}$$

which can be found by integration and differentiation. The mean is defined for $\nu > 4$ and the variance for $\nu > 6$. Note the purposeful use of "$\nu - 2 - 2$" and "$\nu - 2 - 4$" which will become clear with the introduction of the Inverted Wishart distribution.

This parameterization will be followed in this text. Note that the less familiar Inverted Chi-squared distribution results when

$$\alpha = \frac{\nu_0}{2}, \quad \beta = 2. \tag{2.1.58}$$

2.1.6   Student t

The Scalar Student t-distribution [1, 17, 22, 41] is used to describe continuous real-valued random variables with slightly heavier tails than the Normal distribution. It is derived by taking

$$x \sim N(\mu, \sigma^2) \quad \text{and} \quad g \sim W(\sigma^{-2}, 1, \nu), \tag{2.1.59}$$

transforming variables to

$$t = \nu^{\frac{1}{2}} g^{-\frac{1}{2}} (x - \mu) + t_0 \quad \text{and} \quad w = g, \tag{2.1.60}$$

with Jacobian

$$J(x, g \to t, w) = \nu^{-\frac{1}{2}} w^{\frac{1}{2}}, \tag{2.1.61}$$

and then integrating with respect to w. In the derivation, x could be the average of independent and identically distributed Scalar Normal variates with common mean and variance, while g could be the sum of the squares of deviations of these variates about their average.

A random variable that follows a Scalar Student t-distribution is denoted

$$t | \nu, t_0, \sigma^2, \phi^2 \sim t(\nu, t_0, \sigma^2, \phi^2), \tag{2.1.62}$$

where $(\nu, t_0, \sigma^2, \phi^2)$ are degrees of freedom, location, scale, and spread parameters which parameterize the distribution given by

$$p(t | \nu, t_0, \sigma^2, \phi^2) = \frac{\Gamma(\frac{\nu+1}{2})}{(\nu\pi)^{\frac{1}{2}} \Gamma(\frac{\nu}{2})} \frac{\sigma^{-1} \phi^{-\nu}}{\left[\phi^2 + \frac{1}{\nu} \left(\frac{t-t_0}{\sigma} \right)^2 \right]^{\frac{\nu+1}{2}}}, \tag{2.1.63}$$

with

$$t \in \mathbb{R}, \quad \nu \in \mathbb{R}^+ \quad t_0 \in \mathbb{R}, \quad \sigma \in \mathbb{R}^+, \quad \phi \in \mathbb{R}^+. \tag{2.1.64}$$

Properties

The mean, mode, and variance of the Scalar Student t-distribution are

$$E(t | \nu, t_0, \sigma^2, \phi^2) = t_0, \tag{2.1.65}$$

$$Mode(t | \nu, t_0, \sigma^2, \phi^2) = t_0, \tag{2.1.66}$$

$$var(t | \nu, t_0, \sigma^2, \phi^2) = \frac{\nu}{\nu - 2} \phi^2 \sigma^2, \tag{2.1.67}$$

which can be found by integration and differentiation. Note that this parameterization is a generalization of the typical one used which can be found when $\phi^2 = 1$.

The mean of the Scalar Student t-distribution only exists for $\nu > 1$ and the variance only exists for $\nu > 2$. If $\nu \in (0, 1]$, then neither the mean nor the variance exists. When $\nu = 1$, the Scalar Student t-distribution is the Cauchy distribution whose mean and variance or first and second moments do not exist. As the number of degrees of freedom increases, a random variate which follows the Scalar Student t-distribution $t \sim t(\nu, t_0, \sigma^2, \phi^2)$ approaches a Normal distribution $t \sim N(t_0, \phi^2 \sigma^2)$ [17, 41].

2.1.7 F-Distribution

The F-distribution [1, 22, 66] is used to describe continuous random variables which are strictly positive. It is derived by taking

$$x_1 \sim W(1,1,\nu_1) \quad \text{and} \quad x_2 \sim W(1,1,\nu_2), \qquad (2.1.68)$$

and transforming variables to

$$x = \frac{x_1/\nu_1}{x_2/\nu_2}. \qquad (2.1.69)$$

In the derivation, x_1 and x_2 could be independent sums or squared deviations of standard Normal variates.

A random variable that follows an F-distribution is denoted

$$x|\nu_1,\nu_2 \sim F(\nu_1,\nu_2), \qquad (2.1.70)$$

where (ν_1,ν_2) referred to as the numerator and denominator degrees of freedom respectively, which parameterize the distribution given by

$$p(x|\nu_1,\nu_2) = \frac{\Gamma\left(\frac{\nu_1+\nu_2}{2}\right)}{\Gamma\left(\frac{\nu_1}{2}\right)\Gamma\left(\frac{\nu_2}{2}\right)} \left(\frac{\nu_1}{\nu_2}\right)^{\frac{\nu_1}{2}} x^{\frac{\nu_1}{2}-1} \left(1+\frac{\nu_1}{\nu_2}x\right)^{-\frac{\nu_1+\nu_2}{2}}, \qquad (2.1.71)$$

with

$$x \in \mathbb{R}^+, \quad \nu_1 \in \mathbb{N} \quad \nu_2 \in \mathbb{N}. \qquad (2.1.72)$$

Properties

The mean, mode, and variance of the F-distribution are

$$E(x|\nu_1,\nu_2) = \frac{\nu_2}{\nu_2-2}, \qquad (2.1.73)$$

$$Mode(x|\nu_1,\nu_2) = \frac{\nu_2(\nu_1-2)}{\nu_1(\nu_2+2)}, \qquad (2.1.74)$$

$$var(x|\nu_1,\nu_2) = \frac{2\nu_2^2(\nu_1+\nu_2-2)}{\nu_1(\nu_2-2)^2(\nu_2-4)}, \qquad (2.1.75)$$

which can be found by integration and differentiation.

The mean of the F-distribution only exists for $\nu_2 > 2$, and mode for $\nu_1 > 2$, while the variance only exists for $\nu_2 > 4$. The square of a variate t which follows a Scalar Student t-distribution, $t \sim t(\nu,0,0,1)$ is a variate which follows an F-distribution with $\nu_1 = 1$ and $\nu_2 = \nu$ degrees of freedom. The result of transforming a variate x which follows an F-distribution $x \sim F(\nu_1,\nu_2)$ by $1/[1+(\nu_1/\nu_2)x]$ is a Beta variate with $\alpha = \nu_2/2$ and $\beta = \nu_1/2$.

2.2   Vector Distributions

A p-variate vector observation x is a collection of p scalar observations, say $x_1, \ldots, x_p$, arranged in a column.

2.2.1   Multivariate Normal

The p-variate Multivariate Normal distribution [17, 41] is used to simultaneously describe a collection of p continuous real-valued random variables.

A random variable that follows a p-variate Multivariate Normal distribution with mean vector μ and covariance matrix Σ is denoted

$$x|\mu, \Sigma \sim N(\mu, \Sigma), \tag{2.2.1}$$

where (μ, Σ) parameterize the distribution which is given by

$$p(x|\mu, \Sigma) = (2\pi)^{-\frac{p}{2}} |\Sigma|^{-\frac{1}{2}} e^{-\frac{1}{2}(x-\mu)'\Sigma^{-1}(x-\mu)} \tag{2.2.2}$$

with

$$x \in \mathbb{R}^p, \quad \mu \in \mathbb{R}^p, \quad \Sigma > 0, \tag{2.2.3}$$

where $\mathbb{R}^p$ denotes the set of p-dimensional real numbers and $\Sigma > 0$ that Σ belongs to the set of p-dimensional positive definite matrices.

Properties

The mean, mode, and variance of the Multivariate Normal distribution are

$$E(x|\mu, \Sigma) = \mu, \tag{2.2.4}$$
$$Mode(x|\mu, \Sigma) = \mu, \tag{2.2.5}$$
$$var(x|\mu, \Sigma) = \Sigma, \tag{2.2.6}$$

which can be found by integration and differentiation.

Since x follows a Multivariate Normal distribution, the conditional and marginal distributions of any subset are Multivariate Normal distributions [17, 41].

The p-variate Normal distribution is that distribution, which other with finite first and second moments tend to on average according to the central limit theorem.

2.2.2   Multivariate Student t

The Multivariate Student t-distribution [17, 41] is used to describe continuous real-valued random variables with slightly heavier tails than the Multivariate Normal distribution. It is derived by taking

$$x \sim N(\mu, \phi^{-2}) \quad \text{and} \quad G \sim W(\Sigma, p, \nu), \tag{2.2.7}$$

transforming variables to

$$t = \nu^{\frac{1}{2}} G^{-\frac{1}{2}}(x - \mu) + t_0 \quad \text{and} \quad W = G, \tag{2.2.8}$$

with Jacobian

$$J(x, G \rightarrow t, W) = \nu^{-\frac{p}{2}} W^{\frac{p}{2}}, \tag{2.2.9}$$

and then integrating with respect to W. In the derivation, x could be the average of of independent and identically distributed Vector Normal variates with common mean vector and covariance matrix, while G could be the sum of the squares of deviations of these variates about their average.

A random variable that follows a p-variate Multivariate Student t-distribution [17, 41] is denoted

$$t | \nu, t_0, \Sigma, \phi^2 \sim t(\nu, t_0, \Sigma, \phi^2) \tag{2.2.10}$$

where $(\nu, \mu, \Sigma, \phi^2)$ parameterize the distribution which is given by

$$p(t | \nu, t_0, \Sigma, \phi^2) = \frac{k_t (\phi^2)^{-\frac{\nu}{2}} |\Sigma|^{-\frac{1}{2}}}{[\phi^2 + \frac{1}{\nu}(t - t_0)' \Sigma^{-1}(t - t_0)]^{\frac{\nu+p}{2}}}, \tag{2.2.11}$$

where

$$k_t = \frac{\Gamma\left(\frac{\nu+p}{2}\right)}{(\nu\pi)^{\frac{p}{2}} \Gamma\left(\frac{\nu}{2}\right)} \tag{2.2.12}$$

with

$$t \in \mathbb{R}^p, \quad \nu \in \mathbb{R}^+, \quad t_0 \in \mathbb{R}^p, \quad \Sigma > 0, \quad \phi \in \mathbb{R}^+. \tag{2.2.13}$$

Properties

The mean, mode, and variance of the Multivariate Student t-distribution are

$$E(t | \nu, t_0, \Sigma, \phi^2) = t_0, \tag{2.2.14}$$

$$Mode(t | \nu, t_0, \Sigma, \phi^2) = t_0, \tag{2.2.15}$$

$$var(t | \nu, 0, \Sigma, \phi^2) = \frac{\nu}{\nu - 2} \phi^2 \Sigma, \tag{2.2.16}$$

which can be found by integration and differentiation. Note that this parameterization is a generalization of the typical one used which can be found when $\phi^2 = 1$.

The mean of the Multivariate Student t-distribution exists for $\nu > 1$ and the variance for $\nu > 2$. When $\nu = 1$, the Multivariate Student t-distribution is the Multivariate Cauchy distribution whose mean and variance or first and second moments do not exist.

As the number of degrees of freedom increases, a random variate which follows the Multivariate Student t-distribution $t \sim t(\nu, t_0, \Sigma, \phi^2)$ approaches a Normal distribution $t \sim N(t_0, \phi^2 \Sigma)$ [17, 41].

2.3   Matrix Distributions

2.3.1   Matrix Normal

The $n \times p$ Matrix Normal distribution [17, 31] can be derived as a special case of the np-variate Multivariate Normal distribution when the covariance matrix is separable. Denote an np-dimensional Multivariate Normal distribution with np-dimensional mean μ and $np \times np$ covariance matrix Ω by

$$p(x|\mu, \Omega) = (2\pi)^{-\frac{np}{2}} |\Omega|^{-\frac{1}{2}} e^{-\frac{1}{2}(x-\mu)'\Omega^{-1}(x-\mu)}. \tag{2.3.1}$$

A separable matrix is one of the form $\Omega = \Phi \otimes \Sigma$ where $\otimes$ is the Kronecker product which multiplies every entry of its first matrix argument by its entire second matrix argument.

The Kronecker product of Φ and Σ which are n- and p-dimensional matrices respectively, is

$$\Phi \otimes \Sigma = \begin{pmatrix} \phi_{11}\Sigma & \cdots & \phi_{1n}\Sigma \\ & \vdots & \\ \phi_{n1}\Sigma & \cdots & \phi_{nn}\Sigma \end{pmatrix}. \tag{2.3.2}$$

Substituting the separable covariance matrix into the above distribution yields

$$p(x|\mu, \Sigma, \Phi) = (2\pi)^{-\frac{np}{2}} |\Phi \otimes \Sigma|^{-\frac{1}{2}} e^{-\frac{1}{2}(x-\mu)'(\Phi \otimes \Sigma)^{-1}(x-\mu)} \tag{2.3.3}$$

which upon using the matrix identities

$$|\Phi \otimes \Sigma|^{-\frac{1}{2}} = |\Phi|^{-\frac{p}{2}} |\Sigma|^{-\frac{n}{2}},$$

and

$$(x-\mu)'(\Phi \otimes \Sigma)^{-1}(x-\mu) = tr\Phi^{-1}(X-M)\Sigma^{-1}(X-M)',$$

where $x = (X') = (x_1', ..., x_n')'$, $X' = (x_1, ..., x_n)$, $\mu = vec(M') = (\mu_1', ..., \mu_n')'$, and $M' = (\mu_1, ..., \mu_n)$, then Equation 2.3.3 becomes

$$p(X|M,\Sigma,\Phi) = (2\pi)^{-\frac{np}{2}}|\Phi|^{-\frac{p}{2}}|\Sigma|^{-\frac{n}{2}}e^{-\frac{1}{2}tr\Phi^{-1}(X-M)\Sigma^{-1}(X-M)'}. \qquad (2.3.4)$$

The "vec" operator $vec(\cdot)$ which stacks the columns of its matrix argument from left to right into a single vector has been used as has the trace operator $tr(\cdot)$ which gives the sum of the diagonal elements of its square matrix argument.

A random variable that follows an $n \times p$ Matrix Normal distribution is denoted

$$X|M,\Sigma,\Phi \sim N(M,\Phi \otimes \Sigma) \qquad (2.3.5)$$

where (M,Σ,Φ) parameterize the above distribution with

$$X \in \mathbb{R}^{n \times p}, \quad M \in \mathbb{R}^{n \times p}, \quad \Sigma,\Phi > 0. \qquad (2.3.6)$$

The matrices Σ and Φ are commonly referred to as the within and between covariance matrices. Sometimes they are referred to as the right and left covariance matrices.

Properties

The mean, mode, and variance of the Matrix Normal distribution are

$$E(X|M,\Sigma,\Phi) = M, \qquad (2.3.7)$$
$$Mode(X|M,\Sigma,\Phi) = M, \qquad (2.3.8)$$
$$var(vec(X')|M,\Sigma,\Phi) = \Phi \otimes \Sigma, \qquad (2.3.9)$$

which can be found by integration and differentiation.

Since X follows a Matrix Normal distribution, the conditional and marginal distributions of any row or column subset are Multivariate Normal distributions [17, 41]. It should also be noted that the mean of the ith row of X, x_i' is the corresponding i^{th} row of M, μ_i', and the covariance of the ith row of X is $\phi_{ii}\Sigma$, where ϕ_{ii} is the element in the ith row and i^{th} column of Φ. The covariance between the ith and i'^{th} rows of X is $\phi_{ii'}\Sigma$, where $\phi_{ii'}$ is the element in the ith row and i'^{th} column of Φ. Similarly, the mean of the jth column of X is the jth column of M and the covariance between the jth and j'^{th} columns of X is $\sigma_{jj'}\Phi$.

Simply put, if

$$X = \begin{pmatrix} x_1' \\ \vdots \\ x_n' \end{pmatrix} = (X_1,\ldots,X_p), \qquad (2.3.10)$$

$$M = \begin{pmatrix} \mu_1' \\ \vdots \\ \mu_n' \end{pmatrix} = (M_1,\ldots,M_p), \qquad (2.3.11)$$

$\phi_{ii'}$ denotes the ii'^{th} element of Φ and $\sigma_{jj'}$ denotes the jj'^{th} element of Σ then

$$var(x_i|\mu_i,\phi_{ii},\Sigma) = \phi_{ii}\Sigma, \tag{2.3.12}$$
$$cov(x_i,x_{i'}|\mu_i,\mu_{i'},\phi_{ii'},\Sigma) = \phi_{ii'}\Sigma, \tag{2.3.13}$$
$$var(X_j|M_j,\sigma_{jj},\Phi) = \sigma_{jj}\Phi, \tag{2.3.14}$$
$$cov(X_j,X_{j'}|M_j,M_{j'},\sigma_{jj'},\Phi) = \sigma_{jj'}\Phi. \tag{2.3.15}$$

2.3.2   Wishart

A Wishart variate is found as a variate which is the transpose product $G = (X - M)'(X - M)$, where X is a $\nu_0 \times p$ Matrix Normal variate with mean matrix M and covariance matrix $I_{\nu_0} \otimes \Upsilon$. Note that if $p = 1$, this is the sum of the squares of ν_0 centered independent Normal variates with common mean μ and variance υ^2, $g = (x_1 - \mu)^2 + \cdots + (x_{\nu_0} - \mu)^2$. The covariance matrix Υ enters into the Wishart distribution as follows. A $p \times p$ random symmetric matrix G that follows a Wishart distribution [17, 41] is denoted

$$G|\Upsilon,p,\nu_0 \sim W(\Upsilon,p,\nu_0), \tag{2.3.16}$$

where (Υ,p,ν_0) parameterize the distribution which is given by

$$p(C|\Upsilon,p,\nu_0) - k_W|\Upsilon|^{-\frac{\nu_0}{2}}|G|^{\frac{\nu_0-p-1}{2}}e^{-\frac{1}{2}tr\Upsilon^{-1}G}, \tag{2.3.17}$$

where

$$k_W^{-1} = 2^{\frac{\nu_0 p}{2}}\pi^{\frac{p(p-1)}{4}}\prod_{j=1}^{p}\Gamma\left(\frac{\nu_0+1-j}{2}\right) \tag{2.3.18}$$

with

$$G > 0, \quad \nu_0 \in \mathbb{R}^+, \quad \Upsilon > 0, \tag{2.3.19}$$

and ">0" is used to denote that both G and Υ belong to the set of positive definite matrices. Although the Wishart distribution was derived from ν_0 (an integer valued positive number) vector Normal variates, there is no restriction that ν_0 in the Wishart distribution be integer valued.

Properties

The mean, mode, and variance of the Wishart distribution are

$$E(G|\nu_0,\Upsilon) = \nu_0\Upsilon, \tag{2.3.20}$$
$$Mode(G|\nu_0,\Upsilon) = (\nu_0 - p - 1)\Upsilon, \tag{2.3.21}$$
$$var(g_{ij}|\nu_0,\Upsilon) = \nu_0(\upsilon_{ij}^2 + \upsilon_{ii}\upsilon_{jj}), \tag{2.3.22}$$
$$cov(g_{ij}g_{kl}|\nu_0,\Upsilon) = \nu_0(\upsilon_{ik}\upsilon_{jl} + \upsilon_{il}\upsilon_{jk}), \tag{2.3.23}$$

which can be found by integration and differentiation, where g_{ij} and v_{ij} denote the ij^{th} elements of G and Υ respectively. The mode of the Wishart distribution is defined for $\nu_0 > p+1$.

The Wishart distribution is the Multivariate (Matrix variate) analog of the univariate Gamma distribution.

2.3.3   Inverted Wishart

An Inverted Wishart variate Σ is found as a variate which is the reciprocal of a Wishart variate, $\Sigma = G^{-1}$. A $p \times p$ random matrix Σ that follows an Inverted Wishart distribution [17, 41] is denoted

$$\Sigma | Q, p, \nu \sim IW(Q, p, \nu), \tag{2.3.24}$$

where (Q, p, ν) parameterize the distribution which is given by

$$p(\Sigma | \nu, Q) = k_{IW} |Q|^{\frac{\nu - p - 1}{2}} |\Sigma|^{-\frac{\nu}{2}} e^{-\frac{1}{2} tr \Sigma^{-1} Q}, \tag{2.3.25}$$

where

$$k_{IW}^{-1} = 2^{\frac{(\nu - p - 1)p}{2}} \pi^{\frac{p(p-1)}{4}} \prod_{j=1}^{p} \Gamma\left(\frac{\nu - p - j}{2}\right) \tag{2.3.26}$$

with

$$\Sigma > 0, \quad \nu \in \mathbb{R}^+, \quad Q > 0. \tag{2.3.27}$$

Note: In the transformation of variable from G to Σ,

$$Q = \Upsilon^{-1}, \quad \nu_0 = \nu - p - 1, \tag{2.3.28}$$

and the Jacobian of the transformation is

$$J(G \rightarrow \Sigma) = |\Sigma|^{-(p+1)}. \tag{2.3.29}$$

Although the Inverted Wishart distribution was derived from $\nu - p - 1$ (an integer valued positive number) vector Normal variates, there is no restriction that ν in the Inverted Wishart distribution be integer valued.

Properties

The mean, mode, and variance of the Inverted Wishart distribution are

$$E(\Sigma | \nu, Q) = \frac{Q}{\nu - 2p - 2}, \tag{2.3.30}$$

$$Mode(\Sigma | \nu, Q) = \frac{Q}{\nu}, \tag{2.3.31}$$

$$var(\sigma_{ii}|\nu, Q) = \frac{2q_{ii}^2}{(\nu - 2p - 2)^2(\nu - 2p - 4)}, \tag{2.3.32}$$

$$var(\sigma_{ii'}|\nu, Q) = \frac{q_{ii}q_{i'i'} + \frac{\nu - 2p}{\nu - 2p - 2}q_{ii'}^2}{(\nu - 2p - 1)(\nu - 2p - 2)(\nu - 2p - 4)}, \tag{2.3.33}$$

$$cov(\sigma_{ii'}, \sigma_{ii'}|\nu, Q) = \frac{\frac{2}{\nu - 2p - 2}q_{ii}q_{ii'} + q_{ii}q_{i'i'} + q_{ii'}q_{ii'}}{(\nu - 2p - 1)(\nu - 2p - 2)(\nu - 2p - 4)} \tag{2.3.34}$$

which can be found by integration and differentiation. The mean is defined for $\nu > 2p + 2$ while the variances and covariances are defined for $\nu > 2p + 4$. The variances are defined for $i \neq i'$. Where σ_{ij} and q_{ij} denote the ij^{th} element of Σ and Q respectively.

2.3.4   Matrix T

The Matrix Student T-distribution [17, 41] is used to describe continuous random variables with slightly heavier tails than the Normal distribution. It is derived by taking

$$X \sim N(M, I_n \otimes \Sigma) \quad \text{and} \quad G \sim W(\Phi^{-1}, p, \nu), \tag{2.3.35}$$

transforming variables to

$$T = \nu^{\frac{1}{2}} G^{-\frac{1}{2}}(X - M) + T_0 \quad \text{and} \quad W = G, \tag{2.3.36}$$

with Jacobian

$$J(X, G \rightarrow T, W) = \nu^{-\frac{np}{2}} W^{\frac{p}{2}}, \tag{2.3.37}$$

and then integrating with respect to W. In the derivation, X could be the average of independent and identically distributed Matrix Normal variates with common mean and variance, while G could be the sum of the squares of deviations of these variates about their average.

A random variable T follows a $n \times p$ Matrix Student T-distribution [17, 41] is denoted

$$T|\nu, T_0, \Sigma, \Phi \sim T(\nu, T_0, \Sigma, \Phi), \tag{2.3.38}$$

where (ν, T_0, Σ, Φ) parameterize the distribution which is given by

$$p(T|\nu, T_0, \Sigma, \Phi) = k_T \frac{|\Phi|^{\frac{\nu}{2}}|\Sigma|^{-\frac{n}{2}}}{|\Phi + \frac{1}{\nu}(T - T_0)\Sigma^{-1}(T - T_0)'|^{\frac{\nu + p}{2}}}, \tag{2.3.39}$$

where

$$k_T = \frac{\prod_{j=1}^n \Gamma\left(\frac{\nu + p + 1 - j}{2}\right)}{(\nu\pi)^{\frac{np}{2}} \prod_{j=1}^n \Gamma\left(\frac{\nu + 1 - j}{2}\right)} \tag{2.3.40}$$

with

$$T \in \mathbb{R}^{n \times p}, \quad \nu \in \mathbb{R}^+, \quad T_0 \in \mathbb{R}^{n \times p}, \quad \Sigma, \Phi > 0. \qquad (2.3.41)$$

Properties

The mean, mode, and variance of the Matrix Student T-distribution are

$$E(T|\nu, T_0, \Sigma, \Phi) = T_0, \qquad (2.3.42)$$

$$Mode(T|\nu, T_0, \Sigma, \Phi) = T_0, \qquad (2.3.43)$$

$$var(vec(T')|\nu, T_0, \Sigma, \Phi) = \frac{\nu}{\nu - 2}(\Phi \otimes \Sigma) \qquad (2.3.44)$$

which can be found by integration and differentiation. Note that in typical parameterizations [17, 41], the degrees of freedom ν and the matrix Φ are grouped together as a single matrix.

Since T follows a Matrix Student T-distribution, the conditional and marginal distributions of any row or column of T are Multivariate Student t-distribution [17, 41].

The mean of the Matrix Student T-distribution exists for $\nu > 1$ and the variance exists for $\nu > 2$. When the hyperparameter $\nu = 1$, the Matrix Student T-distribution is the Matrix Cauchy distribution whose mean and variance or first and second moments do not exist. As the number of degrees of freedom ν increases, a random matrix variate which follows the Matrix Student T-distribution, $T \sim T(\nu, T_0, \Sigma, \Phi)$ approaches the Matrix Normal distribution $T \sim N(T_0, \Phi \otimes \Sigma)$ [17].

Multivariate generalizations of the Binomial and Beta distribution also exist. Since they are not used in this text, they have been omitted. For a description of the Multivariate Binomial distribution see [30] or [33] and for the Multivariate Beta see [17].

Exercises

1. Compute the mean, mode, and variance of the Scalar Normal distribution.

2. Compute the mean, mode, and variance of the Scalar Student t-distribution.

3. Compute the mean, mode, and variance of the Gamma distribution.

4. Compute the mean, mode, and variance of the Inverted Gamma distribution.

5. Look at the mean, mode, and covariance matrix for the Multivariate Normal distribution. Reason that the mean, mode, and variance of the Scalar Normal distribution follows by letting $p = 1$.

6. Look at the mean, mode, and covariance matrix of the Multivariate Student t-distribution. Reason that the mean, mode, and variance of the Scalar Student t-distribution follows by letting $p = 1$.

7. Look at the mean, mode, and covariance matrix for the Matrix Normal distribution. Reason that the mean, mode, and variance of the Multivariate Normal distribution follows by letting $n = 1$ (and hence the mean, mode, and variance of the Scalar Student t-distribution by letting $n = 1$ and $p = 1$).

8. Look at the mean, mode, and covariance matrix of the Matrix Student T-distribution. Reason that the mean, mode, and variance of the Multivariate Student t-distribution follows by letting $n = 1$ (and hence the mean, mode, and variance of the Scalar Student t-distribution by letting $n = 1$ and $p = 1$).

9. Look at the mean, mode, variances, and covariances of the Wishart distribution. Reason that the mean, mode, and variance of the Gamma distribution follows by letting $p = 1$.

10. Look at the mean, mode, variances, and covariances of the Inverted Wishart distribution. Reason that the mean, mode, and variance of the Gamma distribution follows by letting $p = 1$.

11. Look at the Multivariate Normal distribution. Reason that the Scalar Normal distribution follows by letting $p = 1$.

12. Look at the Multivariate Student t-distribution. Reason that the Scalar Student t-distribution follows by letting $p = 1$.

13. Look at the Matrix Normal distribution. Reason that the Multivariate Normal distribution follows by letting $n = 1$ (and hence the Scalar Normal distribution by letting $n = 1$ and $p = 1$).

14. Look at the Matrix Student T-distribution. Reason that the Multivariate Student t-distribution follows by letting $n = 1$ (and hence the Scalar Student t-distribution by letting $n = 1$ and $p = 1$).

15. Look at the Wishart distribution. Reason that the Gamma distribution follows by letting $p = 1$.

16. Look at the Inverted Wishart distribution. Reason that the inverse Gamma distribution follows by letting $p = 1$.

3

Introductory Bayesian Statistics

Those persons who have experience with Bayesian Statistics [3, 42] can skip this Chapter. Bayesian Statistics quantifies available prior knowledge either using data from prior experiments or subjective beliefs from substantive experts. This prior knowledge is in the form of prior distributions which quantify beliefs about various parameter values that are formally incorporated into the inferences via Bayes' rule.

3.1   Discrete Scalar Variables

3.1.1   Bayes' Rule and Two Simple Events

Bayesian Statistics is based on Bayes' rule or conditional probability. It is well known that the probability of events A and B both occurring can be written as the probability of A occurring multiplied by the probability of B occurring given that A has occurred. This is written as

$$P(A \text{ and } B) = P(A)P(B|A) \qquad (3.1.1)$$

which is the (general) rule for probability multiplication. If we rearrange the terms then we get the formula for conditional probability

$$P(B|A) = \frac{P(A \text{ and } B)}{P(A)} \qquad (3.1.2)$$

which is Bayes' rule or theorem. Pictorially this is represented in Figure 3.1 in what is called a Venn diagram.

Example:
    Consider a standard deck of 52 cards.
Let A = the event of a King chosen randomly.
Let B = the event of a Heart chosen randomly.
    What is the probability of selecting a Heart given we have selected a King?
$P(B|A) = \frac{P(A \text{ and } B)}{P(A)}$
    We use conditional probability. Then the probability of event B given that event A has occurred is

FIGURE 3.1

Pictorial representation of Bayes' rule.

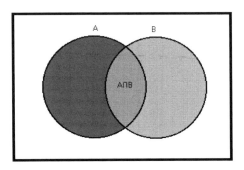

$$P(B|A) = \frac{\frac{1}{52}}{\frac{4}{52}} = \frac{1}{4}.$$

3.1.2 Bayes' Rule and the Law of Total Probability

Bayes' rule can be extended to determine the probability of each of k events given that event A has occurred. The probability of one of the k events occurring, say event B_i, given that event A has occurred is

$$P(B_i|A) = \frac{P(A \text{ and } B_i)}{P(A)} = \frac{P(A \text{ and } B_i)}{\sum_{i=1}^{k} P(A \text{ and } B_i)} = \frac{P(B_i)p(A|B_i)}{\sum_{i=1}^{k} P(B_i)p(A|B_i)},$$
$$(3.1.3)$$

where the B_i's are mutually exclusive events and

$$A \subseteq \bigcup_{i=1}^{k} B_i. \qquad (3.1.4)$$

The probabilities of the $P(B_i)$'s are (prior) probabilities for each of the k events occurring.

The denominator of the above equation is called the Law of Total Probability. This version of Bayes' rule is represented pictorially in Figure 3.2 as a Venn diagram.

The law of total probability is defined to be

$$P(A) = \sum_{i=1}^{k} P(B_i)p(A|B_i) \qquad (3.1.5)$$

FIGURE 3.2
Pictorial representation of the Law of Total Probability.

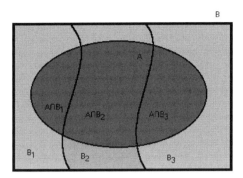

which states that if we sum over the probabilities of event A occurring given that a particular event B_i has occurred multiplied by the probabilities of the event B_i occurring, this results in the probability of event A occurring. Where again, the events B_i are mutually exclusive and exhaustive.

Example:
   To evaluate the effectiveness of a medical testing procedure such as for disease screening or illegal drug use, we will evaluate the probability of a false negative or a false positive using the following notation
   T^+: The test is positive
   T^-: The test is negative
   D^+: The person has the disease
   D^-: The person does not have the disease.
   The "sensitivity" of a test is the probability of a positive result given the person has the disease. We will assume that a particular test has

$$P[T^+|D^+] = 0.99.$$

   The "specificity" of the test is the probability the test is negative given the person does not have the disease is

$$P[T^-|D^-] = 0.99.$$

   If the proportion in the general public infected with the disease is 1 per million or 0.000001, find the probability of a false positive $P[D^-|T^+]$.
   From Bayes' rule

$$P[D^-|T^+] = \frac{P(D^-)P(T^+|D^-)}{P(T^+)} \qquad (3.1.6)$$

and by the Law of Total Probability we get the probability of testing positive

$$\begin{aligned}
P(T^+) &= P(D^+)P(T^+|D^+) + P(D^-)P(T^+|D^-) \\
&= (0.000001)(0.99) + (0.999999)(0.01) \\
&= 0.00000099 + 0.00999999 \\
&= 0.01000098.
\end{aligned}$$

The probability of a false positive or of the test for the disease giving a positive result when the person does not in fact have the disease is

$$P[D^-|T^+] = \frac{0.00999999}{0.01000098} = 0.99990101.$$

3.2 Continuous Scalar Variables

Bayes' rule also applies to continuous random variables. Let's assume that we have a continuous random variable x that is specified to come from a distribution that is indexed by a parameter θ.

Prior

We can quantify our prior knowledge as to the parameter value and assess a prior distribution

$$p(\theta). \tag{3.2.1}$$

Likelihood

That is, the distribution (or likelihood) of the random variable x is

$$p(x|\theta). \tag{3.2.2}$$

Posterior

We now apply Bayes' rule to obtain the posterior distribution

$$p(\theta|x) = \frac{p(\theta)p(x|\theta)}{p(x)}, \tag{3.2.3}$$

where the denominator is given by

$$p(x) = \int p(\theta)p(x|\theta)\, d\theta \tag{3.2.4}$$

which is the continuous version of the discrete Law of Total Probability. Inferences can be made from the posterior distribution of the model parameter θ

given the data x which includes information from both the prior distribution and the likelihood instead of only from the likelihood. From the posterior distribution, mean and modal estimators as well as interval estimates can be obtained for θ by integration or differentiation. This will be described later. (It might help to make an analogy of x to A and θ to B.)

This is generalized to a random sample of size n, $x_1,\ldots,x_n$ from a distribution that depends on J parameters $\theta_1,\ldots,\theta_J$.

Prior

We quantify available prior knowledge regarding the parameters in the form of a joint prior distribution on the parameters (before taking the random sample)

$$p(\theta_1,\ldots,\theta_J), \tag{3.2.5}$$

where they are not necessarily independent.

Likelihood

With an independent sample of size n, the joint distribution of the observations is

$$p(x_1,\ldots,x_n|\theta_1,\ldots,\theta_J) = \prod_{i=1}^{n} p(x_i|\theta_1,\ldots,\theta_J). \tag{3.2.6}$$

Posterior

We apply Bayes' rule to obtain a posterior distribution for the parameters. The posterior distribution is

$$p(\theta_1,\ldots,\theta_J|x_1,\ldots,x_n) = \frac{p(\theta_1,\ldots,\theta_J)p(x_1,\ldots,x_n|\theta_1,\ldots,\theta_J)}{p(x_1,\ldots,x_n)}, \tag{3.2.7}$$

where the denominator is given by

$$p(x_1,\ldots,x_n) = \int p(\theta_1,\ldots,\theta_J)p(x_1,\ldots,x_n|\theta) \, d\theta_1 \ldots d\theta_J, \tag{3.2.8}$$

where $\theta = (\theta_1,\ldots,\theta_J)$. (It might help to make an analogy of $x_1,\ldots,x_n$ to A and $\theta_1,\ldots,\theta_J$ to B.)

Example:

Let's consider a random sample of size n that is specified to come from a population that is Normally distributed with mean μ and variance σ^2. This is denoted as $x_i \sim N(\mu,\sigma^2)$, where $i = 1,\ldots,n$.

Prior

Before seeing the data, we quantify available prior knowledge regarding the parameters μ and σ^2 in the form of a joint prior distribution on the parameters

$$p(\mu, \sigma^2), \tag{3.2.9}$$

where μ and σ^2 are not necessarily independent.

Likelihood
    The likelihood of all the n observations is

$$p(x_1, \ldots, x_n | \mu, \sigma^2) = (2\pi\sigma^2)^{-\frac{n}{2}} e^{-\frac{\sum_{i=1}^{n}(x_i - \mu)^2}{2\sigma^2}}, \tag{3.2.10}$$

where $x_1, \ldots, x_n$ are the data.

Posterior
    We apply Bayes' rule to obtain a joint posterior distribution

$$p(\mu, \sigma^2 | x_1, \ldots, x_n) = \frac{p(\mu, \sigma^2)p(x_1, \ldots, x_n | \mu, \sigma^2)}{p(x_1, \ldots, x_n)} \tag{3.2.11}$$

for the mean and variance. From this joint posterior distribution which contains information from the prior distribution and the likelihood, we can obtain estimates of the parameters. This will be described later.

    Bayesian statisticians usually neglect the denominator of the posterior distribution, as alluded to earlier, to get

$$p(\theta_1, \ldots, \theta_J | x_1, \ldots, x_n) = kp(\theta_1, \ldots, \theta_J)p(x_1, \ldots, x_n | \theta_1, \ldots, \theta_J)$$
$$\propto p(\theta_1, \ldots, \theta_J)p(x_1, \ldots, x_n | \theta_1, \ldots, \theta_J), \tag{3.2.12}$$

where "$\propto$" denotes proportionality and the constant k, which does not depend on the variates $\theta_1, \ldots, \theta_J$, can be found by integration.

3.3 Continuous Vector Variables

    We usually have several variables measured on an individual; thus, we are rarely interested in a single random variable. We are interested in p-dimensional vector observations $x_1, \ldots, x_n$ where $x_i = (x_{1i}, \ldots, x_{pi})'$ for $i = 1, \ldots, n$. The observations are specified to come from a distribution with parameters $\theta_1, \ldots, \theta_J$ where the θ's may be scalars, vectors, or matrices.

Prior
    We quantify available prior knowledge (before performing the experiment and obtaining the data) in the form of a joint prior distribution for the parameters

$$p(\theta_1, \ldots, \theta_J), \tag{3.3.1}$$

where they are not necessarily independent.

Likelihood

With an independent sample of size n, the joint distribution (likelihood) of the observation vectors is the product of the individual distributions (likelihoods) and is given by

$$p(x_1,\ldots,x_n|\theta_1,\ldots,\theta_J) = \prod_{i=1}^{n} p(x_i|\theta_1,\ldots,\theta_J). \qquad (3.3.2)$$

Posterior

We apply Bayes' rule to obtain a posterior distribution for the parameters. The posterior distribution is

$$p(\theta_1,\ldots,\theta_J|x_1,\ldots,x_n) = \frac{p(\theta_1,\ldots,\theta_J)p(x_1,\ldots,x_n|\theta_1,\ldots,\theta_J)}{p(x_1,\ldots,x_n)}, \qquad (3.3.3)$$

where the denominator is given by

$$p(x_1,\ldots,x_n) = \int p(\theta_1,\ldots,\theta_J)p(x_1,\ldots,x_n|\theta) \, d\theta_1 \ldots d\theta_J, \qquad (3.3.4)$$

where $\theta = (\theta_1,\ldots,\theta_J)$. (It might help to make an analogy of $x_1,\ldots,x_n$ to A and $\theta_1,\ldots,\theta_J$ to B.)

Remember we neglect the denominator to get

$$p(\theta_1,\ldots,\theta_J|x_1,\ldots,x_n) \propto p(\theta_1,\ldots,\theta_J)p(x_1,\ldots,x_n|\theta_1,\ldots,\theta_J) \qquad (3.3.5)$$

which states that the posterior distribution is proportional to the product of the prior times the likelihood.

3.4   Continuous Matrix Variables

Just as we are able to observe scalar and vector valued variables, we can also observe matrix valued variables X.

Prior

We can quantify available prior knowledge regarding the parameters with the use of a joint prior distribution

$$p(\theta_1,\ldots,\theta_J) \qquad (3.4.1)$$

where the θ's are possibly matrix valued and are not necessarily independent.

Likelihood

With an independent sample of size n, from the joint distribution $p(X|\theta_1,\ldots,\theta_J)$ of the observation matrices is

$$p(X_1,\ldots,X_n|\theta_1,\ldots,\theta_J) = \prod_{i=1}^{n} p(X_i|\theta_1,\ldots,\theta_J). \qquad (3.4.2)$$

Posterior

We apply Bayes' rule just as we have done for scalar and vector valued variates to obtain a joint posterior distribution for the parameters. The joint posterior distribution is

$$p(\theta_1,\ldots,\theta_J|X_1,\ldots,X_n) = \frac{p(\theta_1,\ldots,\theta_J)p(X_1,\ldots,X_n|\theta_1,\ldots,\theta_J)}{p(X_1,\ldots,X_n)}, \qquad (3.4.3)$$

where the denominator is given by

$$p(X_1,\ldots,X_n) = \int p(\theta_1,\ldots,\theta_J)p(X_1,\ldots,X_n|\theta)\, d\theta_1\ldots d\theta_J. \qquad (3.4.4)$$

(It might help to make an analogy of $X_1,\ldots,X_n$ to A and $\theta_1,\ldots,\theta_J$ to B.)

Remember we neglect the denominator of the joint posterior distribution to get

$$p(\theta_1,\ldots,\theta_J|X_1,\ldots,X_n) \propto p(\theta_1,\ldots,\theta_J)p(X_1,\ldots,X_n|\theta_1,\ldots,\theta_J) \qquad (3.4.5)$$

in which the joint posterior distribution is proportional to the product of the prior distribution and the likelihood distribution.

From the posterior distribution, estimates of the parameters are obtained. Estimation of the parameters is described later.

Exercises

1. State Bayes' rule for the probability of event B occurring given that event A has occurred.

2. Assume that we select a Beta prior distribution

$$p(\varrho) \propto \varrho^{\alpha-1}(1-\varrho)^{\beta-1}$$

   for the probability of success in a Binomial experiment with likelihood

$$p(x|\varrho) \propto \varrho^{x}(1-\varrho)^{n-x}.$$

   Write the posterior distribution $p(\varrho|x)$ of ϱ.

3. Assume that we select the joint prior distribution

$$p(\mu,\sigma^2) \propto p(\mu|\sigma^2)p(\sigma^2),$$

   where

$$p(\mu|\sigma^2) \propto (\sigma^2)^{-\frac{1}{2}}e^{-\frac{(\mu-\mu_0)^2}{2\sigma^2}},$$
$$p(\sigma^2) \propto (\sigma^2)^{-\frac{\nu}{2}}e^{-\frac{q}{2\sigma^2}},$$

   and have a likelihood given by

$$p(x_1,\ldots,x_n|\mu,\sigma^2) \propto (\sigma^2)^{-\frac{n}{2}}e^{-\sum_{i=1}^{n}\frac{(x_i-\bar{x})^2}{2\sigma^2}}.$$

   Write the joint posterior distribution $p(\mu,\sigma^2|x_1,\ldots,x_n)$ of μ and σ^2.

4

Prior Distributions

In this Chapter, we discuss the specification of the form of the prior distributions for our parameters which we are using to quantify our available prior knowledge. We can specify any prior distribution that we like but our choice is guided by the range of values of the parameters.

For example, the probability of success in a Binomial experiment has $(0,1)$ as its range of possible values, the mean of a Normal distribution has $(-\infty, +\infty)$ as its range of values, and the variance of a Normal distribution has $(0, +\infty)$ as its range of values.

There are three common types of prior distributions.

1. Vague (uninformative or diffuse),

2. Conjugate, and

3. Generalized Conjugate.

Even though our choice is guided by the range of values for the parameters, any distribution that is defined solely in that range of values can be used. However, the choice of Conjugate prior distributions have natural updating properties and can simplify the estimation procedure. Further, Conjugate prior distributions are usually rich enough to quantify our available prior information.

4.1 Vague Priors

4.1.1 Scalar Variates

The vague prior distribution can be placed on either a parameter that is bounded (has a finite range of values) or unbounded (has an infinite range of values).

If a vague prior is placed on a parameter θ that has a finite range of values, over the interval (a,b), then the prior distribution is a Uniform distribution over (a,b) indicating that all values in this range are a priori equally likely. That is,

$$p(\theta) = \begin{cases} \frac{1}{b-a}, & \text{if } a < \theta < b \\ 0, & \text{otherwise} \end{cases}, \qquad (4.1.1)$$

the Uniform distribution and we write

$$p(\theta) \propto (\text{a constant}). \qquad (4.1.2)$$

A vague prior is a little different when we place it on a parameter that is unbounded.

Consider $\theta = \mu$ to be the mean of a Normal distribution. If we wish to place a Uniform prior on it, then we have

$$p(\mu) = \begin{cases} \frac{1}{2a}, & \text{if } -a < \mu < a \\ 0, & \text{otherwise,} \end{cases} \qquad (4.1.3)$$

where $a \to \infty$ and again we write

$$p(\mu) \propto (\text{a constant}). \qquad (4.1.4)$$

Principle of Stable Estimation

As previously stated [41], in 1962 it was noted [10] that the posterior distribution is proportional to the product of the prior and the likelihood. Therefore, to be vague about a parameter θ, we only have to place a Uniform prior over the range of values for θ where the likelihood is non-negligible.

If the parameter $\theta = \sigma^2$ is the variance of a Normal distribution, then we take $\log(\sigma^2)$ to be uniform over the entire real line and by transforming back to a distribution on σ^2 we have

$$p(\sigma^2) \propto \frac{1}{\sigma^2}. \qquad (4.1.5)$$

A description of this can be found in [9]. Another justification [25] is based on a prior distribution which expresses minimal information. The vague prior distribution in Equation 4.1.5 is an "improper" prior distribution. That is,

$$\int_0^\infty p(\sigma^2) \, d\sigma^2 \qquad (4.1.6)$$

is not finite.

It should be noted that estimability problems may arise when using vague prior distributions.

4.1.2   Vector Variates

A vague prior distribution for a vector-valued mean such as for a Multivariate Normal distribution is the same as that for a Scalar Normal distribution

$$p(\mu) \propto \text{(a constant)}, \qquad (4.1.7)$$

where $\mu = (\mu_1, \ldots, \mu_p)$.

4.1.3   Matrix Variates

A vague prior distribution for a matrix-valued mean such as for a Matrix Normal distribution is the same as for the scalar and vector versions

$$p(M) \propto \text{(a constant)}, \qquad (4.1.8)$$

where the matrix M is $M = (\mu_1, \ldots, \mu_n)'$. The rows of M are individual μ vectors.

The generalization of the univariate vague prior distribution on a variance to a covariance matrix is

$$p(\Sigma) \propto |\Sigma|^{-\frac{p+1}{2}} \qquad (4.1.9)$$

Which is often refered to as Jeffreys invariant prior distribution. Note that this reduces to the scalar version when $p = 1$.

4.2   Conjugate Priors

Conjugate prior distributions are informative prior distributions. Conjugate prior distributions follow naturally from classical statistics. It is well known that if a set of data were taken in two parts, then an analysis which takes the first part as a prior for the second part is equivalent to an analysis which takes both parts together. The Conjugate prior distribution for a parameter is of great utility and is obtained by writing down the likelihood, interchanging the roles of the random variable and the parameter, and "enriching" the distribution so that it does not depend on the data set [41, 42]. The Conjugate prior distribution has the property that when combined with the likelihood, the resulting posterior is in the same "family" of distributions.

4.2.1　Scalar Variates

Beta

The number of heads x when a coin is flipped n_0 independent times with y tails $(n_0 = x + y)$, follows a Binomial distribution

$$p(x|\varrho) \propto \varrho^x (1-\varrho)^{n_0-x}. \tag{4.2.1}$$

We now implement the Conjugate procedure in order to obtain the prior distribution for the probability of heads, $\theta = \varrho$. First, interchange the roles of x and ϱ

$$p(\varrho|x) \propto \varrho^x (1-\varrho)^{n_0-x}, \tag{4.2.2}$$

and now "enrich" it so that it does not depend on the current data set to obtain

$$p(\varrho) \propto \varrho^{\alpha-1}(1-\varrho)^{\beta-1}. \tag{4.2.3}$$

This is the Beta distribution. This Conjugate procedure implies that a good choice is to use the Beta distribution to quantify available prior information regarding the probability of success in a Binomial experiment. The quantities α and β are hyperparameters to be assessed. Hyperparameters are parameters of the prior distribution.

As previously mentioned, the use of the Conjugate prior distribution has the extra advantage that the resulting posterior distribution is in the same family.

Example:

The prior distribution for the probability of heads when flipping a certain coin is

$$p(\varrho) \propto \varrho^{\alpha-1}(1-\varrho)^{\beta-1}. \tag{4.2.4}$$

and the likelihood for a random sample subsequently taken is

$$p(x|\varrho) \propto \varrho^x (1-\varrho)^{n_0-x}. \tag{4.2.5}$$

When these are combined to form the posterior distribution of ϱ, the result is

$$\begin{aligned} p(\varrho|x) &\propto p(\varrho)p(x|\varrho) \\ &\propto \varrho^{(\alpha+x)-1}(1-\varrho)^{(\beta+n_0-x)-1}. \end{aligned} \tag{4.2.6}$$

The prior distribution belongs to the family of Beta distributions as does the posterior distribution. This is a feature of Conjugate prior distributions.

Normal

The observation can be specified to have come from a Scalar Normal distribution, $x|\mu,\sigma^2 \sim N(\mu,\sigma^2)$ with σ^2 either known or unknown. The likelihood is

$$p(x|\mu,\sigma^2) \propto (\sigma^2)^{-\frac{1}{2}} e^{-\frac{(x-\mu)^2}{2\sigma^2}}, \qquad (4.2.7)$$

which is often called the "kernel" of a Normal distribution.

If we interchange the roles of x and μ, then we obtain

$$p(\mu) \propto (\sigma^2)^{-\frac{1}{2}} e^{-\frac{(\mu-x)^2}{2\sigma^2}} \qquad (4.2.8)$$

thus implying that we should select our prior distribution for μ from the Normal family.

We then select as our prior distribution for μ to be

$$p(\mu|\sigma^2) \propto (\sigma^2)^{-\frac{1}{2}} e^{-\frac{(\mu-\mu_0)^2}{2\sigma^2}}, \qquad (4.2.9)$$

where we have "enriched" the prior distribution with the use of μ_0 so that it does not depend on the data. The quantity μ_0 is a hyperparameter to be assessed. By specifying scalar quantities μ_0 and σ^2, the Normal prior distribution is completely determined.

Inverted Gamma

If we interchange the roles of x and σ^2 in the likelihood, then

$$p(\sigma^2) \propto (\sigma^2)^{-\frac{1}{2}} e^{-\frac{(x-\mu)^2}{2\sigma^2}} \qquad (4.2.10)$$

thus implying that we should select our prior distribution for σ^2 from the Inverted Gamma family.

We then select as our prior distribution for σ^2

$$p(\sigma^2) \propto (\sigma^2)^{-\frac{\nu}{2}} e^{-\frac{q}{2\sigma^2}}, \qquad (4.2.11)$$

where we have "enriched" the prior distribution with the use of q so that it does not depend on the data. The quantity q is a hyperparameter to be assessed. By specifying scalar quantities q and ν, the Inverted Gamma prior distribution is completely determined.

Using the Conjugate procedure to obtain prior distributions, we obtain Table 4.1.

4.2.2 Vector Variates

Normal

The observation can be specified to have come from a Multivariate or Vector

TABLE 4.1
Scalar variate Conjugate priors.

Likelihood	Parameter(s)	Prior Family
Scalar Binomial	p	Beta
Scalar Normal σ^2 known	μ	Normal
Scalar Normal μ known	σ^2	Inverted Gamma
Scalar Normal	(μ, σ^2)	Normal-Inverted Gamma

Normal distribution, $x|\mu, \Sigma \sim N(\mu, \Sigma)$ with Σ either known or unknown. The likelihood is

$$p(x|\mu, \Sigma) \propto |\Sigma|^{-\frac{1}{2}} e^{-\frac{1}{2}(x-\mu)'\Sigma^{-1}(x-\mu)}. \qquad (4.2.12)$$

If we interchange the roles of x and μ, then

$$p(\mu) \propto |\Sigma|^{-\frac{1}{2}} e^{-\frac{1}{2}(\mu-x)'\Sigma^{-1}(\mu-x)} \qquad (4.2.13)$$

thus implying that we should select our prior distribution for μ from the Normal family.

We then select as our prior distribution for μ to be

$$p(\mu|\Sigma) \propto |\Sigma|^{-\frac{1}{2}} e^{-\frac{1}{2}(\mu-\mu_0)'\Sigma^{-1}(\mu-\mu_0)}, \qquad (4.2.14)$$

where we have "enriched" the prior distribution with the use of μ_0 so that it does not depend on the data. The vector quantity μ_0 is a hyperparameter to be assessed. By specifying the vector quantity μ_0 and the matrix quantity Σ, the Vector or Multivariate Normal prior distribution is completely determined.

Inverted Wishart

If we interchange the roles of x and Σ in the Vector or Multivariate Normal likelihood and use the property of the trace operator, then

$$p(\Sigma) \propto |\Sigma|^{-\frac{1}{2}} e^{-\frac{1}{2}tr\Sigma^{-1}(x-\mu)(x-\mu)'} \qquad (4.2.15)$$

thus implying that we should select our prior distribution for Σ from the Inverted Wishart family.

We then select as our prior distribution for Σ

$$p(\Sigma) \propto |\Sigma|^{-\frac{\nu}{2}} e^{-\frac{1}{2}tr\Sigma^{-1}Q}, \qquad (4.2.16)$$

where we have "enriched" the prior distribution with the use of Q and ν so that it does not depend on the data. The quantities ν and Q are hyperparameters to be assessed. By specifying the matrix quantity Q and the scalar quantity ν, the Inverted Wishart prior distribution is completely determined.

Using the Conjugate procedure to obtain prior distributions, we obtain Table 4.2 where "IW" is used to denote Inverted Wishart.

TABLE 4.2
Vector variate Conjugate priors.

Likelihood	Parameter(s)	Prior Family
Multivariate Normal Σ known	μ	Multivariate Normal
Multivariate Normal μ known	Σ	Inverted Wishart
Multivariate Normal	(μ, Σ)	Normal-IW

4.2.3   Matrix Variates

Normal

The observation can be specified to have come from a Matrix Normal Distribution, $X|M, \Phi, \Sigma \sim N(M, \Phi \otimes \Sigma)$ with Φ and Σ either known or unknown. The Matrix Normal likelihood is

$$p(X|M, \Sigma, \Phi) \propto |\Phi|^{-\frac{p}{2}} |\Sigma|^{-\frac{n}{2}} e^{-\frac{1}{2} tr \Phi^{-1}(X-M)\Sigma^{-1}(X-M)'}. \qquad (4.2.17)$$

If we interchange the roles of X and M, then

$$p(M) \propto |\Phi|^{-\frac{p}{2}} |\Sigma|^{-\frac{n}{2}} e^{-\frac{1}{2} tr \Phi^{-1}(M-X)\Sigma^{-1}(M-X)'} \qquad (4.2.18)$$

thus implying that we should select our prior distribution for M from the Matrix Normal family.

We then select as our prior distribution for M

$$p(M|\Sigma, \Phi) \propto |\Phi|^{-\frac{p}{2}} |\Sigma|^{-\frac{n}{2}} e^{-\frac{1}{2} tr \Phi^{-1}(M-M_0)\Sigma^{-1}(M-M_0)'}, \qquad (4.2.19)$$

where we have "enriched" the prior distribution with the use of M_0 so that it does not depend on the data. The quantity M_0 is a hyperparameter to be assessed. By specifying the matrix quantity M_0 and the matrix quantities Φ and Σ, the Matrix Normal prior distribution is completely determined.

Inverted Wishart

If we interchange the roles of X and Σ in the Matrix Normal likelihood and use the property of the trace operator, then

$$p(\Sigma) \propto |\Phi|^{-\frac{p}{2}} |\Sigma|^{-\frac{n}{2}} e^{-\frac{1}{2} tr \Sigma^{-1}(X-M)'\Phi^{-1}(X-M)} \qquad (4.2.20)$$

thus implying that we should select our prior distribution for Σ from the Inverted Wishart family.

We then select as our prior distribution for Σ

$$p(\Sigma) \propto |\Sigma|^{-\frac{\nu}{2}} e^{-\frac{1}{2} tr \Sigma^{-1} Q}, \qquad (4.2.21)$$

where we have "enriched" the prior distribution with the use of Q and ν so that it does not depend on the data. The quantities ν and Q are hyperparameters to be assessed. By specifying the matrix quantity Q and the scalar quantity ν, the Inverted Wishart prior distribution is completely determined.

Taking the same Matrix Normal likelihood, and interchanging the role for Φ,

$$p(\Phi|X, M, \Sigma) \propto |\Phi|^{-\frac{p}{2}}|\Sigma|^{-\frac{n}{2}}e^{-\frac{1}{2}tr\Phi^{-1}(X-M)\Sigma^{-1}(X-M)'} \qquad (4.2.22)$$

thus implying that we should select our prior distribution for Σ from the Inverted Wishart family

$$p(\Phi|\kappa, \Psi) \propto |\Phi|^{-\frac{\eta}{2}}e^{-\frac{1}{2}tr\Phi^{-1}\Psi}, \qquad (4.2.23)$$

where we have "enriched" the prior distribution with the use of Ψ and κ so that it does not depend on the data. The quantities κ and Ψ are hyperparameters to be assessed. By specifying the matrix quantity Ψ and the scalar quantity κ, the Inverted Wishart prior distribution is completely determined.

Using the Conjugate procedure to obtain prior distributions, we obtain Table 4.3 where "IW" is used to denote Inverted Wishart.

TABLE 4.3
Matrix variate Conjugate priors.

Likelihood	Parameter(s)	Prior Family
Normal (Φ, Σ) known	M	Matrix Normal
Matrix Normal (M, Φ) known	Σ	Inverted Wishart
Matrix Normal (M, Σ) known	Φ	Inverted Wishart
Matrix Normal	(M, Φ, Σ)	Normal-IW-IW

4.3   Generalized Priors

At times, Conjugate prior distributions are not sufficient to quantify the prior knowledge we have about the parameter values [49]. When this is the case, generalized Conjugate prior distributions can be used. Generalized Conjugate prior distributions are found by writing down the likelihood, interchanging the roles of the random variable and the parameter, "enriching" the distribution so that it does not depend on the data set, and assuming that the priors on each of the parameters are independent [41].

4.3.1   Scalar Variates

Normal
   The observation can be specified to have come from a Scalar Normal dis-

tribution, $x|\mu,\sigma^2 \sim N(\mu,\sigma^2)$ with σ^2 either known or unknown. The Normal likelihood is given by

$$p(x|\mu,\sigma^2) \propto (\sigma^2)^{-\frac{1}{2}} e^{-\frac{(x-\mu)^2}{2\sigma^2}}. \qquad (4.3.1)$$

If we interchange the roles of x and μ, then we obtain

$$p(\mu) \propto (\sigma^2)^{-\frac{1}{2}} e^{-\frac{(\mu-x)^2}{2\sigma^2}} \qquad (4.3.2)$$

thus implying that we should select our prior distribution for μ from the Normal family.

We then select as our prior distribution for μ

$$p(\mu) \propto (\delta^2)^{-\frac{1}{2}} e^{-\frac{(\mu-\mu_0)^2}{2\delta^2}}, \qquad (4.3.3)$$

where we have "enriched" the prior distribution with the use of μ_0 so that it does not depend on the data and made it independent of the other parameter σ^2 through δ^2. The quantities μ_0 and δ^2 are hyperparameters to be assessed. By specifying scalar quantities μ_0 and δ^2, the Normal prior distribution is completely determined.

Inverted Gamma

If we interchange the roles of x and σ^2 in the Normal likelihood then

$$p(\sigma^2) \propto (\sigma^2)^{-\frac{1}{2}} e^{-\frac{(x-\mu)^2}{2\sigma^2}} \qquad (4.3.4)$$

thus implying that we should select our prior distribution for σ^2 from the Inverted Gamma family.

We then select our prior distribution for σ^2 to be

$$p(\sigma^2) \propto (\sigma^2)^{-\frac{\nu}{2}} e^{-\frac{q}{2\sigma^2}}, \qquad (4.3.5)$$

where we have "enriched" the prior distribution with the use of q so that it does not depend on the data. The quantities q and ν are hyperparameters to be assessed. By specifying the scalar quantities q and ν, the Inverted Gamma prior distribution is completely determined. The generalized Conjugate procedure yields the same prior distribution for the variance σ^2 as the Conjugate procedure.

Using the generalized Conjugate procedure to obtain prior distributions, we obtain Table 4.4 where "IG" is used to denote Inverted Gamma.

4.3.2   Vector Variates

Normal

The observation can be specified to have come from a Multivariate or vector

TABLE 4.4
Scalar variate generalized Conjugate priors.

Likelihood	Parameter(s)	Prior Family
Scalar Normal σ^2 known	μ	Generalized Normal
Scalar Normal μ known	σ^2	Inverted Gamma
Scalar Normal	(μ, σ^2)	Generalized Normal-IG

Normal distribution, $x|\mu, \Sigma \sim N(\mu, \Sigma)$ with Σ either known or unknown. The Multivariate Normal likelihood is

$$p(x|\mu, \Sigma) \propto |\Sigma|^{-\frac{1}{2}} e^{-\frac{1}{2}(x-\mu)'\Sigma^{-1}(x-\mu)}. \qquad (4.3.6)$$

If we interchange the roles of x and μ in the likelihood, then

$$p(\mu) \propto |\Sigma|^{-\frac{1}{2}} e^{-\frac{1}{2}(\mu-x)'\Sigma^{-1}(\mu-x)} \qquad (4.3.7)$$

thus implying that we should select our prior distribution for μ from the Normal family.

We then select as our prior distribution for μ

$$p(\mu) \propto |\Delta|^{-\frac{1}{2}} e^{-\frac{1}{2}(\mu-\mu_0)'\Delta^{-1}(\mu-\mu_0)}, \qquad (4.3.8)$$

where we have "enriched" the prior distribution with the use of μ_0 so that it does not depend on the data and made it independent of the other parameter Σ through Δ. The quantities μ_0 and Δ are hyperparameters to be assessed. By specifying the vector quantity μ_0 and the matrix quantity Δ, the Multivariate Normal prior distribution is completely determined.

Inverted Wishart

If we interchange the roles of x and Σ and use the property of the trace operator, then

$$p(\Sigma) \propto |\Sigma|^{-\frac{1}{2}} e^{-\frac{1}{2}tr\Sigma^{-1}(x-\mu)(x-\mu)'} \qquad (4.3.9)$$

thus implying that we should select our prior distribution for Σ from the Inverted Wishart family.

We then select as our prior distribution for Σ

$$p(\Sigma) \propto |\Sigma|^{-\frac{\nu}{2}} e^{-\frac{1}{2}tr\Sigma^{-1}Q}, \qquad (4.3.10)$$

where we have "enriched" the prior distribution with the use of Q and ν so that it does not depend on the data. The quantities ν and Q are hyperparameters to be assessed. By specifying the matrix quantity Q and the scalar quantity ν, the Inverted Wishart prior distribution is completely determined. The generalized Conjugate procedure yields the same prior distribution for Σ as the Conjugate procedure.

Using the generalized Conjugate procedure to obtain prior distributions, we obtain Table 4.5 where "IW" is used to denote Inverted Wishart and "GMN," Generalized Matrix Normal.

TABLE 4.5
Vector variate generalized Conjugate priors.

Likelihood	Parameter(s)	Prior Family
Multivariate Normal Σ known	μ	GMN
Multivariate Normal μ known	Σ	Inverted Wishart
Multivariate Normal	(μ, Σ)	GMN-IW

4.3.3 Matrix Variates

Normal

The observation can be specified to have come from a Matrix Normal Distribution, $X|M, \Phi, \Sigma \sim N(M, \Phi \otimes \Sigma)$ with Φ and Σ either known or unknown. The Matrix Normal likelihood is

$$p(X|M, \Sigma, \Phi) \propto |\Phi|^{-\frac{p}{2}} |\Sigma|^{-\frac{n}{2}} e^{-\frac{1}{2} tr \Phi^{-1}(X-M)\Sigma^{-1}(X-M)'}. \qquad (4.3.11)$$

If we interchange the roles of X and M, then

$$p(M) \propto |\Phi|^{-\frac{p}{2}} |\Sigma|^{-\frac{n}{2}} e^{-\frac{1}{2} tr \Phi^{-1}(M-X)\Sigma^{-1}(M-X)'} \qquad (4.3.12)$$

thus implying that we should select our prior distribution for M from the Matrix Normal family.

We then select our prior distribution for M to be the Matrix Normal distribution

$$p(M) \propto |\chi|^{-\frac{p}{2}} |\Xi|^{-\frac{n}{2}} e^{-\frac{1}{2} tr \chi^{-1}(M-M_0)\Xi^{-1}(M-M_0)'}, \qquad (4.3.13)$$

where we have "enriched" the prior distribution with the use of M_0 so that it does not depend on the data X and made it independent of the other parameters Φ and Σ through Ξ and χ. The quantities M_0, Ξ, and χ are hyperparameters to be assessed. By specifying the matrix quantity M_0 and the matrix quantities Ξ and χ, the Matrix Normal prior distribution is completely determined.

Inverted Wishart

If we interchange the roles of X and Σ in the Matrix Normal likelihood and use the property of the trace operator, then

$$p(\Sigma) \propto |\Phi|^{-\frac{p}{2}} |\Sigma|^{-\frac{n}{2}} e^{-\frac{1}{2} tr \Sigma^{-1}(X-M)'\Phi^{-1}(X-M)} \qquad (4.3.14)$$

thus implying that we should select our prior distribution for Σ from the Inverted Wishart family.

We then select our prior distribution for Σ to be the Inverted Wishart distribution

$$p(\Sigma) \propto |\Sigma|^{-\frac{\nu}{2}} e^{-\frac{1}{2} tr \Sigma^{-1} Q}, \qquad (4.3.15)$$

where we have "enriched" the prior distribution with the use of Q and ν so that it does not depend on the data. The quantities ν and Q are hyperparameters to be assessed. By specifying the matrix quantity Q and the scalar quantity ν, the Inverted Wishart prior distribution is completely determined.

Taking the same Matrix Normal likelihood, and interchanging the role for Φ, we have

$$p(\Phi|X, M, \Sigma) \propto |\Phi|^{-\frac{p}{2}} |\Sigma|^{-\frac{n}{2}} e^{-\frac{1}{2} \Phi^{-1}(X-M) tr \Sigma^{-1}(X-M)'} \qquad (4.3.16)$$

thus implying that we should select our prior distribution for Φ from the Inverted Wishart family

$$p(\Phi|\kappa, \Psi) \propto |\Phi|^{-\frac{\kappa}{2}} e^{-\frac{1}{2} tr \Phi^{-1} \Psi}, \qquad (4.3.17)$$

where we have "enriched" the prior distribution with the use of Ψ and κ so that it does not depend on the data. The quantities κ and Ψ are hyperparameters to be assessed. By specifying the matrix quantity Ψ and the scalar quantity κ, the Inverted Wishart prior distribution is completely determined.

Using the generalized Conjugate procedure to obtain prior distributions, we obtain Table 4.6 where "IW" is used to denote Inverted Wishart and "GMN," Generalized Matrix Normal.

TABLE 4.6
Matrix variate generalized Conjugate priors.

Likelihood	Parameter(s)	Prior Family
Matrix Normal (Φ, Σ) known	M	GMN
Matrix Normal (M, Φ) known	Σ	Inverted Wishart
Matrix Normal (M, Σ) known	Φ	Inverted Wishart
Matrix Normal	(M, Φ, Σ)	GMN-IW-IW

4.4   Correlation Priors

In this section, Conjugate prior distributions are derived for the correlation coefficient between observation vectors. In the context of Bayesian Factor

Analysis, a Generalized Beta distribution has been used for the correlation coefficient ρ in the between vector correlation matrix Φ [50]. This was done when Φ was either the intraclass correlation matrix

$$\Phi = \begin{pmatrix} 1 & \rho & \rho & \cdots & \rho \\ & 1 & \rho & \cdots & \rho \\ & & \ddots & & \vdots \\ & & & & \rho \\ & & & & 1 \end{pmatrix} = (1-\rho)I_n + \rho e_n e_n', \qquad (4.4.1)$$

where e_n is a column vector of ones and $-\frac{1}{n-1} < \rho < 1$ or the first order Markov correlation matrix

$$\Phi = \begin{pmatrix} 1 & \rho & \rho^2 & \cdots & \rho^{n-1} \\ \rho & 1 & \rho & \cdots & \rho^{n-2} \\ \vdots & \vdots & \vdots & & \vdots \\ \rho^{n-1} & \rho^{n-2} & & \cdots & 1 \end{pmatrix}, \qquad (4.4.2)$$

where $0 < |\rho| < 1$.

When the Generalized Beta prior distribution and the likelihood are combined, the result is a posterior distribution which is unfamiliar. This unfamiliar posterior distribution required a rejection sampling technique to generate random variates as outlined in Chapter 6. A Conjugate prior distribution can be derived and the rejection sampling avoided.

Given X which follows a Matrix Normal distribution

$$p(X|M, \Sigma, \Phi) = (2\pi)^{-\frac{np}{2}} |\Phi|^{-\frac{p}{2}} |\Sigma|^{-\frac{n}{2}} e^{-\frac{1}{2} tr \Phi^{-1}(X-M)\Sigma^{-1}(X-M)'}, \qquad (4.4.3)$$

where $x = (X') = (x_1', ..., x_n')'$, $X' = (x_1, ..., x_n)$, $\mu = vec(M') = (\mu_1', ..., \mu_n')'$, $M' = (\mu_1, ..., \mu_n)$, and Φ is either an intraclass or first order Markov correlation matrix; the Conjugate prior distribution for ρ in Φ is found as follows.

4.4.1   Intraclass

If we determine the intraclass structure in Equation 4.4.1 that has the correlation between any two observations being the same, then we can use the result that the determinant of Φ has the form

$$|\Phi| = (1-\rho)^{n-1}[1+\rho(n-1)] \qquad (4.4.4)$$

and the result that the inverse of Φ has the form

$$\Phi^{-1} = \frac{I_n}{1-\rho} - \frac{\rho e_n e_n'}{(1-\rho)[1+(n-1)\rho]} \qquad (4.4.5)$$

which is again a matrix with intraclass correlation structure [41].

With these results, the Matrix Normal distribution can be written as

$$p(X|M,\Sigma,\Phi) \propto |\Phi|^{-\frac{p}{2}}|\Sigma|^{-\frac{n}{2}}e^{-\frac{1}{2}tr\Phi^{-1}(X-M)\Sigma^{-1}(X-M)'}$$

$$\propto |\Phi|^{-\frac{p}{2}}e^{-\frac{1}{2}tr\Phi^{-1}\Psi}$$

$$p(X|M,\Sigma,\rho) \propto (1-\rho)^{-\frac{(n-1)p}{2}}[1+(n-1)\rho]^{-\frac{p}{2}}$$

$$\times e^{-\frac{1}{2(1-\rho)}\left[k_1 - \frac{\rho k_2}{1+(n-1)\rho}\right]}, \qquad (4.4.6)$$

where $k_1 = tr(\Psi)$, $k_2 = tr(e_n e'_n \Psi)$, and $\Psi = (X-M)\Sigma^{-1}(X-M)'$. It should be noted that k_1 and k_2 can be written as

$$k_1 = \sum_{i=1}^{n} \Psi_{ii} \quad \text{and} \quad k_2 = \sum_{i'=1}^{n}\sum_{i=1}^{n} \Psi_{ii'}. \qquad (4.4.7)$$

We now implement the Conjugate procedure in order to obtain the prior distribution for the between vector correlation coefficient ρ. If we interchange the roles of X and ρ, then we obtain

$$p(\rho) \propto (1-\rho)^{-\frac{\alpha\beta}{2}}[1+\alpha\rho]^{-\frac{\beta}{2}}e^{-\frac{1}{2(1-\rho)}\left[k_1 - \frac{\rho k_2}{1+\alpha\rho}\right]}, \qquad (4.4.8)$$

where α, β, and Ψ for $k_1 = tr(\Psi)$, $k_2 = tr(e_n e'_n \Psi)$ are hyperparameters to be assessed. With appropriate choices of α and β, for example, $\alpha = n-1$ and $\beta = p$, this prior distribution is Conjugate for ρ.

4.4.2　Markov

If we determine the first order Markov structure in Equation 4.4.2 that has the correlation between observations decrease with the power of the difference between the observation numbers, then we can use the results [41] that the determinant of Φ has the form

$$|\Phi| = (1-\rho^2)^{n-1} \qquad (4.4.9)$$

and that the inverse of such a patterned matrix has the form

$$\Phi^{-1} = \frac{1}{1-\rho^2}\begin{pmatrix} 1 & -\rho & & & 0 \\ -\rho & (1+\rho^2) & -\rho & & \\ & \ddots & \ddots & \ddots & \\ & & (1+\rho^2) & -\rho \\ 0 & & & -\rho & 1 \end{pmatrix}. \qquad (4.4.10)$$

With these results, the Matrix Normal distribution can be written as

$$p(X|M,\Sigma,\Phi) \propto |\Phi|^{-\frac{p}{2}}|\Sigma|^{-\frac{n}{2}}e^{-\frac{1}{2}tr\Phi^{-1}(X-M)\Sigma^{-1}(X-M)'},$$

$$\propto |\Phi|^{-\frac{p}{2}}e^{-\frac{1}{2}tr\Phi^{-1}\Psi}$$

$$p(X|M,\Sigma,\rho) \propto (1-\rho^2)^{-\frac{(n-1)p}{2}}e^{-\frac{k_1-\rho k_2+\rho^2 k_3}{2(1-\rho^2)}}, \tag{4.4.11}$$

where

$$\Psi_1 = I_n, \quad \Psi_2 = \begin{pmatrix} 0 & 1 & & & 0 \\ 1 & 0 & 1 & & \\ & \ddots & \ddots & \ddots & \\ & & 0 & 1 \\ 0 & & & 1 & 0 \end{pmatrix}, \quad \Psi_3 = \begin{pmatrix} 0 & & & & 0 \\ & 1 & & & \\ & & \ddots & & \\ & & & 1 & \\ 0 & & & & 0 \end{pmatrix}, \tag{4.4.12}$$

$$k_1 = tr(\Psi_1\Phi) = \sum_{i=1}^{n}\Psi_{ii}, \tag{4.4.13}$$

$$k_2 = tr(\Phi_2\Psi) = \sum_{i=1}^{n-1}(\Psi_{i,i+1}+\Psi_{i+1,i}), \tag{4.4.14}$$

and

$$k_3 = tr(\Phi_3\Psi) = \sum_{i=2}^{n-1}\Psi_{ii}. \tag{4.4.15}$$

We now implement the Conjugate procedure in order to obtain the prior distribution for the between vector correlation coefficient ρ. If we interchange the roles of X and ρ, then we obtain

$$p(\rho) \propto (1-\rho^2)^{-\frac{\alpha\beta}{2}}e^{-\frac{k_1-\rho k_2+\rho^2 k_3}{2(1-\rho^2)}}, \tag{4.4.16}$$

where α, β, and Ψ for k_1, k_2, and k_3 defined above are hyperparameters to be assessed. With appropriate choices of α and β, for example, $\alpha = n-1$ and $\beta = p$, this prior distribution is Conjugate for ρ.

The vague priors are used when there is little or no specific knowledge as to various parameter values. The Conjugate prior distributions are used when we have specific knowledge as to parameter values either in the form of a previous similar experiment or from substantive expert beliefs. The generalized

Conjugate prior distributions are used in the rare situation where we believe that the Conjugate prior distributions are too restrictive to correctly assess prior information. Vague prior distributions are not used in this text because they may lead to nonunique solutions for every model except for the Bayesian Regression model.

Exercises

1. What is the family of Conjugate prior distributions for ϱ corresponding to a Scalar Binomial likelihood $p(x|\varrho)$?

2. What are the families of Conjugate prior distributions for μ and σ^2 corresponding to a Scalar Normal likelihood $p(x|\mu, \sigma^2)$?

3. What are the families of Conjugate prior distributions for μ and Σ corresponding to a Multivariate Normal likelihood $p(x|\mu, \Sigma)$?

5

Hyperparameter Assessment

5.1 Introduction

This Chapter describes methods for assessing the hyperparameters for prior distributions used to quantify available prior knowledge regarding parameter values. When observed data arise from Binomial, Scalar Normal, Multivariate Normal, and Matrix Normal distributions as in this text, the prior distributions of the parameters of these distributions contain parameters themselves termed hyperparameters. These prior distributions are quite often the Scalar Beta, Scalar Normal, Multivariate Normal, Matrix Normal, Inverted Gamma, and Inverted Wishart distributions. The hyperparameters of these prior distributions need to be assessed so that the prior distribution can be identified. There are two ways the hyperparameters can be assessed, either in a pure subjective way which expresses expert knowledge and beliefs or by use of data from a previous similar experiment.

Throughout this chapter, we will be in the predata acquisition stage of an experiment. We will quantify available prior knowledge regarding values of parameters of the model which is specified with a likelihood. We will quantify how likely the values of the parameters in the likelihood are, prior to seeing any current data. This can be accomplished by using data from a previous similar experiment or by using subjective expert opinion in the form of a virtual set of data.

5.2 Binomial Likelihood

Before performing a Binomial experiment and gathering data, we have foresight in knowing that a similar experiment has been carried out and data exist in the form of n_0 observations $x_1, \ldots, x_{n_0}$. The likelihood of these n_0 random variates is

$$p(x_1, \ldots, x_{n_0} | \varrho) \propto \varrho^{\sum_{i=1}^{n_0} x_i} (1 - \varrho)^{n_0 - \sum_{i=1}^{n_0} x_i}$$

$$\propto \varrho^x (1-\varrho)^{n_0-x}, \tag{5.2.1}$$

where the new variable $x = \sum_{i=1}^{n_0} x_i$ is the number of successes (heads) and the new variable $y = n_0 - x$ is the number of failures (tails). With the number of successes x in n_0 Bernoulli trials known, we now view ϱ, the probability of success, as the random variable with known parameters x and n_0.

It can be recognized that the random variable ϱ is (Scalar) Beta distributed.

5.2.1   Scalar Beta

The probability of success ϱ has the Beta distribution

$$p(\varrho) \propto \varrho^x (1-\varrho)^{n_0-x}, \tag{5.2.2}$$

which is compared to its typical parameterization

$$p(\varrho) \propto \varrho^{\alpha-1}(1-\varrho)^{\beta-1} \tag{5.2.3}$$

and it is seen that the hyperparameters α and β of the prior distribution for the probability of success ϱ are $\alpha = x+1$ and $\beta = y+1$.

The scalar hyperparameters α and β can also be assessed by purely subjective means. A substantive field expert can assess them in the following way.

Imagine that we are not able to visually inspect and have not observed any realizations from the coin being flipped. As described in [50], if we imagine a virtual flipping of the coin n_0 times, then let $x = \alpha - 1$ denote the number of virtual heads and $y = \beta - 1$ the number of virtual tails such that $x+y = n_0$. If $\alpha = \beta = 1$, then $n_0 = 0$, implying the absence of virtual coin flipping or the absence of specific prior information. This corresponds to a vague or Uniform prior distribution with mean $\frac{1}{2}$. The parameter values $\alpha = 200$ and $\beta = 100$ imply 199 heads and 99 tails which is strong prior information with a mean of $\frac{2}{3}$. The larger the virtual sample size n_0, the stronger the prior information we have and the more peaked the prior is around its mean.

5.3   Scalar Normal Likelihood

Before performing an experiment in which Scalar Normal random variates will result, we have foresight in knowing that a similar experiment has been carried out and data exist in the form of n_0 observations $x_1, \ldots, x_{n_0}$. The likelihood of these n_0 random variates is

$$p(x_1, \ldots, x_{n_0} | \mu, \sigma^2) \propto (\sigma^2)^{-\frac{n_0}{2}} e^{\frac{-\sum_{i=1}^{n_0}(x_i-\mu)^2}{2\sigma^2}}. \tag{5.3.1}$$

With the numerical values of the observations $x_1, \ldots, x_{n_0}$ known, we now view the parameters μ and σ^2 of this distribution as being the random variables with known parameters involving n_0 and $x_1, \ldots, x_{n_0}$.

By rearranging and performing some algebra on the above distribution, it can be seen that μ and σ^2 are Scalar Normal and Inverted Gamma distributed.

5.3.1   Scalar Normal

The random parameter μ from a sample of Scalar Normal random variates has a Scalar Normal distribution

$$p(\mu|\sigma^2) \propto (\sigma^2)^{-\frac{n_0}{2}} e^{-\frac{(\mu - \bar{x})^2}{2\sigma^2/n_0}} \qquad (5.3.2)$$

which is compared to its typical parameterization (where σ is used generically)

$$p(\mu|\sigma^2) \propto (\sigma^2)^{-\frac{n_0}{2}} e^{-\frac{(\mu - \mu_0)^2}{2\sigma^2}} \qquad (5.3.3)$$

and it is seen that the hyperparameter μ_0 of the prior distribution for the mean is $\mu_0 = \bar{x}$.

5.3.2   Inverted Gamma or Scalar Inverted Wishart

The random parameter σ^2 from a sample of Scalar Normal random variates has an Inverted Gamma distribution

$$p(\sigma^2) \propto (\sigma^2)^{-\frac{n_0}{2}} e^{-\frac{1}{2}\frac{\sum_{i=1}^{n_0}(x_i - \bar{x})^2}{\sigma^2}} \qquad (5.3.4)$$

which is compared to its typical parameterization

$$p(\sigma^2) \propto (\sigma^2)^{-\frac{\nu}{2}} e^{-\frac{1}{2}\frac{q}{\sigma^2}} \qquad (5.3.5)$$

and it is seen that the hyperparameters ν and q of the Inverted Gamma distribution given in Chapter 2, which here is a prior distribution for the variance σ^2, are $\nu = n_0$ and $q = \sum_{i=1}^{n_0}(x_i - \bar{x})^2$.

The scalar hyperparameters μ_0, ν, and q can also be assessed by purely subjective means. A substantive field expert can assess them in the following way.

If we imagine a virtual sample of size n_0, $x_1, \ldots, x_{n_0}$, then a substantive expert can determine a value of the mean of the sample data $\mu_0 = \bar{x}$ which would represent the most probable value to be the average (also a value he would expect since the mean and mode of the Scalar Normal distribution are identical). The substantive expert can also determine the most probable value for the variance of this virtual sample σ_0^2 and the hyperparameters ν and q are $\nu = n_0$ and $q = n_0 \sigma_0^2$.

5.4   Multivariate Normal Likelihood

Before performing an experiment in which Multivariate Normal random variates will result, we have foresight in knowing that a similar experiment has been carried out and data exist in the form of n_0 vector valued observations $x_1, \ldots, x_{n_0}$. The likelihood of these n_0 random variates is

$$p(x_1, \ldots, x_{n_0} | \mu, \Sigma) \propto |\Sigma|^{-\frac{n_0}{2}} e^{-\frac{1}{2} \sum_{i=1}^{n_0} (x_i - \mu)' \Sigma^{-1} (x_i - \mu)} \tag{5.4.1}$$

With the numerical values of the observations $x_1, \ldots, x_{n_0}$ known, we now view the parameters μ and Σ of this distribution as being the random variables with known parameters involving n_0 and $x_1, \ldots, x_{n_0}$.

By rearranging and performing some algebra on the above distribution, it can be seen that μ and Σ are Multivariate Normal and Inverted Wishart distributed.

5.4.1   Multivariate Normal

The random parameter μ from a sample of Multivariate Normal random variates of dimension p has a Multivariate Normal distribution

$$p(\mu | \Sigma) \propto |\Sigma|^{-\frac{n_0}{2}} e^{-\frac{1}{2} (\mu - \bar{x})' (\Sigma/n_0)^{-1} (\mu - \bar{x})} \tag{5.4.2}$$

which is compared to its typical parameterization (where Σ is used generically)

$$p(\mu | \Sigma) \propto |\Sigma|^{-\frac{n_0}{2}} e^{-\frac{1}{2} (\mu - \mu_0)' \Sigma^{-1} (\mu - \mu_0)} \tag{5.4.3}$$

and it is seen that the hyperparameter μ_0 of the prior distribution for the mean is $\mu_0 = \bar{x}$.

5.4.2   Inverted Wishart

The random parameter Σ from a sample of Multivariate Normal random variates has an Inverted Wishart distribution

$$p(\Sigma) \propto |\Sigma|^{-\frac{n_0}{2}} e^{-\frac{1}{2} tr \Sigma^{-1} \sum_{i=1}^{n_0} (x_i - \bar{x})(x_i - \bar{x})'} \tag{5.4.4}$$

which is compared to its typical parameterization

$$p(\Sigma) \propto |\Sigma|^{-\frac{\nu}{2}} e^{-\frac{1}{2} tr \Sigma^{-1} Q} \tag{5.4.5}$$

and it is seen that the hyperparameters ν and Q of the prior distribution for the covariance matrix Σ are $\nu = n_0$ and $Q = \sum_{i=1}^{n_0} (x_i - \bar{x})(x_i - \bar{x})'$.

The vector, scalar, and matrix hyperparameters μ_0, ν, and Q can also be assessed by purely subjective means. A substantive field expert can assess them in the following way.

If we imagine a virtual sample of size n_0, $x_1, \ldots, x_{n_0}$, then a substantive expert can determine a value of the mean of the sample data $\mu_0 = \bar{x}$ which would represent the most probable value to be the average (also a value he would expect since the mean and mode of the Multivariate Normal distribution are identical). The substantive expert can also determine the most probable value for the covariance matrix of this virtual sample Σ_0 and the hyperparameters ν and Q are $\nu = n_0$ and $Q = n_0 \Sigma_0$.

5.5 Matrix Normal Likelihood

Before performing an experiment in which Matrix Normal random variates will result, we have foresight in knowing that a similar experiment has been carried out and data exist in the form of n_0 matrix valued observations $X_1, \ldots, X_{n_0}$ of dimension n_1 by p_1. The likelihood of these n_0 random variates is

$$p(X_1, \ldots, X_{n_0} | M, \Sigma, \Phi) \propto |\Sigma|^{-\frac{n_0 n_1}{2}} |\Phi|^{-\frac{n_0 p_1}{2}} e^{-\frac{1}{2} \sum_{i=1}^{n_0} tr \Phi^{-1}(X_i - M)\Sigma^{-1}(X_i - M)'}.$$

(5.5.1)

With the numerical values of the observations $X_1, \ldots, X_{n_0}$ known, we now view the parameters M, Σ, and Φ of this distribution as being the random variables with known parameters involving n_0 and $X_1, \ldots, X_{n_0}$.

By rearranging and performing some algebra on the above distribution, it can be seen that M, Σ, and Φ are Matrix Normal, Inverted Wishart, and Inverted Wishart distributed.

5.5.1 Matrix Normal

The random parameter M from a sample of Matrix Normal random variates has a Matrix Normal distribution

$$p(M | \Sigma, \Phi) \propto e^{-\frac{1}{2} tr \Phi^{-1}(M - \bar{X})(\Sigma/n_0)^{-1}(M - \bar{X})'}$$

(5.5.2)

which is compared to its typical parameterization (where Σ is used generically)

$$p(M | \Sigma, \Phi) \propto e^{-\frac{1}{2} tr \Phi^{-1}(M - M_0)\Sigma^{-1}(M - M_0)'}$$

(5.5.3)

and it is seen that the hyperparameter M_0 of the prior distribution for the mean is $M_0 = \bar{X}$.

5.5.2   Inverted Wishart

The random parameter Σ from a sample of Matrix Normal random variates has an Inverted Wishart distribution

$$p(\Sigma) \propto |\Sigma|^{-\frac{n_0 n_1}{2}} |\Phi|^{-\frac{n_0 p_0}{1}} e^{-\frac{1}{2} tr \Sigma^{-1} \sum_{i=1}^{n_0} (X_i - \bar{X})' \Phi^{-1}(X_i - \bar{X})} \qquad (5.5.4)$$

which is compared to its typical parameterization

$$p(\Sigma) \propto |\Sigma|^{-\frac{\nu}{2}} e^{-\frac{1}{2} tr \Sigma^{-1} Q} \qquad (5.5.5)$$

and it is seen that the hyperparameters ν and Q of the prior distribution for the covariance matrix Σ are $\nu = n_0 n_1$ and $Q = \sum_{i=1}^{n_0} (X_i - \bar{X})' \Phi^{-1}(X_i - \bar{X})$.

Similarly, the random parameter Φ from a sample of Matrix Normal random variates has an Inverted Wishart distribution

$$p(\Phi) \propto |\Sigma|^{-\frac{n_0 n_1}{2}} |\Phi|^{-\frac{n_0 p_0}{1}} e^{-\frac{1}{2} tr \Phi^{-1} \sum_{i=1}^{n_0} (X_i - \bar{X}) \Sigma^{-1}(X_i - \bar{X})'} \qquad (5.5.6)$$

which is compared to its typical parameterization

$$p(\Phi) \propto |\Phi|^{-\frac{\kappa}{2}} e^{-\frac{1}{2} tr \Phi^{-1} \Psi} \qquad (5.5.7)$$

and it is seen that the hyperparameters κ and Ψ of the prior distribution for the covariance matrix Φ are $\kappa = n_0 p_1$ and $\Psi = \sum_{i=1}^{n_0} (X_i - \bar{X}) \Sigma^{-1}(X_i - \bar{X})'$.

Note that the equations for Q and Ψ are coupled. This means that there is not a closed form analytic solution for estimating Φ and Σ. Their values must be computed in an iterative fashion with an initial value similar to the ICM algorithm which will be presented in Chapter 6.

The matrix, scalar, matrix, scalar, and matrix hyperparameters M_0, ν, Q, κ, Ψ, can also be assessed by purely subjective means. A substantive field expert can assess them in the following way.

If we imagine a virtual sample of size n_0, $X_1, \ldots, X_{n_0}$, then a substantive expert can determine a value of the mean of the sample data $M_0 = \bar{X}$ which would represent the most probable value to be the average (also a value he would expect since the mean and mode of the Matrix Normal distribution are identical). The substantive expert can also determine the most probable value for the covariance matrix of this virtual sample Σ_0 and the hyperparameters ν and Q are $\nu = n_0 n_1$ and $Q = n_0 \Sigma_0$. The substantive expert can also determine the most probable value for the covariance matrix of this virtual sample Φ_0 and the hyperparameters κ and Ψ are $\kappa = n_0 p_1$ and $\Psi = n_0 \Phi_0$.

Exercises

1. Assume that we have $n_0 = 25$ (virtual or actual data) observations from a Binomial experiment with $x = 15$ successes. What values would you assess for the hyperparameters α and β for a Conjugate Beta prior distribution on ϱ, the probability of success?

2. Assume that we have $n_0 = 30$ (virtual or actual data) observations for a Scalar Normal variate with sample mean and sample sum of square deviates given by

$$\bar{x} = 50, \qquad \sum_{i=1}^{30}(x_i - \bar{x})^2 = 132.$$

What value would you assess for the hyperparameter μ_0 of a Conjugate Scalar Normal prior distribution for the mean μ and what hyperparameters ν and q of the Conjugate Inverse Gamma prior distribution for the variance σ^2?

3. Assume that we have $n_0 = 50$ (virtual or actual data) observations for a Multivariate Normal variate with sample mean vector and sample sum of square deviates matrix given by

$$\bar{x} = \begin{pmatrix} 50 \\ 100 \\ 75 \end{pmatrix}, \qquad \sum_{i=1}^{30}(x_i - \bar{x})(x_i - \bar{x})' = \begin{pmatrix} 50.000 & 12.500 & 3.125 \\ 12.500 & 50.000 & 12.500 \\ 3.125 & 12.500 & 50.000 \end{pmatrix}.$$

What value would you assess for the vector hyperparameter μ_0 of a Conjugate Multivariate Normal prior distribution for the mean vector μ and what scalar and matrix hyperparameters ν and Q of the Conjugate Inverse Gamma prior distribution for the covariance matrix Σ?

6

Bayesian Estimation Methods

In this Chapter we define the two methods of parameter estimation which are used in this text, namely, marginal posterior mean and joint maximum a posteriori estimators. Typically these estimators are found by integration and differentiation to arrive at explicit equations for their computation. There are often instances where explicit closed form equations are not possible. In these instances, numerical integration and maximization estimation procedures are required. The typical explicit integration and differentiation procedures are discussed as are the numerical estimation procedures used. The numerical estimation procedures are Gibbs sampling for sampling based marginal posterior means and the iterated conditional modes algorithm (ICM) for joint maximum posterior (joint posterior modal) estimates.

6.1 Marginal Posterior Mean

Often we have a set of parameters, $\theta = (\theta_1, \ldots, \theta_J)$ in our posterior distribution $p(\theta|X)$ where X represents the data which may be a collection of scalar, vector, or matrix observations. The marginal posterior distribution of any of the parameters, say θ_j, can be obtained by integrating $p(\theta|X)$ with respect to all parameters except θ_j. That is, the marginal posterior distribution of θ_j is

$$p(\theta_j|X) = \int p(\theta_1, \ldots, \theta_J|X) \, d\theta_1 \ldots d\theta_{j-1} \, d\theta_{j+1} \ldots d\theta_J \qquad (6.1.1)$$

where the integral is evaluated over the appropriate range of the set parameters. After calculating the marginal posterior distribution for each of the parameters, marginal posterior estimators such as

$$\hat{\theta}_j = E(\theta_j|X) = \int \theta_j p(\theta_j|X) d\theta_j \qquad (6.1.2)$$

can be calculated which is the marginal mean estimator.

6.1.1   Matrix Integration

When computing marginal distributions [42], joint posterior distributions are integrated with respect to scalar, vector, or matrix variates. Let's consider the variates to be of the matrix form and perform integration. Scalar and vector analogs follow as special cases. Integration of a posterior distribution with respect to a matrix variate is typically carried out first by algebraic manipulation of the integrand and finally by recognition.

To motivate the integration with respect to matrices, consider the problem of estimating the p dimensional mean vector μ and covariance matrix Σ from a Multivariate Normal distribution $p(x|\mu, \Sigma)$. Available prior knowledge regarding the mean vector and covariance matrix is quantified in the form of the joint (Conjugate) prior distribution $p(\mu, \Sigma)$ and a random sample $x_1, \dots, x_n$ is taken with likelihood

$$p(x_1, \dots, x_n | \mu, \Sigma) = \prod_{i=1}^{n} p(x_i | \mu, \Sigma). \tag{6.1.3}$$

The posterior distribution is given by

$$p(\mu, \Sigma | X) \propto p(\mu, \Sigma) p(X | \mu, \Sigma) \tag{6.1.4}$$

and marginal posterior distributions

$$p(\mu | X) = \int p(\mu, \Sigma) p(X | \mu, \Sigma) \, d\Sigma, \tag{6.1.5}$$

$$p(\Sigma | X) = \int p(\mu, \Sigma) p(X | \mu, \Sigma) \, d\mu, \tag{6.1.6}$$

where the random sample denoted by $X' = (x_1, \dots, x_n)$. The first integral is taken over the set of all p-dimensional positive definite symmetric matrices and the second over p-dimensional real space.

In general, if we were presented with a joint posterior distribution $p(\theta | X)$ that was a function of $\theta = (\theta_1, \theta_2)$, the marginal posterior distribution of θ_1 is found by integration with respect to θ_2 as

$$p(\theta_1 | X) = \int p(\theta | X) \, d\theta_2. \tag{6.1.7}$$

This integration is often carried out by algebraically manipulating the terms in $p(\theta | X)$ to write it as

$$p(\theta | X) = g(\theta_1 | X) h(\theta_2 | \theta_1, X), \tag{6.1.8}$$

where $h(\theta_2 | \theta_1, X)$ is recognized as being a known distribution except for a multiplicative normalizing constant with respect to θ_2. The multiplicative normalizing constant $k(\theta_1|X)$ can depend on the parameter θ_1 and the data X but not on θ_2. The posterior distribution is such that

$$p(\theta|X) = \frac{g(\theta_1|X)}{k(\theta_1|X)} k(\theta_1|X) h(\theta_2|\theta_1, X) \qquad (6.1.9)$$

and the integration is carried out by taking those terms that do not depend on θ_2 out of the integrand and then recognizing that the integrand is unity. Mathematically this procedure is described as

$$p(\theta_1|X) = \int \frac{g(\theta_1|X)}{k(\theta_1|X)} k(\theta_1|X) h(\theta_2|\theta_1, X) \, d\theta_2 \qquad (6.1.10)$$

$$= \frac{g(\theta_1|X)}{k(\theta_1|X)} \int k(\theta_1|X) h(\theta_2|\theta_1, X) \, d\theta_2 \qquad (6.1.11)$$

$$= \frac{g(\theta_1|X)}{k(\theta_1|X)}. \qquad (6.1.12)$$

The integral will be the integral of a probability distribution function that we recognize as unity.

Integration is similarly performed when integrating with respect to θ_1 to determine the marginal posterior distribution of θ_2. This method also applies when $\theta = (\theta_1, \theta_2, \ldots, \theta_J)$.

In computing marginal posterior distributions for the mean vector and covariance matrix of a Multivariate Normal distribution with Conjugate priors, integration of the joint posterior distribution will be carried out with respect to the error covariance matrix Σ to find the marginal posterior distribution $p(\mu|X)$ of the mean vector μ. The integration is as follows.

The joint posterior distribution of the mean vector and covariance matrix is

$$p(\mu, \Sigma|X) \propto |\Sigma|^{-\frac{(n+\nu+1)}{2}} e^{-\frac{1}{2} tr \Sigma^{-1}[(X-e_n\mu')'(X-e_n\mu')+(\mu-\mu_0)(\mu-\mu_0)'+Q]},$$

$$(6.1.13)$$

which upon inspection is an Inverted Wishart distribution except for a normalizing constant

$$k(\mu|X) = |(X - e_n\mu')'(X - e_n\mu') + (\mu - \mu_0)(\mu - \mu_0)' + Q|^{\frac{\nu_* - p - 1}{2}},$$

$$(6.1.14)$$

where $\nu_* = n + \nu + 1$.

The joint posterior distribution is written as

$$p(\mu, \Sigma | X) \propto |(X - e_n \mu')'(X - e_n \mu') + (\mu - \mu_0)(\mu - \mu_0)' + Q|^{-\frac{\nu_* - p - 1}{2}}$$
$$\times |(X - e_n \mu')'(X - e_n \mu') + (\mu - \mu_0)(\mu - \mu_0)' + Q|^{\frac{\nu_* - p - 1}{2}}$$
$$\times |\Sigma|^{-\frac{\nu_*}{2}} e^{-\frac{1}{2} tr \Sigma^{-1} [(X - e_n \mu')'(X - e_n \mu') + (\mu - \mu_0)(\mu - \mu_0)' + Q]}.$$

(6.1.15)

Upon integrating this joint posterior distribution, the marginal posterior distribution of the mean vector μ is

$$p(\mu | X) \propto \int p(\mu, \Sigma | X) \, d\Sigma$$
$$\propto |(X - e_n \mu')'(X - e_n \mu') + (\mu - \mu_0)(\mu - \mu_0)' + Q|^{-\frac{\nu_* - p - 1}{2}}$$
$$\times \int |(X - e_n \mu')'(X - e_n \mu') + (\mu - \mu_0)(\mu - \mu_0)' + Q|^{\frac{\nu_* - p - 1}{2}}$$
$$\times |\Sigma|^{-\frac{\nu_*}{2}} e^{-\frac{1}{2} tr \Sigma^{-1} [(X - e_n \mu')'(X - e_n \mu') + (\mu - \mu_0)(\mu - \mu_0)' + Q]} d\Sigma$$

(6.1.16)

$$\propto \frac{1}{|(X - e_n \mu')'(X - e_n \mu') + (\mu - \mu_0)(\mu - \mu_0)' + Q|^{\frac{\nu_* - p - 1}{2}}},$$

(6.1.17)

where the integral was recognized as being an Inverted Wishart distribution except for its proportionality constant which did not depend on Σ. Upon performing some matrix algebra, the above yields a marginal posterior distribution which is recognized as being a Multivariate Student t-distribution.

Similarly, integration is performed with respect to the mean vector μ.

6.1.2   Gibbs Sampling

Gibbs sampling [13, 14] is a stochastic integration method that draws random variates from the posterior conditional distribution for each of the parameters conditional on fixed values of all the other parameters and the data X. Let $p(\theta | X)$ be the posterior distribution of the parameters where θ is the set of parameters and X is the data. Let θ be partitioned as $\theta = (\theta_1, \theta_2, \ldots, \theta_J)$ into J groups of parameters. Ideally, we would like to perform the integration of the joint posterior distribution to obtain marginal posterior distributions

$$p(\theta_j | X) = \int p(\theta_1, \ldots, \theta_J | X) \, d\theta_1 \ldots d\theta_{j-1} \, d\theta_{j+1} \ldots d\theta_J$$

(6.1.18)

and marginal posterior mean estimates

$$E(\theta_j | X) = \int \theta_j p(\theta_j | X) d\theta_j.$$

(6.1.19)

Unfortunately, these integrations are usually of very high dimension and not always available in a closed form. This is why we need the Gibbs sampling procedure. With the random variates drawn from the posterior conditional distributions

$$p(\theta_j|\theta_1,\ldots,\theta_{j-1},\theta_{j+1},\ldots,\theta_J,X) = \frac{p(\theta_1,\ldots,\theta_{j-1},\theta_j,\theta_{j+1},\ldots,\theta_J,|X)}{p(\theta_1,\ldots,\theta_{j-1},\theta_{j+1},\ldots,\theta_J|X)}$$
$$\propto p(\theta_1,\ldots,\theta_{j-1},\theta_j,\theta_{j+1},\ldots,\theta_J,|X)$$

$$(6.1.20)$$

we can determine the marginal posterior distributions (Equation 6.1.18) and any marginal posterior quantities such as the marginal posterior means (Equation 6.1.19).

For the Gibbs sampling, we begin with an initial value for the parameters

$$\bar{\theta}^{(0)} = (\bar{\theta}_1^{(0)},\bar{\theta}_2^{(0)},\ldots,\bar{\theta}_J^{(0)}),$$

and at the lth iteration define

$$\bar{\theta}^{(l+1)} = (\bar{\theta}_1^{(l+1)},\bar{\theta}_2^{(l+1)},\ldots,\bar{\theta}_J^{(l+1)})$$

by the values from

$$\bar{\theta}_1^{(l+1)} = \text{a random variate from } p(\bar{\theta}_1|\bar{\theta}_2^{(l)},\bar{\theta}_3^{(l)},\ldots,\bar{\theta}_J^{(l)},X), \qquad (6.1.21)$$
$$\bar{\theta}_2^{(l+1)} = \text{a random variate from } p(\bar{\theta}_2|\bar{\theta}_1^{(l+1)},\bar{\theta}_3^{(l)},\ldots,\bar{\theta}_J^{(l)},X), \qquad (6.1.22)$$
$$\vdots$$
$$\bar{\theta}_J^{(l+1)} = \text{a random variate from } p(\bar{\theta}_J|\bar{\theta}_1^{(l+1)},\bar{\theta}_2^{(l+1)},\ldots,\bar{\theta}_{J-1}^{(l+1)},X), \quad (6.1.23)$$

that is, at each step l drawing a random variate from the associated conditional posterior distribution. To apply this method we need to determine the posterior conditional of each θ_j, the posterior distribution of each θ_j conditional on the fixed values of all the other elements of θ and X from $p(\theta|X)$.

After drawing $s + L$ random variates of each we will have $\bar{\theta}^{(1)},\bar{\theta}^{(2)},\ldots,$ $\bar{\theta}^{(s+1)},\ldots,\bar{\theta}^{(s+L)}$. The first s random variates called the "burn in" are discarded and the remaining L variates are kept.

It has been shown [14] that under mild conditions the L randomly sampled variates for each of the parameters constitute a random sample from the corresponding marginal posterior distribution given the data and that for any measurable function of the sample values whose expectation exists, the average of the function of the sample values converges almost surely to the expected value of the population parameter values.

The marginal posterior distributions (Equation 6.1.18) are computed to be

$$\bar{p}(\theta_j|X) = \frac{1}{L}\sum_{l=1}^{L}\delta\left(\theta_j - \bar{\theta}_j^{(s+l)}\right),\ j=1,\ldots,J, \qquad (6.1.24)$$

where $\delta(\cdot)$ denotes the Kronecker delta function

$$\delta\left(\theta_j - \bar{\theta}_j^{(s+l)}\right) = \begin{cases} 1, & \theta_j = \bar{\theta}_j^{(s+l)} \\ 0, & \text{otherwise} \end{cases} \qquad (6.1.25)$$

and the marginal posterior mean estimators of the parameters (Equation 6.1.19) are computed to be $\bar{\theta} = (\bar{\theta}_1,\ldots,\bar{\theta}_J)$ where

$$\bar{\theta}_j = \bar{E}(\theta_j|X)$$
$$= \frac{1}{L}\sum_{l=1}^{L}\bar{\theta}_j^{(s+l)},\quad j=1,\ldots,J. \qquad (6.1.26)$$

The marginal posterior estimators of the variances of the parameter can be similarly found as

$$\overline{var}(\theta_j|X) = \frac{1}{L}\sum_{l=1}^{L}\left(\bar{\theta}_j^{(s+l)}\right)^2 - \left(\frac{1}{L}\sum_{l=1}^{L}\bar{\theta}_j^{(s+l)}\right)^2 \qquad (6.1.27)$$

if θ_j is a scalar variate,

$$\overline{var}(\theta_j|X) = \frac{1}{L}\sum_{l=1}^{L}\left(\bar{\theta}_j^{(s+l)}\right)\left(\bar{\theta}_j^{(s+l)}\right)' - \bar{\theta}_j\bar{\theta}_j' \qquad (6.1.28)$$

if θ_j is a vector variate, and

$$\overline{var}(\theta_j|X) = \frac{1}{L}\sum_{l=1}^{L}vec\left(\bar{\theta}_j^{(s+l)}\right)vec\left(\bar{\theta}_j^{(s+l)}\right)' - vec\left(\bar{\theta}_j\right)vec\left(\bar{\theta}_j\right)' \qquad (6.1.29)$$

if θ_j is a matrix variate. In fact, the posterior estimate of any function of the parameters can be found.

Credibility interval estimates can also be found with the use of nonparametric techniques and all of the retained sample variates. In practice, a distributional specification can be used with appropriate marginal posterior moments to define it. For example, instead of retaining L random matrix variates of dimension $p \times (q+1)$ for the matrix of regression coefficients, the marginal posterior mean and covariance matrices are used in a Normal distribution. This is reasonable since the Conjugate prior and the posterior conditional distributions for the matrix of Regression coefficients are both Normal.

6.1.3 Gibbs Sampling Convergence

The Gibbs sampling procedure in the current form was developed [14, 19] as a way to avoid direct multidimensional integration.

It is well known that the full posterior conditional distributions uniquely determine the full joint distribution when the random variables have a joint distribution whose distribution function is strictly positive over the sample space [13]. Since the posterior conditionals uniquely determine the full joint distribution, they also uniquely determine the posterior marginals. It was shown that under mild conditions, the following results are true [14].

Result 1 (Convergence)
The randomly generated variates from the posterior conditional distributions, $(\bar{\theta}_1^{(l)}, \bar{\theta}_2^{(l)}, \ldots, \bar{\theta}_J^{(l)})$ converge in distribution to the true parameter values $(\theta_1, \theta_2, \ldots, \theta_J)$. This convergence is denoted by

$$(\bar{\theta}_1^{(l)}, \bar{\theta}_2^{(l)}, \ldots, \bar{\theta}_J^{(l)}) \xrightarrow{d} (\theta_1, \theta_2, \ldots, \theta_J)$$

and hence for each j, the average of the random variates $\bar{\theta}_j^{(l)}$ converges to its corresponding parameter value θ_j which has the distribution $p(\theta_j)$, written as $\bar{\theta}_j^{(l)} \xrightarrow{d} \theta_j \sim p(\theta_j)$ as $l \to \infty$.

Result 2 (Rate)
Using the sup norm, the joint posterior distribution of $(\bar{\theta}_1^{(l)}, \bar{\theta}_2^{(l)}, \ldots, \bar{\theta}_J^{(l)})$ converges to the true joint posterior distribution $p(\theta_1, \theta_2, \ldots, \theta_J)$ at a geometric rate in l, when visiting in order.

Result 3 (Ergodic Theorem)
For any measurable function T of the parameter values $(\bar{\theta}_1, \bar{\theta}_2, \ldots, \bar{\theta}_J)$ whose expectation exists, the average of the measurable functions of the sample variates, as the number of sample variates tends toward infinity, converges almost surely to its expected value. This is expressed as

$$\lim_{L \to \infty} \frac{1}{L} \sum_{l=1}^{L} T(\bar{\theta}_1^{(l)}, \bar{\theta}_2^{(l)}, \ldots, \bar{\theta}_J^{(l)}) \xrightarrow{a.s.} E(T(\theta_1, \theta_2, \ldots, \theta_J)).$$

converges almost surely [47] to its expectation.

With these results, we are guaranteed convergence of the Gibbs sampling estimation method.

6.1.4 Normal Variate Generation

The generation of random variates from Normal distributions is described in terms of the Matrix Normal distribution with vector and scalar distributions as special cases. The mathematical symbols here are used generically.

An $n \times p$ random Matrix Normal variate X with mean matrix M_X and covariance matrix $\Phi \otimes \Sigma$ can be generated from np independent scalar standard Normal variates with mean zero and variance one. This is performed with the following Matrix Normal property.

If Y_X is an $n \times p$ random matrix whose elements are independent standard Normal scalar variates, written

$$Y_X \sim N(0, I_n \otimes I_p) \tag{6.1.30}$$

and then using a transformation of variable result [17]

$$X = A_X Y_X B'_X + M_X \sim N(M_X, A_X A'_X \otimes B_X B'_X), \tag{6.1.31}$$

where M is the mean matrix, $\Phi = A_X A'_X$, and $\Sigma = B_X B'_X$. The covariance matrices Σ and Φ have been factorized using a method such as a Cholesky factorization also called decomposition [32] or a factorization by eigenvalues and eigenvectors [53].

It was assumed that a method is available to generate standard Normal scalar variates. If a Normal random variate generation method is not available, these variates may be generated as follows [5, 22]. Generate two variates y_1 and y_2 which are Uniform on the unit interval. Define

$$x_1 = (-2 \log y_1)^{\frac{1}{2}} \cos(2\pi y_2) \tag{6.1.32}$$

$$x_2 = (-2 \log y_1)^{\frac{1}{2}} \sin(2\pi y_2). \tag{6.1.33}$$

Then, x_1 and x_2 are independent Scalar Normal variates with mean 0 and variance 1. Now, only a method to generate Uniform random variates on the unit interval is needed.

6.1.5   Wishart and Inverted Wishart Variate Generation

A $p \times p$ random Wishart matrix variate G or a $p \times p$ random Inverse Wishart matrix variate Σ can be generated as follows. By generating a $\nu_0 \times p$ standard Matrix Normal variate Y_G as above and then using the transformation of variable result [17], a Wishart distributed matrix variate

$$Y'_G Y_G \sim W(I_p, p, \nu_0) \tag{6.1.34}$$

can be generated. Upon using the transformation of variable result [41]

$$G = A_G (Y'_G Y_G) A'_G \sim W(A_G A'_G, p, \nu_0), \tag{6.1.35}$$

where $\Upsilon = A_G A'_G$ has been factorized using technique such as the Cholesky factorization [32] or using eigenvalues and eigenvectors [53].

Inverted Wishart matrix variates can be generated by first generating a $\nu_0 \times p$ standard Matrix Normal variate Y_Σ as above and then using the transformation of variable result [17]

$$(Y_\Sigma'Y_\Sigma)^{-1} \sim IW(I_p, p, \nu) \qquad (6.1.36)$$

and finally the transformation of variable result [41]

$$\Sigma = A_\Sigma(Y_\Sigma'Y_\Sigma)^{-1}A_\Sigma' \sim IW(A_\Sigma A_\Sigma', p, \nu), \qquad (6.1.37)$$

where $\nu_0 = \nu - p - 1$ and $Q = A_\Sigma A_\Sigma'$ has been factorized.

If the degrees of freedom ν_0 is not an integer, then random Wishart variates can be generated with the use of independent Gamma (Scalar Wishart) distributed variates with real valued parameters.

6.1.6   Factorization

In implementing the Gibbs sampling algorithm, matrix factorizations have to be computed. Two possibilities are the Cholesky and Eigen factorizations. In the following, assume that we wish to factor the $p \times p$ covariance matrix Σ.

6.1.6.1   Cholesky Factorization

Cholesky's method for factorizing a symmetric positive definite matrix Σ of dimension p is very straightforward. This factorization $\Sigma = A_\Sigma A_\Sigma'$ has the property that A_Σ be a lower triangular matrix. Denote the ij^{th} element of Σ and A_Σ to be σ_{ij} and a_{ij} respectively. Simple formulas [32] for the method are

$$a_{11} = \sqrt{\sigma_{11}} \qquad (6.1.38)$$

$$a_{ii} = \sqrt{\sigma_{ii} - \sum_{k=1}^{i-1} a_{ik}^2} \qquad i = 2, \ldots, n \qquad (6.1.39)$$

$$a_{i1} = \frac{\sigma_{i1}}{a_{11}} \qquad i = 1, \ldots, n \qquad (6.1.40)$$

$$a_{ij} = \frac{1}{a_{jj}}\left(\sigma_{ij} - \sum_{k=1}^{j-1} a_{ik}a_{jk}\right) \qquad i = j+1, \ldots, n; j \geq 2. \qquad (6.1.41)$$

6.1.6.2   Eigen Factorization

The matrix Σ can also be factorized using eigenvalues and eigenvectors as

$$\Sigma = (WD_\theta^{\frac{1}{2}})(WD_\theta^{\frac{1}{2}})' \qquad (6.1.42)$$

$$= A_\Sigma A_\Sigma', \qquad (6.1.43)$$

where the columns of W are the orthonormal eigenvectors which sequentially maximize the percent of variation and D_θ is a diagonal matrix with elements θ_j which are the eigenvalues.

The vector w_1 is now determined to be that vector that maximizes the variance subject to $w_1'w_1 = 1$. The method of Lagrange multipliers is applied

$$\frac{\partial}{\partial w_1}[w_1'\Sigma w_1 - \theta_1(w_1'w_1 - 1)] = 2(\Sigma - \theta_1 I_p)w_1 = 0$$

and since $w_1 \neq 0$, there can only be a solution if

$$|\Sigma - \theta_1 I_p| = 0.$$

It is apparent that θ_1 must be an eigenvalue of Σ, and w_1 is a normalized eigenvector of Σ. There are p such eigenvalues that satisfy the equation. The largest is selected. The other rows of W are found in a similar fashion with the additional constraints that they are orthogonal to the previous ones. For a more detailed account of the procedure refer to [41]. Previous work [53] used this Eigen factorization in the context of factorizing separable matrices $\Omega = \Phi \otimes \Sigma$ in which the covariance matrices Φ and Σ were patterned with exact known formulas for computing the eigenvectors and eigenvalues. Only the eigenvalues and eigenvectors of Φ and Σ were needed and not of Ω.

Occasionally a matrix factorization will be represented as

$$\Sigma = (WD_\theta^{\frac{1}{2}})(WD_\theta^{\frac{1}{2}})' \qquad (6.1.44)$$

$$= (\Sigma^{\frac{1}{2}})(\Sigma^{\frac{1}{2}})' \qquad (6.1.45)$$

or the factorization of an inverse as

$$\Sigma^{-1} = (WD_\theta^{-\frac{1}{2}})(WD_\theta^{-\frac{1}{2}})' \qquad (6.1.46)$$

$$= (\Sigma^{-\frac{1}{2}})(\Sigma^{-\frac{1}{2}})'. \qquad (6.1.47)$$

These are refered to as the square root matrices.

6.1.7   Rejection Sampling

Random variates can also be generated from an arbitrary distribution function by using the rejection sampling method [16, 40, 48]. Occasionally the (posterior conditional) distribution of a model parameter is not recognized as one of the well-known standard distributions from which we can easily generate random variates. When this is the case, a rejection sampling technique can be employed.

Assume that we are able to generate a random variate from the distribution $f(x)$ whose support (range of x values for which the distribution function is nonzero) is the same as $p(x)$. If the support of $f(x)$ is not the same as that of $p(x)$, then a truncated distribution function which restricts the range of values of $f(x)$ to that of $p(x)$ can be used. Further assume that $p(x) \leq cf(x)$ for all x and a positive constant c.

The random variate from $f(x)$ can be used to generate a random variate from $p(x)$. We generate a random variate y_0 from $f(y)$ and can retain this generated variate as being from $p(y)$. The rejection sampling is a simple two step process [48] which proceeds as follows.

Step 1: Generate a random variate y_0 from convenient distribution $f(y)$ and independently generate u_0 from a Uniform distribution on the unit interval.

Step 2: If the Uniform random variate $u_0 \leq \frac{p(y_0)}{cf(y_0)}$ where c is a constant, then let $x_0 = y_0$. If not repeat step 1.

The retained random variate x_0 generated by the above rejection sampling process is a random variate from $p(x)$. This can be shown to be true in the following manner.

Denote the retained variate by x_0. Let N denote the number of iterations of the above steps required to retain x_0, and y_N denote a variate which took N iterations to be retained. The probability of the retained value x_0 being less than another value x is

$$
\begin{aligned}
P[x_0 \leq x] &= P[y_N \leq x] \\
&= P\left[y_0 \leq x \middle| u_0 \leq \frac{p(y_0)}{cf(y_0)}\right] \\
&= \frac{1}{k} P\left[y_0 \leq x, u_0 \leq \frac{p(y_0)}{cf(y_0)}\right] \\
&= \frac{1}{k} \int_{-\infty}^{x} f(y) \int_{0}^{\frac{p(y)}{cf(y)}} du\, dy \\
&= \frac{1}{k} \int_{-\infty}^{x} \frac{p(y)}{cf(y)} f(y)\, dy \\
&= \frac{1}{kc} \int_{-\infty}^{x} p(y)\, dy,
\end{aligned}
\tag{6.1.48}
$$

where $k = P[u_0 \leq \frac{p(y_0)}{cf(y_0)}]$ (the probability of the event on the right side of the conditioning in the second line of Equation 6.1.48). We can see that by letting $x \to \infty$, $k = \frac{1}{c}$. The constant c does not have to be $\frac{1}{k}$. In fact, it can be any number greater than $\frac{1}{k}$, but the smaller the value of c, the greater the probability of retaining a generated random variate.

The above concept is illustrated in the following example.

Example:

Let's use the rejection sampling technique to generate a random variate from the Beta distribution

$$p(\varrho) = k\varrho^{\alpha-1}(1-\varrho)^{\beta-1}, \tag{6.1.49}$$

where $\alpha > 1$, $\beta > 1$, $0 < \varrho < 1$, and k is the proportionality constant. Note that when α and β are integers, $k = \frac{(\alpha+\beta-1)!}{(\alpha-1)!(\beta-1)!}$.

Since the Beta distribution is confined to the unit interval, a good choice for the convenient distribution from which we can generate random variates is the Uniform distribution

$$f(\varrho) = 1, \tag{6.1.50}$$

where $0 < \varrho < 1$.

Without knowing the proportionality constant for $p(\varrho)$, we can determine the rejection criteria $u_0 \leq \frac{p(y_0)}{cf(y_0)}$ of step 2. This is done by finding the maximum value of

$$\frac{p(\varrho)}{f(\varrho)} = k\varrho^{\alpha-1}(1-\varrho)^{\beta-1} \tag{6.1.51}$$

by differentiation with respect to ϱ which yields

$$\frac{d}{d\varrho}\frac{p(\varrho)}{f(\varrho)} = k[(\alpha-1)\varrho^{\alpha-2}(1-\varrho)^{\beta-1} - (\beta-1)\varrho^{\alpha-1}(1-\varrho)^{\beta-2}]. \tag{6.1.52}$$

Upon setting this derivative equal to zero, the maximum is seen to be the mode of the Beta distribution

$$\frac{\alpha-1}{\alpha+\beta-2} \tag{6.1.53}$$

which gives

$$\frac{p(\varrho)}{f(\varrho)} \leq k\left(\frac{\alpha-1}{\alpha+\beta-2}\right)^{\alpha-1}\left(1 - \frac{\alpha-1}{\alpha+\beta-2}\right)^{\beta-1} = c. \tag{6.1.54}$$

Therefore the ratio,

$$\frac{p(\varrho)}{cf(\varrho)} = \left[\left(\frac{\alpha-1}{\alpha+\beta-2}\right)^{\alpha-1}\left(1 - \frac{\alpha-1}{\alpha+\beta-2}\right)^{\beta-1}\right]^{-1} \varrho^{\alpha-1}(1-\varrho)^{\beta-1} \tag{6.1.55}$$

and the rejection sampling procedure proceeds as follows.

Step 1: Generate random Uniform variates u_1 and u_2.
Step 2: If

$$u_2 \leq \left[\left(\frac{\alpha-1}{\alpha+\beta-2}\right)^{\alpha-1}\left(1 - \frac{\alpha-1}{\alpha+\beta-2}\right)^{\beta-1}\right]^{-1} u_1^{\alpha-1}(1-u_1)^{\beta-1}, \tag{6.1.56}$$

then let the retained random variate from $p(\varrho)$ be $x_0 = u_1$. If not repeat step 1.

The average number of times that step 1 will be performed is c given above. Thus, the smaller the value of c, the fewer times step 1 will be performed on average.

The rejection sampling procedure can be generalized to a Multivariate formulation to generate random vectors [40].

6.2 Maximum a Posteriori

The joint posterior distribution may also be jointly maximized with respect to the parameters. If again $\theta = (\theta_1, \ldots, \theta_J)$, then maximum a posteriori (MAP) estimators (the analog of maximum likelihood estimators for posterior distributions) can be found by differentiation.

To determine joint maximum a posteriori estimators, differentiate the joint posterior distribution

$$p(\theta_1, \ldots, \theta_J | X) \tag{6.2.1}$$

with respect to each of the parameters, set the result equal to zero as

$$\frac{\partial}{\partial \theta_1} p(\theta_1, \ldots, \theta_J | X) \Big|_{\theta_1 = \hat{\theta}_1, \ldots, \theta_J = \hat{\theta}_J} = \cdots \frac{\partial}{\partial \theta_J} p(\theta_1, \ldots, \theta_J | X) \Big|_{\theta_1 = \hat{\theta}_1, \ldots, \theta_J = \hat{\theta}_J} = 0, \tag{6.2.2}$$

and solve the resulting system of J equations with J unknowns to find the joint maximum a posteriori estimators

$$\hat{\theta}_j = \underset{\theta_j}{\text{Arg Max}} \, p(\theta_j | \hat{\theta}_1, \ldots, \hat{\theta}_J, X) \tag{6.2.3}$$

$$= \hat{\theta}_j(\hat{\theta}_1, \ldots, \hat{\theta}_J, X) \tag{6.2.4}$$

for each of the parameters. In addition to the joint posterior modal estimates, conditional maximum a posteriori quantities such as variances can also be found as

$$var(\theta_j | \hat{\theta}_1, \ldots, \hat{\theta}_J, X) = \int (\theta_j - \bar{\theta}_j)^2 p(\theta_j | \hat{\theta}_1, \ldots, \hat{\theta}_J, X) \, d\theta_j, \tag{6.2.5}$$

where $\bar{\theta}_j$ is the conditional posterior mean which is often the mode.

6.2.1 Matrix Differentiation

When determining maxima of scalar functions of matrices by differentiation, the logarithm and trace are often used because maximizing a function
is equivalent to maximizing its logarithm due to monotonicity. Some matrix
(vector) derivative results that are often used [18, 41, 46] are

1. For general θ, $|\theta| > 0$,
   $\frac{\partial}{\partial \theta} \log |\theta| = (\theta^{-1})'$.

2. For symmetric $\theta = \theta'$, $|\theta| > 0$,
   $\frac{\partial}{\partial \theta} \log |\theta| = [2\theta^{-1} - diag(\theta^{-1})]$.

3. For general A and $\theta, |\theta| > 0$,
   $\frac{\partial}{\partial \theta} tr\left(\theta^{-1} A\right) = -\theta^{-1} A \theta^{-1}$.

4. For symmetric $A = A'$ and $\theta = \theta', |\theta| > 0$,
   $\frac{\partial}{\partial \theta} tr\left(\theta A\right) = -[2A^{-1} - diag(A^{-1})]$.

5. For symmetric $A = A'$, $A : q \times q$, $B = B'$, $B : p \times p$, and general $\theta, \theta_0 : p \times q$,
   then $\frac{\partial}{\partial \theta} tr\left[A(\theta - \theta_0)'B(\theta - \theta_0)\right] = 2B(\theta - \theta_0)A$.

where $diag(\cdot)$ denotes the diagonalization operator which forms a diagonal
matrix consisting only of the diagonal elements of its matrix argument and θ
has been used generically to denote either a matrix, a vector, or a scalar.

To verify that these estimators are maxima and not minima, the Hessian
matrix of second derivatives is computed. If the Hessian matrix is negative
definite, then the estimators are maxima and not minima.

6.2.2 Iterated Conditional Modes (ICM)

Iterated Conditional Modes [36, 40] is a deterministic optimization method
that finds the joint posterior modal estimators also known as the maximum a
posteriori estimates of $p(\theta|X)$ where θ denotes the collection of scalar, vector,
or matrix parameters, and X denotes the data. It is useful when the system of
J equations from differentiation does not yield closed form analytic equations
for each of the maxima.

Assume that $\theta = (\theta_1, \theta_2)$ where θ_1 and θ_2 are scalars and the posterior
distribution of θ is $p(\theta_1, \theta_2|X)$. We have a surface in 3-dimensional space.
We have θ_1 along one axis and θ_2 along the other with $p(\theta_1, \theta_2|X)$ being the
height of the surface or hill.

We want to find the top of the hill which is the same as finding the peak
or maximum of the function $p(\theta_1, \theta_2|X)$ with respect to both θ_1 and θ_2. As
usual, the maximum of a surface is found by differentiating with respect to
each variable (direction) and setting the result equal to zero.

The maximum of the function $p(\theta_1, \theta_2|X)$ with respect to each of the vari-
ables are

$$\tilde{\theta}_1^{(l+1)} = \overset{\text{Arg Max}}{\theta_1} \, p(\theta_1|\tilde{\theta}_2^{(l)}, X) \qquad (6.2.6)$$

$$= \tilde{\theta}_1(\tilde{\theta}_2^{(l)}, X) \qquad (6.2.7)$$

$$\tilde{\theta}_2^{(l+1)} = \overset{\text{Arg Max}}{\theta_2} \, p(\theta_2|\tilde{\theta}_1^{(l+1)}, X) \qquad (6.2.8)$$

$$= \tilde{\theta}_2(\tilde{\theta}_1^{(l+1)}, X) \qquad (6.2.9)$$

which satisfies

$$\frac{\partial}{\partial\theta_1} p(\theta_1, \theta_2|X)\Big|_{\theta_1=\tilde{\theta}_1} = \frac{\partial}{\partial\theta_2} p(\theta_1, \theta_2)\Big|_{\theta_2=\tilde{\theta}_2} = 0, \qquad (6.2.10)$$

which is the same as

$$\frac{\partial}{\partial\theta_1} p(\theta_1|\theta_2, X)p(\theta_2|X)\Big|_{\theta_1=\tilde{\theta}_1} = \frac{\partial}{\partial\theta_2} p(\theta_2|\theta_1, X)p(\theta_1|X)\Big|_{\theta_2=\tilde{\theta}_2} = 0 \quad (6.2.11)$$

or

$$p(\theta_2|X)\frac{\partial}{\partial\theta_1} p(\theta_1|\theta_2, X)\Big|_{\theta_1=\tilde{\theta}_1} = p(\theta_1|X)\frac{\partial}{\partial\theta_2} p(\theta_2|\theta_1, X)\Big|_{\theta_2=\tilde{\theta}_2} = 0 \quad (6.2.12)$$

assuming that $p(\theta_1|X) \neq 0$ and $p(\theta_2|X) \neq 0$.

We can obtain the posterior conditionals (functions) $p(\theta_1|\theta_2, X)$ and $p(\theta_2|\theta_1, X)$ along with their respective modes (maximum) $\tilde{\theta}_1 = \tilde{\theta}_1(\theta_2, X)$ and $\tilde{\theta}_2 = \tilde{\theta}_2(\theta_1, X)$.

We have the maximum of θ_1, $\tilde{\theta}_1$ for a given value of (conditional on) θ_2, and the maximum of θ_2, $\tilde{\theta}_2$ for a given value of (conditional on) θ_1.

The optimization procedure consists of

1. Select an initial value for θ_2; call it $\tilde{\theta}_2^{(0)}$.

2. Calculate the modal (maximal) value of $p(\theta_1|\tilde{\theta}_2^{(0)}, X)$, $\tilde{\theta}_1^{(1)}$.

3. Calculate the modal (maximal) value of $p(\theta_2|\tilde{\theta}_1^{(1)}, X)$, $\tilde{\theta}_2^{(1)}$.

4. Continue to calculate the remainder of the sequence $\tilde{\theta}_1^{(1)}, \tilde{\theta}_2^{(1)}, \tilde{\theta}_1^{(2)}, \tilde{\theta}_2^{(2)}, \ldots$ until convergence is reached.

If the posterior conditional distributions are not unimodal, we may converge to a local maximum and not the global maximum. If each of the posterior conditionals are unimodal, then the Hessian matrix is negative definite and the converged maximum is always the global maximum.

When convergence is reached, the point estimators $(\tilde{\theta}_1, \tilde{\theta}_2)$ are the maximum a posteriori estimators.

This method can be generalized to more than two parameters [40]. If θ is partitioned by $\theta = (\theta_1, \theta_2, \ldots, \theta_J)$ into J groups of parameters, we begin with a starting point $\tilde{\theta}^{(0)} = (\tilde{\theta}_1^{(0)}, \tilde{\theta}_2^{(0)}, \ldots, \tilde{\theta}_J^{(0)})$ and at the lth iteration define $\tilde{\theta}^{(l+1)}$ by

$$\tilde{\theta}_1^{(l+1)} = \underset{\theta_1}{\text{Arg Max}} \; p(\theta_1 | \tilde{\theta}_2^{(l)}, \ldots, \tilde{\theta}_J^{(l)}, X) \tag{6.2.13}$$

$$= \tilde{\theta}_1(\tilde{\theta}_2^{(l)}, \tilde{\theta}_3^{(l)}, \ldots, \tilde{\theta}_J^{(l)}, X) \tag{6.2.14}$$

$$\tilde{\theta}_2^{(l+1)} = \underset{\theta_2}{\text{Arg Max}} \; p(\theta_2 | \tilde{\theta}_1^{(l+1)}, \tilde{\theta}_3^{(l)}, \ldots, \tilde{\theta}_J^{(l)}, X) \tag{6.2.15}$$

$$= \tilde{\theta}_2(\tilde{\theta}_1^{(l+1)}, \tilde{\theta}_3^{(l)}, \ldots, \tilde{\theta}_J^{(l)}, X) \tag{6.2.16}$$

$$\vdots$$

$$\tilde{\theta}_J^{(l+1)} = \underset{\theta_J}{\text{Arg Max}} \; p(\theta_J | \tilde{\theta}_1^{(l+1)}, \ldots, \tilde{\theta}_{J-1}^{(l+1)}, X) \tag{6.2.17}$$

$$= \tilde{\theta}_1(\tilde{\theta}_2^{(l+1)}, \tilde{\theta}_3^{(l+1)}, \ldots, \tilde{\theta}_{J-1}^{(l+1)}, X) \tag{6.2.18}$$

at each step computing the maximum or mode. To apply this method we need to determine the functions $\tilde{\theta}_j$ which give the maximum of $p(\theta | X)$ with respect to θ_j, conditional on the fixed values of all the other elements of θ.

6.3   Advantages of ICM over Gibbs Sampling

We will show that when Φ is a general symmetric covariance matrix (or known), each of the posterior conditional distributions are unimodal. Thus we do not have to worry about local maxima; we will find global maxima. The reason one would use a stochastic procedure like Gibbs sampling over a deterministic procedure like ICM is to eliminate the possibility of converging to a local mode when the conditional posterior distribution is multimodal.

ICM is slightly simpler to implement than Gibbs and less computationally intensive because Gibbs sampling requires generation of random variates from the conditionals which includes matrix factorizations. ICM simply has to cycle through the posterior conditional modes and convergence is not uncertain as it is with Gibbs sampling. With ICM, we can check for convergence, say every 1000 iterations, by computing the difference between $\theta_j^{(1000l)}$ and $\theta_j^{(1000(l+1))}$ for every j, and if each element is the same to the third decimal, we can claim convergence and stop. This reduces computation time.

ICM should be implemented cautiously when Φ is a correlation matrix such as a first order Markov or an intraclass matrix with a single unknown parameter. The posterior conditional is not necessarily unimodal. ICM might converge to a local maxima. This could however be combated by an exhaustive search over the entire interval for which the single parameter is defined.

6.4 Advantages of Gibbs Sampling over ICM

When the posterior conditionals are not recognizable as unimodal distributions, we might prefer to use a stochastic procedure like Gibbs sampling to eliminate the possibility of converging to a local maxima. Although Gibbs sampling is more computationally intensive than ICM, it is a more general method and gives us more information such as marginal posterior point and interval estimates. Gibbs sampling allows us to make inferences regarding a parameter unconditional of the other parameters.

Exercises

1. Use the following 12 independent random Uniform variates in the interval $(0,1)$ to generate 12 independent random $N(0,1)$ variates.

   TABLE 6.1
   Twelve independent random Uniform variates.

0.9501	0.4565	0.2311	0.0185	0.6068	0.8214
0.4860	0.4447	0.8913	0.6154	0.7621	0.7919

2. Compute the Cholesky factorization $A_\Sigma A'_\Sigma$ of the positive definite symmetric covariance matrix

$$\Sigma = \begin{pmatrix} 2 & 0 & 1 \\ 0 & 4 & 2 \\ 1 & 2 & 3 \end{pmatrix}.$$

3. Using the first three independent random Scalar Normal variates ($N(0,1)$'s) from Exercise 1, generate a 3-dimensional random Multivariate Normal vector valued variate x with mean vector and covariance matrix

$$\mu = \begin{pmatrix} 5 \\ 3 \\ 7 \end{pmatrix} \qquad \Sigma = A_\Sigma A'_\Sigma,$$

   where A_Σ is computed from Exercise 2 above.

4. Compute the Eigen factorization $A_\Sigma A'_\Sigma$ of

$$Q = \begin{pmatrix} 100 & 10 & 10 \\ 10 & 100 & 10 \\ 10 & 10 & 100 \end{pmatrix}.$$

5. Using the last nine independent random Scalar Normal variates ($N(0,1)$'s) from Exercise 1, generate a random 3×3 Inverse Wishart matrix variate Σ with scale matrix $Q = A_\Sigma A'_\Sigma$ given in Exercise 4 above and $\nu = 5$ degrees of freedom.

6. Using the first two independent Uniformly distributed random variates in Exercise 1, generate a random variate from a $B(\alpha = 2, \beta = 2)$ distribution by using the rejection sampling technique.

7. Assume that we have a posterior distribution

$$p(\mu, \sigma^2 | x_1, \ldots, x_n) \propto (\sigma^2)^{-\frac{n+\nu+1}{2}} e^{-\frac{g}{2\sigma^2}},$$

where

$$g = \sum_{i=1}^{n} (x_i - \bar{x})^2 + (\mu - \mu_0)^2 + q.$$

Derive an ICM algorithm for obtaining joint maximum a posteriori estimates of μ and σ^2.

7

Regression

7.1   Introduction

The purpose of this Chapter is to quickly review the Classical Multivariate Regression model before the presentation of Multivariate Bayesian Regression in Chapter 8. This is accomplished by a build up which starts with the Scalar Normal Samples model through to the Simple and Multiple Regression models and finally to the Multivariate Regression model. For each model, estimation and inference is discussed. This is fundamental knowledge which is essential to successfully understanding Part II of the text.

7.2   Normal Samples

Consider the following independent and identically distributed random variables $x_1, \ldots, x_n$ from a Scalar Normal distribution with mean μ and variance σ^2 denoted by $N(\mu, \sigma^2)$. This model can be written in terms of a linear model similar to Regression. This is identical to the linear model

$$x_i = \mu + \epsilon_i, \qquad (7.2.1)$$

where ϵ_i is the Scalar Normally distributed random error with mean zero and variance σ^2 denoted

$$\epsilon_i \sim N(0, \sigma^2) \qquad (7.2.2)$$

for $i = 1, \ldots, n$. The joint distribution of the variables or likelihood is

$$p(x_1, \ldots, x_n | \mu, \sigma^2) = \prod_{i=1}^{n} p(x_i | \mu, \sigma^2)$$

$$= \prod_{i=1}^{n} (2\pi\sigma^2)^{-\frac{1}{2}} e^{-\frac{(x_i - \mu)^2}{2\sigma^2}}$$

$$= (2\pi\sigma^2)^{-\frac{n}{2}} e^{-\frac{1}{2\sigma^2}\sum_{i=1}^{n}(x_i-\mu)^2}. \tag{7.2.3}$$

This model can also be written in terms of vectors

$$x = \begin{pmatrix} x_1 \\ \vdots \\ x_n \end{pmatrix}, \quad e_n = \begin{pmatrix} 1 \\ \vdots \\ 1 \end{pmatrix}, \quad \epsilon = \begin{pmatrix} \epsilon_1 \\ \vdots \\ \epsilon_n \end{pmatrix}, \tag{7.2.4}$$

so that the model is

$$\begin{array}{ccccc} x & = & e_n & \mu & + & \epsilon, \\ n \times 1 & & n \times 1 & 1 \times 1 & & n \times 1 \end{array} \tag{7.2.5}$$

and likelihood is

$$p(x|\mu,\sigma^2) = (2\pi\sigma^2)^{-\frac{n}{2}} e^{-\frac{(x-e_n\mu)'(x-e_n\mu)}{2\sigma^2}}, \tag{7.2.6}$$

where " $'$ " denotes the transpose.

The natural logarithm of $p(x|\mu,\sigma^2)$ can be taken and differentiated with respect to scalars μ and σ^2 in order to obtain values which maximize the likelihood. These are maximum likelihood estimates. However, some algebra on the exponent can also be performed to find the value of μ which yields the maximal value of this likelihood.

$$\begin{aligned}
(x-e_n\mu)'(x-e_n\mu) &= x'x - x'\mu e_n - e_n'\mu x + \mu e_n' e_n \mu \\
&= \mu[e_n'e_n\mu - e_n'x] - x'e_n\mu + x'x \\
&= \mu(e_n'e_n)[\mu - (e_n'e_n)^{-1}e_n'x] - x'e_n\mu + x'x \\
&= \mu(n)[\mu - (n)^{-1}e_n'x] - x'e_n\mu + x'x \\
&= \mu(n)[\mu - \hat{\mu}] - (n)\hat{\mu}\mu + x'x \\
&= (\mu - \hat{\mu})(n)(\mu - \hat{\mu}) + (n)\hat{\mu}\mu - \hat{\mu}(n)\hat{\mu} - (n)\hat{\mu}\mu + x'x \\
&= (\mu - \hat{\mu})(n)(\mu - \hat{\mu}) - \hat{\mu}(n)\hat{\mu} + x'x \\
&= (\mu - \hat{\mu})(n)(\mu - \hat{\mu}) - (n)^{-1}x'e_n e_n'x + x'x \\
&= (\mu - \hat{\mu})(n)(\mu - \hat{\mu}) + x'(I_n - e_n e_n'/n)x, \tag{7.2.7}
\end{aligned}$$

where $\hat{\mu} = (n)^{-1}x'e_n = \bar{x}$. This could have been written with the vector generalization μ' but was not since $\mu' = \mu$.

The likelihood is now

$$p(x|\mu,\sigma^2) = (2\pi\sigma^2)^{-\frac{n}{2}} e^{-\frac{1}{2\sigma^2}[n(\mu-\hat{\mu})^2 + x'\left[I_n - \frac{e_n e_n'}{n}\right]x]}. \tag{7.2.8}$$

It is now obvious that the value of the mean μ which maximizes the likelihood or minimizes the exponent in the likelihood is $\hat{\mu} = \bar{x}$.

Upon differentiating the natural logarithm of the likelihood denoted by $LL = \log[p(x|\mu,\sigma^2)]$ with respect to σ^2, then evaluating it at the maximal values of the parameters, and setting it equal to zero

$$\left.\frac{\partial}{\partial\sigma^2}LL\right|_{\mu=\hat\mu,\sigma^2=\hat\sigma^2} = 0, \tag{7.2.9}$$

we obtain the maximum likelihood estimate of the variance σ^2,

$$\hat\sigma^2 = \frac{(x-e_n\hat\mu)'(x-e_n\hat\mu)}{n}. \tag{7.2.10}$$

Consider the numerator of the estimate $\hat\sigma^2$.

$$\begin{aligned}
NUM &= (x-e_n\hat\mu)'(x-e_n\hat\mu) \\
&= x'x - x'e_n\hat\mu - (e_n\hat\mu)'x + (e_n\hat\mu)'(e_n\hat\mu) \\
&= \hat\mu'[e_n'e_n\hat\mu - e_n'x] - x'e_n\hat\mu + x'x \\
&= \hat\mu'(e_n'e_n)[\hat\mu - (e_n'e_n)^{-1}e_n'x] - x'e_n\hat\mu + x'x \\
&= \hat\mu'(n)[\hat\mu - (n)^{-1}e_n'x] - x'e_n\hat\mu + x'x \\
&= [\hat\mu - (n)^{-1}e_n'x]'(n)[\hat\mu - (n)^{-1}e_n'x] + (n)^{-1}x'e_n(n)\hat\mu \\
&\quad -(n)^{-1}x'e_n(n)(n)^{-1}e_n'x - x'e_n\hat\mu + x'x \\
&= x'\left[I_n - \frac{e_n e_n'}{n}\right]x. \tag{7.2.11}
\end{aligned}$$

This is exactly the g term in the likelihood. Now the likelihood can be written as

$$p(x|\mu,\sigma^2) = (2\pi\sigma^2)^{-\frac{n}{2}}e^{-\frac{1}{2\sigma^2}[n(\mu-\hat\mu)^2+g]}. \tag{7.2.12}$$

In the likelihood for Normal observations as written immediately above, if the normalization coefficients are ignored by using proportionality, it can be partitioned and viewed as a joint distribution of $\hat\mu$ and g. This then becomes

$$\begin{aligned}
p(\hat\mu,g|\mu,\sigma^2) &\propto (\sigma^2)^{-\frac{1}{2}}e^{-\frac{(\hat\mu-\mu)^2}{2\sigma^2/n}}(\sigma^2)^{-\frac{n-1}{2}}g^{\frac{n-1-2}{2}}e^{-\frac{g}{2\sigma^2}} \\
&= \underbrace{\frac{e^{-\frac{(\hat\mu-\mu)^2}{2\sigma^2/n}}}{(2\pi\sigma^2/n)^{\frac{1}{2}}}}_{\hat\mu|\mu,\sigma^2\sim N(\mu,\sigma^2/n)}\underbrace{\frac{(\sigma^2)^{-\frac{n-1}{2}}g^{\frac{n-1-2}{2}}e^{-\frac{g}{2\sigma^2}}}{\Gamma\left(\frac{n-1}{2}\right)2^{\frac{n-1}{2}}}}_{g|\sigma^2\sim W(\sigma^2,1,n-1)}. \tag{7.2.13}
\end{aligned}$$

It should be noted that the estimates of the mean $\hat{\mu}$ and the variance $\hat{\sigma}^2$ are independent by the Neyman factorization criterion [22, 47]. This is proven by showing that

$$p(\hat{\mu}, g | \mu, \sigma^2) = p(\hat{\mu} | \mu, \sigma^2) p(g | \mu, \sigma^2). \tag{7.2.14}$$

If the joint distribution $p(\hat{\mu}, g | \mu, \sigma^2)$ is integrated with respect to g, then

$$
\begin{aligned}
p(\hat{\mu} | \mu, \sigma^2) &= \int p(\hat{\mu}, g | \mu, \sigma^2) \, dg \\
&= \int \frac{e^{-\frac{(\hat{\mu}-\mu)^2}{2\sigma^2/n}}}{(2\pi\sigma^2/n)^{\frac{1}{2}}} \frac{(\sigma^2)^{-\frac{n-1}{2}} g^{\frac{n-1-2}{2}} e^{-\frac{g}{2\sigma^2}}}{\Gamma\left(\frac{n-1}{2}\right) 2^{\frac{n-1}{2}}} \, dg \\
&= \frac{e^{-\frac{(\hat{\mu}-\mu)^2}{2\sigma^2/n}}}{(2\pi\sigma^2/n)^{\frac{1}{2}}} \int \frac{(\sigma^2)^{-\frac{n-1}{2}} g^{\frac{n-1-2}{2}} e^{-\frac{g}{2\sigma^2}}}{\Gamma\left(\frac{n-1}{2}\right) 2^{\frac{n-1}{2}}} \, dg \\
&= \frac{e^{-\frac{(\hat{\mu}-\mu)^2}{2\sigma^2/n}}}{(2\pi\sigma^2/n)^{\frac{1}{2}}}, \tag{7.2.15}
\end{aligned}
$$

and thus,

$$\hat{\mu} | \mu, \sigma^2 \sim N\left(\mu, \sigma^2/n\right). \tag{7.2.16}$$

If the joint distribution $p(\hat{\mu}, g | \mu, \sigma^2)$ is integrated with respect to $\hat{\mu}$, then

$$
\begin{aligned}
p(g | \sigma^2) &= \int p(\hat{\mu}, g | \mu, \sigma^2) \, d\hat{\mu} \\
&= \int \frac{e^{-\frac{(\hat{\mu}-\mu)^2}{2\sigma^2/n}}}{(2\pi\sigma^2/n)^{\frac{1}{2}}} \frac{(\sigma^2)^{-\frac{n-1}{2}} g^{\frac{n-1-2}{2}} e^{-\frac{g}{2\sigma^2}}}{\Gamma\left(\frac{n-1}{2}\right) 2^{\frac{n-1}{2}}} \, d\hat{\mu} \\
&= \frac{(\sigma^2)^{-\frac{n-1}{2}} g^{\frac{n-1-2}{2}} e^{-\frac{g}{2\sigma^2}}}{\Gamma\left(\frac{n-1}{2}\right) 2^{\frac{n-1}{2}}} \int \frac{e^{-\frac{(\hat{\mu}-\mu)^2}{2\sigma^2/n}}}{(2\pi\sigma^2/n)^{\frac{1}{2}}} \, d\hat{\mu} \\
&= \frac{(\sigma^2)^{-\frac{n-1}{2}} g^{\frac{n-1-2}{2}} e^{-\frac{g}{2\sigma^2}}}{\Gamma\left(\frac{n-1}{2}\right) 2^{\frac{n-1}{2}}}, \tag{7.2.17}
\end{aligned}
$$

and thus,

$$g | \sigma^2 \sim W(\sigma^2, 1, n-1). \tag{7.2.18}$$

The joint distribution of μ and g is equal to the product of their marginal distributions. Note that g/σ^2 has the familiar Chi-squared distribution with $n-1$ degrees of freedom.

By changing variables from $\hat{\mu}$ and g to $t = (n-1)^{\frac{1}{2}} g^{-1/2} n^{\frac{1}{2}} (\hat{\mu} - \mu)$ and $w = g$, the joint distribution of t and w becomes

$$p(t, w | \mu, \sigma^2) = \frac{(\sigma^2)^{-\frac{n}{2}} w^{\frac{n-2}{2}} e^{-\frac{w}{2\sigma^2} \left(\frac{t^2}{n-1} + 1 \right)}}{[(n-1)\pi]^{\frac{1}{2}} \Gamma \left(\frac{n-1}{2} \right) 2^{\frac{n}{2}}} \tag{7.2.19}$$

where the Jacobian of the transformation was

$$J(\hat{\mu}, g \rightarrow t, w) = (n-1)^{-\frac{1}{2}} w^{\frac{1}{2}} n^{-\frac{1}{2}}. \tag{7.2.20}$$

Now integrate with respect to w to find the distribution of t (unconditional on w). This is done by making the integrand look like the Scalar Wishart distribution. It is seen that the integrand needs to be multiplied and divided by the same factor.

$$p(t|\mu) = \int p(t, w | \mu, \sigma^2) \, dw$$

$$\propto \int (\sigma^2)^{-\frac{n}{2}} w^{\frac{n-2}{2}} e^{-\frac{w}{2\sigma^2} \left(\frac{t^2}{n-1} + 1 \right)} \, dw$$

$$\propto \left(\frac{t^2}{n-1} + 1 \right)^{-\frac{n}{2}} \int \left(\frac{t^2}{n-1} + 1 \right)^{\frac{n}{2}} (\sigma^2)^{-\frac{n}{2}} w^{\frac{n-2}{2}} e^{-\frac{w}{2\sigma^2} \left(\frac{t^2}{n-1} + 1 \right)} \, dw$$

$$\propto \left(\frac{t^2}{n-1} + 1 \right)^{-\frac{n}{2}} \tag{7.2.21}$$

and now transforming back while taking g to be known

$$p(\hat{\mu} | \mu, g) \propto \left[1 + ng^{-1} (\hat{\mu} - \mu)^2 \right]^{-\frac{n}{2}}$$

$$\propto \left[1 + \frac{1}{n-1} \left(\frac{\hat{\mu} - \mu}{\hat{\sigma}/\sqrt{(n-1)}} \right)^2 \right]^{-\frac{n}{2}}$$

$$\propto \left[1 + \frac{1}{n-1} \left(\frac{\hat{\mu} - \mu}{\hat{\sigma}/\sqrt{(n-1)}} \right)^2 \right]^{-\frac{(n-1)+1}{2}}. \tag{7.2.22}$$

Recall the Scalar Student t-distribution in the statistical distribution Chapter. The estimate of the mean $\hat{\mu}$ given the estimated variance has the familiar Scalar Student t-distribution with $n-1$ degrees of freedom.

7.3   Simple Linear Regression

The Simple Linear Regression model is

$$x_i = \beta_0 + \beta_1 u_{i1} + \epsilon_i \tag{7.3.1}$$

where ϵ_i is the Scalar Normally distributed random error term with mean zero and variance σ^2,

$$\epsilon_i \sim N(0, \sigma^2) \tag{7.3.2}$$

for $i = 1, \ldots, n$. The joint distribution of the variables or likelihood is

$$
\begin{aligned}
p(x_1, \ldots, x_n | \beta_0, \beta_1, \sigma^2, u_{11}, \ldots, u_{n1}) &= \prod_{i=1}^{n} p(x_i | \beta_0, \beta_1, \sigma^2, u_{i1}) \\
&= \prod_{i=1}^{n} (2\pi\sigma^2)^{-\frac{1}{2}} e^{-\frac{(x_i - \beta_0 - \beta_1 u_{i1})^2}{2\sigma^2}} \\
&= (2\pi\sigma^2)^{-\frac{n}{2}} e^{-\frac{1}{2\sigma^2} \sum_{i=1}^{n} (x_i - \beta_0 - \beta_1 u_{i1})^2}.
\end{aligned}
\tag{7.3.3}
$$

This can be written in terms of vectors and matrices

$$
x = \begin{pmatrix} x_1 \\ \vdots \\ x_n \end{pmatrix}, \quad
u_i = \begin{pmatrix} 1 \\ u_{i1} \end{pmatrix}, \quad
U = \begin{pmatrix} u_1' \\ \vdots \\ u_n' \end{pmatrix}, \quad
\beta = \begin{pmatrix} \beta_0 \\ \beta_1 \end{pmatrix}, \quad
\epsilon = \begin{pmatrix} \epsilon_1 \\ \vdots \\ \epsilon_n \end{pmatrix}, \tag{7.3.4}
$$

so that the model is

$$
\begin{array}{ccccc}
x & = & U & \beta & + & \epsilon \\
n \times 1 & & n \times 2 & 2 \times 1 & & n \times 1
\end{array}
\tag{7.3.5}
$$

and the likelihood is

$$p(x | \beta, \sigma^2, U) = (2\pi\sigma^2)^{-\frac{n}{2}} e^{-\frac{(x - U\beta)'(x - U\beta)}{2\sigma^2}}. \tag{7.3.6}$$

The natural logarithm of $p(x | \beta, \sigma^2, U)$ can be taken and differentiated with respect to the vector β and the scalar σ^2 in order to obtain values of β and σ^2 which maximize the likelihood. These are maximum likelihood estimates.

However, some algebra on the exponent can also be performed to find the value of β which yields the maximal value of this likelihood.

$$
\begin{aligned}
(x-U\beta)'(x-U\beta) &= x'x - x'U\beta - \beta'U'x + \beta'U'U\beta \\
&= \beta'[U'U\beta - U'x] - x'U\beta + x'x \\
&= \beta'(U'U)[\beta - (U'U)^{-1}U'x] - x'U\beta + x'x \\
&= \beta'(U'U)(\beta - \hat{\beta}) - x'U\beta + x'x \\
&= (\beta' - \hat{\beta}')(U'U)(\beta - \hat{\beta}) + \hat{\beta}'(U'U)(\beta - \hat{\beta}) \\
&\quad -x'U\beta + x'x \\
&= (\beta - \hat{\beta})'(U'U)(\beta - \hat{\beta}) + \hat{\beta}'(U'U)\beta - \hat{\beta}'(U'U)\hat{\beta} \\
&\quad -x'U\beta + x'x \\
&= (\beta - \hat{\beta})'(U'U)(\beta - \hat{\beta}) + x'U(U'U)^{-1}(U'U)\beta \\
&\quad -x'U(U'U)^{-1}(U'U)(U'U)^{-1}U'x - x'U\beta + x'x \\
&= (\beta - \hat{\beta})'(U'U)(\beta - \hat{\beta}) \\
&\quad -x'U(U'U)^{-1}(U'U)(U'U)^{-1}U'x + x'x \\
&= (\beta - \hat{\beta})'(U'U)(\beta - \hat{\beta}) \\
&\quad +x'[I_n - U(U'U)^{-1}U']x, \qquad (7.3.7)
\end{aligned}
$$

where $\hat{\beta} = (U'U)^{-1}U'x$.

The likelihood is now

$$
p(x|\beta,\sigma^2,U) = (2\pi\sigma^2)^{-\frac{n}{2}} e^{-\frac{1}{2\sigma^2}\left\{(\beta-\hat{\beta})'(U'U)(\beta-\hat{\beta})+x'[I_n-U(U'U)^{-1}U']x\right\}}.
$$
$$(7.3.8)$$

It is now obvious that the value of β which maximizes the likelihood or minimizes the exponent in the likelihood is $\hat{\beta} = (U'U)^{-1}U'x$.

Upon differentiating the natural logarithm of the likelihood denoted by $LL = \log[p(x|\beta,\sigma^2,U)]$ with respect to σ^2, then evaluating it at the maximal values of the parameters, and setting it equal to zero

$$
\left.\frac{\partial}{\partial\sigma^2}LL\right|_{\beta=\hat{\beta},\sigma^2=\hat{\sigma}^2} = 0, \qquad (7.3.9)
$$

we obtain

$$
\hat{\sigma}^2 = \frac{(x-U\hat{\beta})'(x-U\hat{\beta})}{n}. \qquad (7.3.10)
$$

Consider the numerator of the estimate $\hat{\sigma}^2$.

$$
\begin{aligned}
NUM &= (x - U\hat{\beta})'(x - U\hat{\beta}) \\
&= x'x - x'U\hat{\beta} - \hat{\beta}'U'x + \hat{\beta}'U'U\hat{\beta} \\
&= \hat{\beta}'(U'U\hat{\beta} - U'x) - x'U\hat{\beta} + x'x \\
&= \hat{\beta}'(U'U)(\hat{\beta} - (U'U)^{-1}U'x) - x'U\beta\hat{B}eta + x'x \\
&= (\hat{\beta} - (U'U)^{-1}U'x)'(U'U)(\hat{\beta} - (U'U)^{-1}U'x) \\
&\quad + x'U(U'U)^{-1}(U'U)\hat{\beta} - x'U(U'U)^{-1}(U'U)(U'U)^{-1}U'x \\
&\quad - x'U\hat{\beta} + x'x \\
&= x'x - x'U(U'U)^{-1}U'x \\
&= x'[I_n - U(U'U)^{-1}U']x, \hspace{3cm} (7.3.11)
\end{aligned}
$$

This is exactly the g term in the likelihood. Now let's write the likelihood as

$$
p(x|\beta,\sigma^2,U) = (2\pi\sigma^2)^{-\frac{n}{2}} e^{-\frac{1}{2\sigma^2}[(\beta-\hat{\beta})'(U'U)(\beta-\hat{\beta})+g]}, \hspace{1cm} (7.3.12)
$$

where $g = (x - U\hat{\beta})'(x - U\hat{\beta})$.

In the likelihood for the simple linear regression model as written immediately above, if the normalization coefficients are ignored by using proportionality, it can be partitioned and viewed as a joint distribution of $\hat{\beta}$ and g. This then becomes

$$
\begin{aligned}
p(\hat{\beta},g|\beta,\sigma^2) &\propto (\sigma^2)^{-\frac{(q+1)}{2}} e^{-\frac{(\beta-\hat{\beta})'(U'U)(\beta-\hat{\beta})}{2\sigma^2}} (\sigma^2)^{-\frac{n-(q+1)}{2}} g^{\frac{n-(q+1)-2}{2}} e^{-\frac{g}{2\sigma^2}} \\
&= \underbrace{\frac{|\sigma^2(U'U)^{-1}|^{\frac{1}{2}}}{(2\pi)^{\frac{(q+1)}{2}}} e^{-\frac{(\beta-\hat{\beta})'(U'U)(\beta-\hat{\beta})}{2\sigma^2}}}_{\hat{\beta}|\beta,\sigma^2 \sim N(\beta,\sigma^2(U'U)^{-1})} \\
&\quad \times \underbrace{\frac{(\sigma^2)^{-\frac{n-(q+1)}{2}} g^{\frac{n-(q+1)-2}{2}} e^{-\frac{g}{2\sigma^2}}}{\Gamma\left(\frac{n-(q+1)}{2}\right) 2^{\frac{n-(q+1)}{2}}}}_{g|\sigma^2 \sim W(\sigma^2,1,n-(q+1))}, \hspace{1cm} (7.3.13)
\end{aligned}
$$

where q which is unity is generically used.

It should be noted that the estimates of the vector of regression coefficients $\hat{\beta}$ and the variance $\hat{\sigma}^2$ are independent by the Neyman factorization criterion [22, 47]. This is proven by showing that

$$
p(\hat{\beta},g|\beta,\sigma^2) = p(\hat{\beta}|\beta,\sigma^2)p(g|\beta,\sigma^2). \hspace{1cm} (7.3.14)
$$

If the joint distribution $p(\hat{\beta}, g|\beta, \sigma^2)$ is integrated with respect to g, then

$$
\begin{aligned}
p(\hat{\beta}|\beta, \sigma^2) &= \int p(\hat{\beta}, g|\beta, \sigma^2)\, dg \\
&= \int \frac{|\sigma^2(U'U)^{-1}|^{\frac{1}{2}}}{(2\pi)^{\frac{(q+1)}{2}}} e^{-\frac{(\beta-\hat{\beta})'(U'U)(\beta-\hat{\beta})}{2\sigma^2}} \\
&\qquad \times \frac{(\sigma^2)^{-\frac{n-(q+1)}{2}} g^{\frac{n-(q+1)-2}{2}} e^{-\frac{g}{2\sigma^2}}}{\Gamma\left(\frac{n-(q+1)}{2}\right) 2^{\frac{n-(q+1)}{2}}} \, dg \\
&= \frac{|\sigma^2(U'U)^{-1}|^{\frac{1}{2}}}{(2\pi)^{\frac{(q+1)}{2}}} e^{-\frac{(\beta-\hat{\beta})'(U'U)(\beta-\hat{\beta})}{2\sigma^2}} \\
&\qquad \times \int \frac{(\sigma^2)^{-\frac{n-(q+1)}{2}} g^{\frac{n-(q+1)-2}{2}} e^{-\frac{g}{2\sigma^2}}}{\Gamma\left(\frac{n-(q+1)}{2}\right) 2^{\frac{n-(q+1)}{2}}} \, dg \\
&= \frac{|\sigma^2(U'U)^{-1}|^{\frac{1}{2}}}{(2\pi)^{\frac{(q+1)}{2}}} e^{-\frac{(\beta-\hat{\beta})'(U'U)(\beta-\hat{\beta})}{2\sigma^2}},
\end{aligned} \tag{7.3.15}
$$

and thus,

$$
\hat{\beta}|\beta, \sigma^2 \sim N\left(\beta, \sigma^2(U'U)^{-1}\right). \tag{7.3.16}
$$

If the joint distribution $p(\hat{\beta}, g|\beta, \sigma^2)$ is integrated with respect to $\hat{\beta}$, then

$$
\begin{aligned}
p(g|\sigma^2) &= \int p(\hat{\beta}, g|\beta, \sigma^2)\, d\hat{\beta} \\
&= \int \frac{|\sigma^2(U'U)^{-1}|^{\frac{1}{2}}}{(2\pi)^{\frac{(q+1)}{2}}} e^{-\frac{(\beta-\hat{\beta})'(U'U)(\beta-\hat{\beta})}{2\sigma^2}} \\
&\qquad \times \frac{(\sigma^2)^{-\frac{n-(q+1)}{2}} g^{\frac{n-(q+1)-2}{2}} e^{-\frac{g}{2\sigma^2}}}{\Gamma\left(\frac{n-(q+1)}{2}\right) 2^{\frac{n-(q+1)}{2}}} \, d\hat{\beta} \\
&= \frac{(\sigma^2)^{-\frac{n-(q+1)}{2}} g^{\frac{n-(q+1)-2}{2}} e^{-\frac{g}{2\sigma^2}}}{\Gamma\left(\frac{n-(q+1)}{2}\right) 2^{\frac{n-(q+1)}{2}}} \\
&\qquad \times \int \frac{|(U'U)^{-1}|^{\frac{1}{2}}}{(2\pi\sigma^2)^{\frac{(q+1)}{2}}} e^{-\frac{(\beta-\hat{\beta})'(U'U)(\beta-\hat{\beta})}{2\sigma^2}} \, d\hat{\beta} \\
&= \frac{(\sigma^2)^{-\frac{n-(q+1)}{2}} g^{\frac{n-(q+1)-2}{2}} e^{-\frac{g}{2\sigma^2}}}{\Gamma\left(\frac{n-(q+1)}{2}\right) 2^{\frac{n-(q+1)}{2}}},
\end{aligned} \tag{7.3.17}
$$

and thus,

$$g|\sigma^2 \sim W\left(\sigma^2, 1, n-(q+1)\right). \tag{7.3.18}$$

The joint distribution of $\hat{\beta}$ and g is equal to the product of their marginal distributions. Note that g/σ^2 has the familiar Chi-squared distribution with $n-(q+1)$ degrees of freedom.

By changing variables from $\hat{\beta}$ and g to $t = [n-(q+1)]^{\frac{1}{2}} g^{-1/2} (U'U)^{\frac{1}{2}} (\hat{\beta}-\beta)$ and $w = g$, the joint distribution of t and w becomes

$$p(t,w|\beta,\sigma^2) = \frac{(\sigma^2)^{-\frac{n}{2}} w^{\frac{n-2}{2}} e^{-\frac{w}{2\sigma^2} \left[\frac{t't}{n-(q+1)}+1\right]}}{[(n-(q+1))\pi]^{\frac{1}{2}} \Gamma\left(\frac{n-(q+1)}{2}\right) 2^{\frac{n}{2}}}, \tag{7.3.19}$$

where the Jacobian of the transformation was

$$J(\hat{\beta}, g \to t, w) = [n-(q+1)]^{-\frac{(q+1)}{2}} w^{\frac{(q+1)}{2}} |U'U|^{-\frac{1}{2}}. \tag{7.3.20}$$

Now integrate with respect to w to find the distribution of t (unconditional on w). This is done by making the integrand look like the Scalar Wishart distribution. It is seen that the integrand needs to multiplied and divided by the same factor.

$$
\begin{aligned}
p(t|\beta) &= \int p(t,w|\beta,\sigma^2)\, dw \\
&\propto \int (\sigma^2)^{-\frac{n}{2}} w^{\frac{n-(q+1)}{2}} e^{-\frac{w}{2\sigma^2}\left(\frac{t't}{n-(q+1)}+1\right)}\, dw \\
&\propto \left(\frac{t't}{n-(q+1)}+1\right)^{-\frac{n}{2}} \\
&\quad \times \int \left(\frac{t't}{n-(q+1)}+1\right)^{\frac{n}{2}} w^{\frac{n-2}{2}} e^{-\frac{w}{2\sigma^2}\left(\frac{t't}{n-(q+1)}+1\right)}\, dw \\
&\propto \left(\frac{t't}{n-(q+1)}+1\right)^{-\frac{n}{2}} \tag{7.3.21}
\end{aligned}
$$

and now transforming back while taking g to be known

$$p(\hat{\beta}|\beta,g) \propto \frac{\left[\frac{g}{n-(q+1)}(U'U)\right]^{-\frac{1}{2}}}{\left\{1 + \frac{1}{n-(q+1)}\left(\hat{\beta}-\beta\right)'\left[\frac{g}{n-(q+1)}(U'U)^{-1}\right]^{-1}\left(\hat{\beta}-\beta\right)\right\}^{\frac{n}{2}}}. \tag{7.3.22}$$

Recall the Multivariate Student t-distribution in the statistical distribution Chapter. The estimate of the vector of regression coefficients $\hat{\beta}$ has the familiar Multivariate Student t-distribution with $n - (q+1)$ degrees of freedom.

7.4 Multiple Linear Regression

Similarly, the multiple Regression model

$$x_i = \beta_0 + \beta_1 u_{i1} + \cdots + \beta_q u_{iq} + \epsilon_i, \qquad (7.4.1)$$

where ϵ_i is the Scalar Normally distributed random error with mean zero and variance σ^2,

$$\epsilon_i \sim N(0, \sigma^2) \qquad (7.4.2)$$

for $i = 1, \ldots, n$. The joint distribution of these variables or likelihood is

$$
\begin{aligned}
p(x_1, \ldots, x_n | \beta, \sigma^2, U) &= \prod_{i=1}^{n} f(x_i | \beta, \sigma^2, u_i) \\
&= \prod_{i=1}^{n} (2\pi\sigma^2)^{-\frac{1}{2}} e^{-\frac{(x_i - \beta_0 - \beta_1 u_{i1} - \cdots - \beta_q u_{iq})^2}{2\sigma^2}} \\
&= (2\pi\sigma^2)^{-\frac{n}{2}} e^{-\frac{1}{2\sigma^2} \sum_{i=1}^{n} (x_i - \beta_0 - \beta_1 u_{i1} - \cdots - \beta_q u_{iq})^2}.
\end{aligned}
$$

$$(7.4.3)$$

This model can also be written in terms of vectors and matrices

$$
x = \begin{pmatrix} x_1 \\ \vdots \\ x_n \end{pmatrix}, \quad
u_i = \begin{pmatrix} 1 \\ u_{i1} \\ \vdots \\ u_{iq} \end{pmatrix}, \quad
U = \begin{pmatrix} u_1' \\ \vdots \\ u_n' \end{pmatrix}, \quad
\beta = \begin{pmatrix} \beta_0 \\ \vdots \\ \beta_q \end{pmatrix}, \quad
\epsilon = \begin{pmatrix} \epsilon_1 \\ \vdots \\ \epsilon_n \end{pmatrix}, \quad (7.4.4)
$$

so that the model is

$$
\begin{array}{ccccc}
x & = & U & \beta & + & \epsilon \\
n \times 1 & & n \times (q+1) & (q+1) \times 1 & & n \times 1
\end{array}
\qquad (7.4.5)
$$

and the likelihood is

$$p(x|\beta,\sigma^2,U) = (2\pi\sigma^2)^{-\frac{n}{2}} e^{-\frac{(x-U\beta)'(x-U\beta)}{2\sigma^2}}. \qquad (7.4.6)$$

The value of β which maximizes the likelihood can be found the same way as before, namely, $\hat{\beta} = (U'U)^{-1}U'x$. The value of σ^2 that makes the likelihood a maximum can also be found, namely, $\sigma^2 = \hat{\sigma}^2$ which is defined as before.

By the Neyman factorization criterion [22, 47], $\hat{\beta}$ and $\hat{\sigma}^2$ are independent in a similar way as in the Normal Samples and simple linear Regression models. Also in a similar way as in the Normal Samples model and the simple linear Regression model, the distribution of $\hat{\beta}|\beta,\sigma^2$, $g|\sigma^2$, and $\hat{\beta}|\beta$ can also be similarly found.

7.5 Multivariate Linear Regression

The Multivariate Regression model is a generalization of the multiple Regression model to vector valued dependent variable observations.

Previously for the simple linear Regression, (x_i, u_i) pairs were observed, $i = 1,\ldots,n$, where x_i was a scalar but u_i was a $(1+1)$ dimensional vector containing a 1 and an observable u, u_{i1}. We adopted the model

$$x_i = \beta' u_i + \epsilon_i \qquad (7.5.1)$$

which was adopted where

$$\beta = \begin{pmatrix} \beta_0 \\ \beta_1 \end{pmatrix}, \quad u_i = \begin{pmatrix} 1 \\ u_{i1} \end{pmatrix}, \qquad (7.5.2)$$

and then fit a simple line to the data.

In the multiple Regression model, (x_i, u_i) pairs were observed, $i = 1,\ldots,n$ where x_i was a scalar but u_i was a $(q+1)$ dimensional vector containing a 1 and q observable u's, $u_{i1},\ldots,u_{iq}$. The model is

$$\begin{array}{ccccc} x_i & = & \beta' & u_i & + & \epsilon_i \\ 1 \times 1 & & 1 \times (q+1) & (q+1) \times 1 & & 1 \times 1 \end{array} \qquad (7.5.3)$$

which was adopted where

$$\beta = \begin{pmatrix} \beta_0 \\ \vdots \\ \beta_q \end{pmatrix}, \quad u_i = \begin{pmatrix} 1 \\ u_{i1} \\ \vdots \\ u_{iq} \end{pmatrix}, \qquad (7.5.4)$$

and then fit a line in space to the data.

In the Multivariate Regression model, (x_i, u_i) pairs are observed, $i = 1, \ldots, n$ where x_i is a p-dimensional vector and u_i is a $(q+1)$-dimensional vector containing a 1 and q observable u's, $u_{i1}, \ldots, u_{iq}$. The model

$$\begin{array}{cccc} x_i & = & B & u_i & + & \epsilon_i \\ p \times 1 & & p \times (q+1) & (q+1) \times 1 & & p \times 1 \end{array} \tag{7.5.5}$$

where u_i is as in the multiple Regression model and

$$x_i = \begin{pmatrix} x_{i1} \\ \vdots \\ x_{ip} \end{pmatrix}, \quad \beta_j = \begin{pmatrix} \beta_{j0} \\ \vdots \\ \beta_{jq} \end{pmatrix}, \quad B = \begin{pmatrix} \beta_1' \\ \vdots \\ \beta_p' \end{pmatrix}, \quad \epsilon_i = \begin{pmatrix} \epsilon_{i1} \\ \vdots \\ \epsilon_{ip} \end{pmatrix}. \tag{7.5.6}$$

Taking a closer look at this model,

$$\underset{p \times 1}{\begin{pmatrix} x_{i1} \\ \vdots \\ x_{ip} \end{pmatrix}} = \underset{p \times (q+1)}{\begin{pmatrix} \beta_{10} & \cdots & \beta_{1q} \\ & \vdots & \\ \beta_{p0} & \cdots & \beta_{pq} \end{pmatrix}} \underset{(q+1) \times 1}{\begin{pmatrix} 1 \\ u_{i1} \\ \vdots \\ u_{iq} \end{pmatrix}} + \underset{p \times 1}{\begin{pmatrix} \epsilon_{i1} \\ \vdots \\ \epsilon_{ip} \end{pmatrix}}, \tag{7.5.7}$$

which means that for each observation, which is a row in the left-hand side of the model, there is a Regression. Each row has its own Regression complete with its own set of Regression coefficients.

Just like in the simple and multiple Regression models, we specify that the errors are normally distributed. However, the Multivariate Regression model has vector-valued observations and vector-valued errors. The error vector ϵ_i has a Multivariate Normal distribution with a zero mean vector and positive definite covariance matrix Σ

$$\epsilon_i \sim N(0, \Sigma) \tag{7.5.8}$$

for $i = 1, \ldots, n$. The resulting distribution for a given observation vector has a Multivariate Normal distribution

$$p(x_i | B, \Sigma, u_i) = (2\pi)^{-\frac{p}{2}} |\Sigma|^{-\frac{1}{2}} e^{-\frac{1}{2}(x_i - Bu_i)' \Sigma^{-1}(x_i - Bu_i)}, \tag{7.5.9}$$

because the Jacobian of the transformation is unity.

Note: If each of the elements of vector x_i were independent, then Σ would be diagonal. Since we assume that each variable has its own distinct error term (i.e., its own $\sigma_j^2 = \Sigma_{jj}$), $\Sigma = \text{diag}(\sigma_1^2, \ldots, \sigma_p^2)$ and $p(x_i | B, \Sigma, u_i)$ break down into

$$p(x_i|B,\Sigma,u_i) = p(x_{i1},\ldots,x_{ip}|\beta,\sigma^2,u_i)$$

$$= \prod_{j=1}^{p} p(x_{ij}|\beta_j,\sigma_j^2,u_{ij}). \tag{7.5.10}$$

This is what we would get if we just collected independent observations from the multiple Regression model. This means add a subscript j to the multiple Regression model for each of the $j = 1,\ldots,p$ variables and collect them together.

Returning to the vector-valued observations with dependent elements, we have observations for $i = 1,\ldots,n$ whose likelihood is

$$p(x_1,\ldots,x_n|B,\Sigma,U) = \prod_{i=1}^{n} p(x_i|B,\Sigma,U)$$

$$= \prod_{i=1}^{n}(2\pi)^{-\frac{p}{2}}|\Sigma|^{-\frac{1}{2}}e^{-\frac{1}{2}(x_i-Bu_i)'\Sigma^{-1}(x_i-Bu_i)}$$

$$= (2\pi)^{-\frac{np}{2}}|\Sigma|^{-\frac{n}{2}}e^{-\frac{1}{2}\sum_{i=1}^{n}(x_i-Bu_i)'\Sigma^{-1}(x_i-Bu_i)}. \tag{7.5.11}$$

If all of the n observation and error vectors on p variables are collected together as

$$X = \begin{pmatrix} x_1' \\ \vdots \\ x_n' \end{pmatrix}, \quad U = \begin{pmatrix} u_1' \\ \vdots \\ u_n' \end{pmatrix}, \quad E = \begin{pmatrix} \epsilon_1' \\ \vdots \\ \epsilon_n' \end{pmatrix}, \tag{7.5.12}$$

then the model can be written as

$$\begin{array}{ccccc} X & = & U & B' & + & E \\ n\times p & & n\times(q+1) & (q+1)\times p & & n\times p \end{array} \tag{7.5.13}$$

and the likelihood

$$p(X|B,\Sigma,U) = (2\pi)^{-\frac{np}{2}}|\Sigma|^{-\frac{n}{2}}e^{-\frac{1}{2}tr(X-UB')\Sigma^{-1}(X-UB')'}$$

$$= (2\pi)^{-\frac{np}{2}}|\Sigma|^{-\frac{n}{2}}e^{-\frac{1}{2}tr\Sigma^{-1}(X-UB')'(X-UB')}, \tag{7.5.14}$$

where $tr(\cdot)$ is the trace operator which gives the sum of the diagonal elements of its matrix argument.

The natural logarithm of $p(X|B,\Sigma,U)$ can be taken and differentiated with respect to the matrix B and the matrix Σ. But just as with the simple and multiple Regression models, some algebra on the exponent can be performed to find the value of B which yields the maximal value of this likelihood.

$$
\begin{aligned}
(X - UB')'(X - UB') &= X'X - X'UB' - BU'X + BU'UB' \\
&= B[U'UB' - U'X] - X'UB' + X'X \\
&= B(U'U)[B' - (U'U)^{-1}U'X] - X'UB' + X'X \\
&= B(U'U)(B' - \hat{B}') - X'UB' + X'X \\
&= (B - \hat{B})(U'U)(B - \hat{B})' + \hat{B}(U'U)(B - \hat{B})' \\
&\quad - X'UB' + X'X \\
&= (B - \hat{B})(U'U)(B - \hat{B})' + \hat{B}(U'U)B' - \hat{B}(U'U)\hat{B}' \\
&\quad - X'UB' + X'X \\
&= (B - \hat{B})(U'U)(B - \hat{B})' + X'U(U'U)^{-1}(U'U)B' \\
&\quad - X'U(U'U)^{-1}(U'U)(U'U)^{-1}U'X - X'UB' + X'X \\
&= (B - \hat{B})(U'U)(B - \hat{B})' \\
&\quad - X'U(U'U)^{-1}(U'U)(U'U)^{-1}U'X + X'X \\
&= (B - \hat{B})(U'U)(B - \hat{B})' + X'[I_n - U(U'U)^{-1}U']X.
\end{aligned}
$$

$$(7.5.15)$$

The likelihood is now

$$
\begin{aligned}
p(X|B,\Sigma,U) = \\
(2\pi)^{-\frac{np}{2}}|\Sigma|^{-\frac{n}{2}}e^{-\frac{1}{2}tr\Sigma^{-1}[(B-\hat{B})(U'U)(B-\hat{B})'+X'[I_n-U(U'U)^{-1}U']X]}.
\end{aligned}
$$

$$(7.5.16)$$

It is now obvious that the value of B which maximizes the likelihood or minimizes the exponent in the likelihood is $\hat{B}' = (U'U)^{-1}U'X$.

It is also now evident by making an analogy to the simple and multiple Regression models that the value of Σ which maximizes the likelihood is

$$
\hat{\Sigma} = \frac{(X - U\hat{B}')'(X - U\hat{B}')}{n}.
$$

$$(7.5.17)$$

This could have also been proved by differentiation.

Recall that when the univariate or Scalar Normal distribution was introduced, the estimates of the mean $\hat{\mu}$ and the variance $\hat{\sigma}^2$ were independent. Also in the simple and multiple linear Regression models the estimates of the Regression coefficients and the variance were independent. A similar thing is true in Multivariate Regression, namely, that $\hat{B}$ and $\hat{\Sigma}$ are independent.

Consider the numerator of the estimate of Σ, $\hat{\Sigma}$.

$$
\begin{aligned}
NUM &= (X - U\hat{B}')'(X - U\hat{B}') \\
&= X'X - X'U\hat{B}' - \hat{B}U'X + \hat{B}U'U\hat{B}' \\
&= \hat{B}(U'U\hat{B}' - U'X) - X'U\hat{B}' + X'X \\
&= \hat{B}(U'U)(\hat{B}' - (U'U)^{-1}U'X) - X'U\hat{B}' + X'X \\
&= (\hat{B} - X'U(U'U)^{-1})(U'U)(\hat{B} - X'U(U'U)^{-1})' \\
&\quad + X'U(U'U)^{-1}(U'U)\hat{B} - X'U(U'U)^{-1}(U'U)(U'U)^{-1}U'X \\
&\quad - X'U\hat{B} + X'X \\
&= X'X - X'U(U'U)^{-1}U'X \\
&= X'[I_n - U(U'U)^{-1}U']X.
\end{aligned}
\tag{7.5.18}
$$

This is exactly the G term in the likelihood. Now, the likelihood can be written as

$$
p(X|B,\Sigma,U) = (2\pi)^{-\frac{np}{2}} |\Sigma|^{-\frac{n}{2}} e^{-\frac{1}{2}tr\Sigma^{-1}[(B-\hat{B})(U'U)(B-\hat{B})' + (X-U\hat{B}')'(X-U\hat{B}')]},
\tag{7.5.19}
$$

where $G = (X - U\hat{B}')'(X - U\hat{B}')$.

Just as in the Normal Samples and the multiple linear Regression models, if we ignore the normalization coefficients by using proportionality, partition the likelihood, and view this as a distribution of $\hat{B}$ and G, then

$$
p(\hat{B},G|B,\Sigma,U) \propto \underbrace{|\Sigma|^{-\frac{q+1}{2}} e^{-\frac{1}{2}tr(U'U)(\hat{B}-B)'\Sigma^{-1}(\hat{B}-B)}}_{\hat{B}|B,\Sigma \sim N(B,\Sigma\otimes(U'U)^{-1})}
$$
$$
\times \underbrace{|\Sigma|^{-\frac{n-(q+1)}{2}} |G|^{\frac{n-(q+1)-p-1}{2}} e^{-\frac{1}{2}tr\Sigma^{-1}G}}_{G|\Sigma \sim W(\Sigma,p,n-(q+1))}, \tag{7.5.20}
$$

where the first term involves $\hat{B}$ and the second G.

By the Neyman factorization criterion [22, 41, 47], $\hat{B}|B,\Sigma$ and $G|\Sigma$ are independent. This is proven by showing that

$$
p(\hat{B},G|B,\Sigma) = p(\hat{B}|B,\Sigma)p(G|B,\Sigma).
\tag{7.5.21}
$$

If we integrate $p(\hat{B},G|B,\Sigma)$ with respect to G, then

$$
\begin{aligned}
p(\hat{B}|\Sigma,U) &= \int p(\hat{B},G|B,\Sigma) \, dG \\
&\propto \int |\Sigma|^{-\frac{q+1}{2}} e^{-\frac{1}{2}tr(U'U)(\hat{B}-B)'\Sigma^{-1}(\hat{B}-B)}
\end{aligned}
$$

$$\times |\Sigma|^{-\frac{n-(q+1)}{2}} |G|^{\frac{n-(q+1)-p-1}{2}} e^{-\frac{1}{2}tr\Sigma^{-1}G} \, dG$$

$$\propto |\Sigma|^{-\frac{q+1}{2}} e^{-\frac{1}{2}tr(U'U)(\hat{B}-B)'\Sigma^{-1}(\hat{B}-B)}$$

$$\times \int |\Sigma|^{-\frac{n-(q+1)}{2}} |G|^{\frac{n-(q+1)-p-1}{2}} e^{-\frac{1}{2}tr\Sigma^{-1}G} \, dG$$

$$\propto |\Sigma|^{-\frac{q+1}{2}} e^{-\frac{1}{2}tr(U'U)(\hat{B}-B)'\Sigma^{-1}(\hat{B}-B)}, \tag{7.5.22}$$

where the matrix $\hat{B}$ follows a Matrix Normal distribution

$$\hat{B}|B,\Sigma \sim N(B,\Sigma \otimes (U'U)^{-1}). \tag{7.5.23}$$

If we integrate $p(\hat{B},G|B,\Sigma)$ with respect to $\hat{B}$, then

$$p(G|\Sigma,U) = \int p(\hat{B},G|B,\Sigma) \, d\hat{B}$$

$$\propto \int |\Sigma|^{-\frac{q+1}{2}} e^{-\frac{1}{2}tr(U'U)(\hat{B}-B)'\Sigma^{-1}(\hat{B}-B)}$$

$$\times |\Sigma|^{-\frac{n-(q+1)}{2}} |G|^{\frac{n-(q+1)-p-1}{2}} e^{-\frac{1}{2}tr\Sigma^{-1}G} \, d\hat{B}$$

$$\propto |\Sigma|^{-\frac{n-(q+1)}{2}} |G|^{\frac{n-(q+1)-p-1}{2}} e^{-\frac{1}{2}tr\Sigma^{-1}G}$$

$$\times \int |\Sigma|^{-\frac{q+1}{2}} e^{-\frac{1}{2}tr(U'U)(\hat{B}-B)'\Sigma^{-1}(\hat{B}-B)} \, d\hat{B}$$

$$\propto |\Sigma|^{-\frac{n-(q+1)}{2}} |G|^{\frac{n-(q+1)-p-1}{2}} e^{-\frac{1}{2}tr\Sigma^{-1}G}, \tag{7.5.24}$$

where the matrix $G = n\hat{\Sigma} = (X - U\hat{B}')'(X - U\hat{B}')$ follows a Wishart distribution

$$G|\Sigma \sim W(\Sigma, p, n-(q+1)). \tag{7.5.25}$$

By changing variables from $\hat{B}$ and G to $T = [n-(q+1)]^{\frac{1}{2}} G^{-\frac{1}{2}}(\hat{B}-B)(U'U)^{\frac{1}{2}}$ and $W = G$ in a similar fashion as was done before, the distribution of $p(T,W|B,\Sigma,U)$ is

$$p(T,W|B,\Sigma,U) \propto |\Sigma|^{-\frac{n}{2}} |W|^{\frac{n-p-1}{2}} e^{-\frac{1}{2}tr\Sigma^{-1}[\frac{1}{n-q-1} W^{\frac{1}{2}} TT' W^{\frac{1}{2}} +W]}$$

$$\propto |\Sigma|^{-\frac{n}{2}} |W|^{\frac{n-p-1}{2}} e^{-\frac{1}{2}tr\left\{\Sigma^{-1}[\frac{1}{n-q-1} TT' + I_p]\right\}W}, \tag{7.5.26}$$

where the Jacobian of the transformation [41] was

$$J(\hat{B},G \to T,W) = [n-(q+1)]^{\frac{p(q+1)}{2}} |W|^{\frac{q+1}{2}} |U'U|^{-\frac{p}{2}}. \tag{7.5.27}$$

The distribution of T unconditional of Σ is found by integrating $p(T,W|B,\Sigma,U)$ with respect to W,

$$p(T|B,U) \propto \int p(T,W|B,\Sigma,U)\ dW$$

$$\propto \int |\Sigma|^{-\frac{n}{2}} |W|^{\frac{n-p-1}{2}} e^{-\frac{1}{2}tr\left\{\Sigma^{-1}[\frac{1}{n-q-1}TT'+I_p]\right\}W}\ dW$$

$$\propto \left[\frac{1}{n-q-1}TT'+I_p\right]^{-\frac{n}{2}}$$

$$\times \int |\Sigma^{-1}[\frac{1}{n-q-1}TT'+I_p]|^{-\frac{n}{2}} |W|^{\frac{n-p-1}{2}}$$

$$\times e^{-\frac{1}{2}tr\left\{\Sigma^{-1}[\frac{1}{n-q-1}TT'+I_p]\right\}W}\ dW$$

$$\propto [I_p + \frac{1}{n-q-1}TT']^{-\frac{n}{2}} \tag{7.5.28}$$

and now transforming back while taking G to be known

$$p(\hat{B}|B,G,U) \propto \left|G + \frac{1}{n-(q+1)}(\hat{B}-B)[n-(q+1)](U'U)(\hat{B}-B)'\right|^{-\frac{n}{2}} \tag{7.5.29}$$

which is a Matrix Student T-distribution. That is,

$$\hat{B}|B,G,U \sim T\left(n-q-1, B, [(n-q-1)(U'U)]^{-1}, G\right). \tag{7.5.30}$$

Note that the exponent can be written as $n-(q+1)+q+1$ and compared to the definition of the Matrix Student T-distribution given in Equation 2.3.39. This is the conditional posterior distribution of $\hat{B}$ given G. Conditional maximum a posteriori variance estimates can be found. The conditional modal variance is

$$var(\hat{B}|G,X,U) = \frac{n-(q+1)}{n-(q+1)-2}\ G \otimes [(n-q-1)(U'U)]^{-1} \tag{7.5.31}$$

or equivalently

$$var(\hat{\beta}|G,X,U) = \frac{n-(q+1)}{n-(q+1)-2}\ [(n-q-1)(U'U)]^{-1} \otimes G \tag{7.5.32}$$

$$= \hat{\Delta}, \tag{7.5.33}$$

where $\hat{\beta} = vec(\hat{B})$.

Hypotheses can be performed regarding the entire coefficient matrix, a submatrix, a particular row or column, or a particular element.

Significance for the entire coefficient matrix can be evaluated with the use of

$$|I + G^{-1}\hat{B}(U'U)\hat{B}'|^{-1} \qquad (7.5.34)$$

which follows a distribution of the product of independent Scalar Beta variates [12, 41]. General simultaneous hypotheses (which do not assume independence) can be performed regarding the entire Regression coefficient matrix. A similar result holds for a submatrix.

The above distribution for the matrix of regression coefficients can be written in another form

$$p(\hat{B}|X,G,U,B) \propto |W + (\hat{B} - B)'G^{-1}(\hat{B} - B)|^{-\frac{(n-q-1)+(q+1)}{2}}. \qquad (7.5.35)$$

by using Sylvester's theorem [41]

$$|I_n + Y I_p Y'| = |I_p + Y' I_n Y|, \qquad (7.5.36)$$

where $W = (U'U)^{-1}$ has been defined.

It can be shown [17, 41] that the marginal distribution of any column, say the kth of the matrix of $\hat{B}$, $\hat{B}_k$ is Multivariate Student t-distributed

$$p(\hat{B}_k|B_k,G,X,U) \propto \frac{1}{\left[W_{kk} + (\hat{B}_k - B_k)'G^{-1}(\hat{B}_k - B_k)\right]^{\frac{(n-q-p)+p}{2}}}, \qquad (7.5.37)$$

where W_{kk} is the kth diagonal element of W, $k = 0,\ldots,q$.

It can be also be shown [17, 41] that the marginal distribution of any row, say the jth of the matrix of $\hat{B}$, $\hat{\beta}'_j$ is Multivariate Student t-distributed

$$p(\hat{\beta}_j|\beta_j,G,X,U) \propto \frac{1}{\left[G_{jj} + (\hat{\beta}_j - \beta_j)'W^{-1}(\hat{\beta}_j - \beta_j)\right]^{\frac{(n-q-p)+(q+1)}{2}}}, \qquad (7.5.38)$$

where G_{jj} is the jth diagonal element of G, $j = 1,\ldots,p$.

With the marginal distribution of a column or row of $\hat{B}$, significance can be evaluated for the coefficient of a particular independent variable (column of $\hat{B}$) with the use of the statistic

$$F_{p,n-q-p} = \frac{n-q-p}{p} W_{kk}^{-1} \hat{B}'_k G^{-1} \hat{B}_k, \qquad (7.5.39)$$

and for the coefficients of a particular dependent variable (row of $\hat{B}$) with the use of the statistic

$$F_{q+1,n-q-p} = \frac{n-q-p}{q+1} G_{jj}^{-1} \hat{\beta}'_j W^{-1} \hat{\beta}_j. \qquad (7.5.40)$$

It can be shown that these statistics follow F-distributions with $(p, n-q-p)$ or $(q+1, n-q-p)$ numerator and denominator degrees of freedom respectively. Significance can be determined for a subset of variables by determining the marginal distribution of the subset within $\hat{B}_k$ or $\hat{\beta}'_j$ which are also Multivariate Student t-distributed.

It should also be noted that

$$b = \left[1 + \frac{\nu_1}{\nu_2} F_{\nu_1, \nu_2}\right]^{-1} \tag{7.5.41}$$

follows the Scalar Beta distribution $B\left(\frac{\nu_2}{2}, \frac{\nu_1}{2}\right)$.

With the subset of variables being a singleton set, significance can be determined for a particular independent variable with the marginal distribution of the scalar coefficient which is

$$p(\hat{B}_{kj} | B_{kj}, G_{jj}, X, U) \propto \frac{1}{\left[W_{kk} + (\hat{B}_{kj} - B_{kj}) G_{jj}^{-1} (\hat{B}_{kj} - B_{kj})\right]^{\frac{(n-q-p)+1}{2}}}, \tag{7.5.42}$$

where $G_{jj} = (X_j - U\hat{\beta}_j)'(X_j - U\hat{\beta}_j)$ is the jth diagonal element of G. The above can be rewritten in the more familiar form

$$p(\hat{B}_{kj} | B_{kj}, G_{jj}, X, U) \propto \frac{1}{\left[1 + \frac{1}{(n-q-p)} \frac{(\hat{B}_{kj} - B_{kj})^2}{W_{kk}[G_{jj}/(n-q-p)]}\right]^{\frac{(n-q-p)+1}{2}}} \tag{7.5.43}$$

which is readily recognizable as a Scalar Student t-distribution. Note that $\hat{B}_{kj} = \hat{\beta}_{jk}$ and that

$$t = \frac{(\hat{B}_{kj} - B_{kj})}{[W_{kk} G_{jj} (n-q-p)^{-1}]^{\frac{1}{2}}} \tag{7.5.44}$$

follows a Scalar Student t-distribution with $n-q-p$ degrees of freedom, and t^2 follows an F distribution with 1 and $n-q-p$ numerator and denominator degrees of freedom, which is commonly used in Regression [1, 68], derived from a likelihood ratio test of reduced and full models when testing a single coefficient. By using a t statistic instead of an F statistic, positive and negative coefficient values can be identified.

It should also be noted that

$$b = \left[1 + \frac{1}{n-q-p} t^2\right]^{-1} \tag{7.5.45}$$

follows the Scalar Beta distribution $B\left(\frac{n-q-p}{2}, \frac{1}{2}\right)$.

Exercises

1. Assume that we perform an experiment in which we have a multiple Regression, $p = 1$, $n = 8$, with the data vector and design matrix as given in Table 7.1.

TABLE 7.1
Regression data vector and design matrix.

x	1	U	e_n	1	2
1	12.0719	1	1	1	1
2	15.7134	2	1	2	1
3	20.2977	3	1	3	1
4	24.2973	4	1	4	1
5	15.9906	5	1	5	-1
6	20.0818	6	1	6	-1
7	24.0437	7	1	7	-1
8	27.9533	8	1	8	-1

We wish to fit the multiple Regression model

$$x_i = \beta_0 + \beta_1 u_{i1} + \beta_2 u_{i2} + \epsilon_i$$
$$x = U\beta + \epsilon.$$

What is the estimate $\hat{\beta}$ of β? What is the estimate $\hat{\sigma}^2$ of σ^2?

2. What is the distribution and associated parameters of the vector of Regression coefficients?

3. Compute

$$F_{q+1, n-(q+1)} = \frac{\hat{\beta}'(U'U)\hat{\beta}/(q+1)}{g/(n-q-1)}$$

and

$$t_k = \frac{\hat{\beta}_k}{[W_{kk}g/(n-q-1)]^{\frac{1}{2}}}$$

for $k = 0, \ldots, q$ from the data in Exercise 1.

4. Assume that we perform an experiment in which $p = 3$, $q = 2$, and $n = 32$ with data matrix X and design U matrix (including a column of ones for the intercept term) as in Table 7.2.

TABLE 7.2
Regression data and design matrices.

X	1	2	3	U	e_n	1	2
1	12.1814	9.3538	20.9415	1	1	1	1
2	15.8529	11.7987	28.0296	2	1	2	1
3	20.5458	15.1322	35.0787	3	1	3	1
4	23.9659	18.0548	42.3609	4	1	4	1
5	28.0285	20.7695	48.9123	5	1	5	1
6	32.2667	23.4573	56.1558	6	1	6	1
7	36.0148	26.9852	63.1998	7	1	7	1
8	39.9761	29.7473	70.2352	8	1	8	1
9	31.7919	23.1536	58.7520	9	1	9	-1
10	36.0736	26.1269	66.0530	10	1	10	-1
11	39.6660	29.4231	73.0595	11	1	11	-1
12	44.1786	32.1478	79.7481	12	1	12	-1
13	48.4059	34.8391	86.8145	13	1	13	-1
14	51.8271	38.0951	94.2706	14	1	14	-1
15	56.2145	40.7477	100.9671	15	1	15	-1
16	60.3135	43.9951	108.0975	16	1	16	-1
17	75.6016	56.9879	133.0220	17	1	17	1
18	79.6398	60.0000	139.8411	18	1	18	1
19	84.1428	62.9205	146.8601	19	1	19	1
20	87.9000	66.2738	154.1109	20	1	20	1
21	92.1725	68.5315	160.7625	21	1	21	1
22	96.2039	72.1070	168.1953	22	1	22	1
23	100.1780	75.2239	175.1422	23	1	23	1
24	104.3226	78.1827	181.7946	24	1	24	1
25	96.1672	71.1445	170.9336	25	1	25	-1
26	100.2977	74.0101	177.7031	26	1	26	-1
27	103.6994	77.1693	184.4494	27	1	27	-1
28	107.9951	80.1422	192.2466	28	1	28	-1
29	111.9608	82.9361	198.8703	29	1	29	-1
30	115.5990	85.9056	206.0818	30	1	30	-1
31	120.0643	88.9260	213.0585	31	1	31	-1
32	123.7359	91.6312	220.0054	32	1	32	-1

We wish to fit the multiple Regression model

$$x_i = Bu_i + \epsilon_i$$
$$X = UB' + E.$$

What is the estimate $\hat{B}$ of B? What is the estimate $\hat{\Sigma}$ of Σ?

5. What is the distribution and associated parameters of the matrix of Regression coefficients in Exercise 4?

6. Compute

$$F_{p,n-q-p} = \frac{n-q-p}{p} W_{kk}^{-1} \hat{B}_k' G^{-1} \hat{B}_k$$

for k=0,…,q,

$$F_{q+1,n-q-p} = \frac{n-q-p}{q+1} G_{jj}^{-1} \hat{\beta}_j' W^{-1} \hat{\beta}_j.$$

for $j = 1,…,p$, and

$$t_{kj} = \frac{\hat{B}_{kj}}{[W_{kk}G_{jj}/(n-q-p)]^{\frac{1}{2}}} \tag{7.5.46}$$

for all kj combinations from the data in Exercise 4.

7. Recompute the statistics in Exercise 6 assuming $p = 1$, i.e. the observations are independent. Compare your results to those in Exercise 6.

Part II

Models

8

Bayesian Regression

8.1 Introduction

The Multivariate Bayesian Regression model [41] is introduced that requires some knowledge of Multivariate as well as Bayesian Statistics. A concise introduction to this material was presented in Part I. This Chapter considers the unobservable sources (the speakers conversations in the cocktail party problem) to be observable or known. This is done so that the Bayesian Regression model may be introduced which will help lead to the Bayesian Source Separation models.

8.2 The Bayesian Regression Model

A Regression model is used when it is believed that a set of dependent variables may be represented as being linearly related to a set of independent variables. The model may be motivated using a Taylor series expansion to represent a functional relationship as was done in Chapter 1 for the Source Separation model.

The p-dimensional recorded values are denoted as x_i's and the q emitted values along with the overall mean by the $(q+1)$-dimensional u_i's. The u_i's are of dimension $(q+1)$ because they contain a 1 as their first element for the overall mean (constant or intercept). The q specified to be observable sources, the u_{ik}'s are used for distinction from the m unobservable sources, the s_{ik}'s in Chapter 1. The mixing coefficients are denoted by β's to denote that they are coefficients for observable variables (sources) and not λ's for unobservable variables (sources).

Given a set of independent and identically distributed vector-valued observations x_i, $i = 1, \ldots, n$, on p possibly correlated random variables, the Multivariate Regression model on the variable u_i is

$$
\begin{array}{cccc}
(x_i|B, u_i) = & B & u_i & + & \epsilon_i, \\
(p \times 1) & [p \times (q+1)] & [(q+1) \times 1] & (p \times 1)
\end{array}
\tag{8.2.1}
$$

where the matrix of Regression coefficients B is given by

$$B = \begin{pmatrix} \beta'_1 \\ \vdots \\ \beta'_p \end{pmatrix}, \tag{8.2.2}$$

which describes the relationship between $u_i = (1, u_{i1}, \ldots, u_{iq})'$ and x_i while ϵ_i is the p-dimensional Normally distributed error.

Note that if the coefficient matrix B were written as $B = (\mu, B_\star)$ and $u_i = (1, u'_{\star i})'$, then

$$\begin{array}{ccccc}
(x_i|\mu, B_\star, u_i) = & \mu & + & B_\star & u_{\star i} & + & \epsilon_i, \\
(p \times 1) & (p \times 1) & & (p \times q) & (q \times 1) & & (p \times 1)
\end{array} \tag{8.2.3}$$

which is the same basic model as in Source Separation.

More specifically for the Regression model, a given element of x_i, say the jth is

$$\begin{array}{cccc}
(x_{ij}|\beta_j, u_i) = & \beta'_j & u_i & + & \epsilon_{ij}, \\
(1 \times 1) & [1 \times (q+1)] & [(q+1) \times 1] & & (1 \times 1)
\end{array} \tag{8.2.4}$$

where $\beta'_j = (\beta_{j0}, \ldots, \beta_{jq})$. This is also represented as

$$(x_{ij}|\beta_j, u_i) = \sum_{t=0}^{q} \beta_{jt} \, u_{it} + \epsilon_{ij}. \tag{8.2.5}$$

Gathering all observed vectors into a matrix, the Regression model may be written as

$$\begin{array}{ccccc}
(X|B, U) = & U & B' & + & E, \\
(n \times p) & [n \times (q+1)] & [(q+1) \times p] & & (n \times p)
\end{array} \tag{8.2.6}$$

where the matrix of dependent variables (observed mixed sources) X, the independent variables (observable sources) U, and the matrix of errors E are defined to be

$$X = \begin{pmatrix} x'_1 \\ \vdots \\ x'_n \end{pmatrix}, \qquad U = \begin{pmatrix} u'_1 \\ \vdots \\ u'_n \end{pmatrix}, \qquad E = \begin{pmatrix} \epsilon'_1 \\ \vdots \\ \epsilon'_n \end{pmatrix}, \tag{8.2.7}$$

while B is as above.

The ij^{th} element of X is the ith row of U multiplied by the jth column of B' plus the ij^{th} element of E. The model may also be written in terms of columns by parameterizing the dependent variables (observed mixed sources) X, the independent variables (observable source) matrix U, the matrix of Regression coefficients B, and the matrix of errors E as

$$X = (X_1,\ldots,X_p), \qquad U = (e_n, U_1,\ldots,U_q),$$
$$B' = (\beta_1, \beta_1,\ldots,\beta_p), \qquad E = (E_1,\ldots,E_p), \qquad (8.2.8)$$

where e_n is an n-dimensional column vector of ones.
This leads to the model

$$\underset{(n\times 1)}{(X_j|\beta_j, U)} = \underset{[n\times (q+1)]}{U} \quad \underset{[(q+1)\times 1]}{\beta_j} \quad + \quad \underset{(n\times 1)}{E_i}, \qquad (8.2.9)$$

which describes all the observations for a single microphone in the cocktail party problem at all n time points.

8.3 Likelihood

In observing variables, there is inherent random variability or error which statistical models quantify. The errors of observation are characterized to have arisen from a particular distribution. It is specified that the errors of observation are independent and Normally distributed random vectors with p-dimensional mean 0 and $p \times p$ covariance matrix Σ. The Multivariate p-dimensional Normal distribution for the errors described in Chapter 2 is denoted by

$$p(\epsilon_i|\Sigma) \propto |\Sigma|^{-\frac{1}{2}} e^{-\frac{1}{2}\epsilon_i'\Sigma^{-1}\epsilon_i}, \qquad (8.3.1)$$

where ϵ_i is the p-dimensional error vector. It is common in Bayesian Statistics to omit the normalization constant and use proportionality.

From this Multivariate Normal error specification, the distribution of the observation vectors is also Multivariate Normally distributed and given by

$$p(x_i|B, u_i, \Sigma) \propto |\Sigma|^{-\frac{1}{2}} e^{-\frac{1}{2}(x_i - Bu_i)'\Sigma^{-1}(x_i - Bu_i)}, \qquad (8.3.2)$$

because the Jacobian of the transformation from ϵ_i to x_i is unity.

With the matrix representation of the model given by Equation 8.2.6, the distribution of the matrix of observations is a Matrix Normal distribution as described in Chapter 2

$$p(X|B, \Sigma, U) \propto |\Sigma|^{-\frac{n}{2}} e^{-\frac{1}{2}tr(X - UB')\Sigma^{-1}(X - UB')'}, \qquad (8.3.3)$$

where "tr" denotes the trace operator which yields the sum of the diagonal elements of its matrix argument.

8.4   Conjugate Priors and Posterior

In the Bayesian approach to statistical inference, available prior information regarding parameter values is quantified in terms of prior distributions to represent the current state of knowledge before an experiment is performed and data taken. In this Section, Conjugate prior distributions are used to characterize our beliefs regarding the parameters, namely, the Normal and the Inverted Wishart which are described in Chapters 2 and 4. Later in this Chapter, generalized Conjugate prior distributions will be used.

Using the Conjugate procedure given in Chapter 4, the joint prior distribution is $p(B, \Sigma)$ for the parameters B, and Σ is the product of the prior distribution of B given Σ, $p(B|\Sigma)$, and the prior distribution of Σ, $p(\Sigma)$. This is expressed as

$$p(B, \Sigma) = p(B|\Sigma)p(\Sigma), \qquad (8.4.1)$$

where the prior distribution for the Regression coefficients $B|\Sigma$ is Matrix Normally distributed as

$$p(B|\Sigma) \propto |D|^{-\frac{p}{2}} |\Sigma|^{-\frac{(q+1)}{2}} e^{-\frac{1}{2} tr D^{-1}(B-B_0)' \Sigma^{-1}(B-B_0)} \qquad (8.4.2)$$

and the prior distribution for the error covariance matrix Σ is Inverse Wishart distributed as

$$p(\Sigma) \propto |\Sigma|^{-\frac{\nu}{2}} e^{-\frac{1}{2} tr \Sigma^{-1} Q}. \qquad (8.4.3)$$

The quantities D, B_0, ν, and Q are hyperparameters to be assessed. By specifying these hyperparameters, the entire joint prior distribution is determined.

The joint posterior distribution of the model parameters B and Σ with specified Conjugate priors is

$$p(B, \Sigma|X, U) \propto p(\Sigma)p(B|\Sigma)p(X|B, \Sigma, U) \qquad (8.4.4)$$

and after inserting the aforementioned prior distributions and likelihood becomes

$$p(B, \Sigma|X, U) \propto |\Sigma|^{-\frac{(n+\nu+q+1)}{2}}$$
$$\times e^{-\frac{1}{2} tr \Sigma^{-1}[(X-UB')'(X-UB')+(B-B_0)D^{-1}(B-B_0)'+Q]},$$
$$(8.4.5)$$

where the property of the trace operator that the trace of the product of two conformable matrices is equal to the trace of their product in the reverse order was used.

8.5 Conjugate Estimation and Inference

The estimation of and inference on parameters as outlined in Chapter 6 in a Bayesian Statistical procedure is often the most difficult part of the analysis. Two different methods are used to estimate parameters and draw inferences. The methods of estimation are marginal posterior mean and maximum a posteriori estimates. Formulas for marginal mean and joint maximum a posteriori estimates are derived.

8.5.1 Marginalization

Marginal posterior mean estimation of the parameters involves computing the marginal posterior distribution for each of the parameters and then computing mean estimates from these marginal posterior distributions.

To find the marginal posterior distribution of the matrix of Regression coefficients B, the joint posterior distribution Equation 8.4.5 must be integrated with respect to Σ. This can be performed easily by recognizing (as described in Chapter 6) that the posterior distribution is exactly of the same form as an Inverted Wishart distribution except for a proportionality constant. Integration can be easily performed using the definition of an Inverted Wishart distribution. The marginal posterior distribution for the matrix of Regression coefficients B after integrating with respect to Σ is given by

$$p(B|X,U) \propto \frac{1}{|G+(B-\bar{B})(D^{-1}+U'U)(B-\bar{B})'|^{\frac{(n+\nu-p-1+q+1)}{2}}}, \quad (8.5.1)$$

where the matrix

$$\bar{B} = (X'U+B_0D^{-1})(D^{-1}+U'U)^{-1}, \quad (8.5.2)$$

and B_0 is the prior mean while the $p \times p$ matrix G (after some algebra) has been written as

$$\begin{aligned} G = Q + X'X + B_0D^{-1}B_0' \\ - (X'U+B_0D^{-1})(D^{-1}+U'U)^{-1}(X'U+B_0D^{-1})' \end{aligned} \quad (8.5.3)$$

have been defined.

The mean and modal estimate of the matrix of Regression coefficients from this exact marginal posterior distribution is $\bar{B}$. The marginal posterior variance of the matrix of Regression coefficients is

$$var(B|\bar{B},X,U) = \frac{n-(q+1)}{n-(q+1)-2} \, G \otimes [(n-q-1)(U'U)]^{-1} \quad (8.5.4)$$

or equivalently

$$var(\beta|\bar{\beta},X,U) = \frac{n+\nu-p-1}{n+\nu-p-1-2} \, [(n-q-1)(U'U)]^{-1} \otimes G, \quad (8.5.5)$$

$$= \hat{\Delta}, \quad (8.5.6)$$

where $\beta = vec(B)$.

This marginal posterior distribution is easily recognized as being a Matrix Student T-distribution as described in Chapter 2. That is,

$$B|\bar{B},U \sim T\left(n+\nu-p-1, \bar{B}, \left[(n+\nu-p-1)(D^{-1}+U'U)\right]^{-1}, G\right). \quad (8.5.7)$$

Inferences such as tests of hypothesis and credibility intervals can be evaluated on B as in the Regression Chapter. Hypotheses can be performed regarding the entire coefficient matrix, a submatrix, a particular row or column, or a particular element.

The above distribution for the matrix of regression coefficients can be written in another form

$$p(B|X,U,\bar{B}) \propto \frac{1}{|W+(B-\bar{B})'G^{-1}(B-\bar{B})|^{\frac{(n_*-q-p)+(q+1)}{2}}} \quad (8.5.8)$$

by using Sylvester's theorem [41]

$$|I_n+YI_pY'| = |I_p+Y'I_nY|, \quad (8.5.9)$$

where $W = \left(D^{-1}+U'U\right)^{-1}$ and $n_* = n+\nu-p+2q+1$ have been defined.

It can be shown that the marginal distribution of any column, say the kth of the matrix of B, B_k is Multivariate Student t-distributed

$$p(B_k|\bar{B}_k,X,U) \propto \frac{1}{\left[W_{kk}+(B_k-\bar{B}_k)'G^{-1}(B_k-\bar{B}_k)\right]^{\frac{(n_*-q-p)+p}{2}}}, \quad (8.5.10)$$

where W_{kk} is the kth diagonal element of W.

It can be also be shown that the marginal distribution of any row, say the jth of the matrix B, β'_j is Multivariate Student t-distributed

$$p(\beta_j|\bar{\beta}_j,X,U) \propto \frac{1}{\left[G_{jj}+(\beta_j-\bar{\beta}_j)'W^{-1}(\beta_j-\bar{\beta}_j)\right]^{\frac{(n_*-q-p)+(q+1)}{2}}}, \quad (8.5.11)$$

where G_{jj} is the jth diagonal element of G.

With the marginal distribution of a column or row of B, significance can be evaluated for the coefficients of a particular independent variable (column of B) with the use of the statistic

$$F_{p,n_*-q-p} = \frac{n_* - q - p}{p} W_{kk}^{-1} \bar{B}_k' G^{-1} \bar{B}_k, \qquad (8.5.12)$$

or for the coefficients of a particular dependent variable (row of B) with the use of the statistic

$$F_{q+1,n_*-q-p} = \frac{n_* - q - p}{q+1} G_{jj}^{-1} \bar{\beta}_j' W^{-1} \bar{\beta}_j. \qquad (8.5.13)$$

These statistics follow F-distributions with either $(p, n_* - q - p)$ or $(q+1, n_* - q - p)$ numerator and denominator degrees of freedom respectively [39, 41]. Significance can be determined for a subset of coefficients by determining the marginal distribution of the subset within B_k or β_j' which is also Multivariate Student t-distributed.

With the subset of coefficients being a singleton set, significance can be determined for a particular coefficient with the marginal distribution of the scalar coefficient which is

$$p(B_{kj}|\bar{B}_{kj}, X, U) \propto \frac{1}{\left[W_{kk} + (B_{kj} - \bar{B}_{kj}) G_{jj}^{-1} (B_{kj} - \bar{B}_{kj}) \right]^{\frac{(n_*-q-p)+1}{2}}}, \qquad (8.5.14)$$

where G_{jj} is the jth diagonal element of G. The above can be rewritten in the more familiar form

$$p(B_{kj}|\bar{B}_{kj}, X, U) \propto \frac{1}{\left[1 + \frac{1}{(n_*-q-p)} \frac{(B_{kj} - \bar{B}_{kj})^2}{W_{kk}[G_{jj}/(n_*-q-p)]} \right]^{\frac{(n_*-q-p)+1}{2}}} \qquad (8.5.15)$$

which is readily recognizable as a Scalar Student t-distribution. Note that $\bar{B}_{kj} = \bar{\beta}_{jk}$ and that

$$t = \frac{(B_{kj} - \bar{B}_{kj})}{[W_{kk} G_{jj}(n_* - q - p)^{-1}]^{\frac{1}{2}}} \qquad (8.5.16)$$

follows a Scalar Student t-distribution with $n_* - q - p$ degrees of freedom, and t^2 follows an F-distribution with 1 and $n_* - q - p$ numerator and denominator degrees of freedom. The F-distribution is commonly used in Regression [1, 68] and derived from a likelihood ratio test of reduced and full models when testing coefficients. By using a t statistic instead of an F statistic, positive and negative coefficient values can be identified. Even for a modest sample size, this Scalar Student t-distribution typically has a large number of degrees of freedom $(n+\nu-q-2p+1)$ so that it is nearly equivalent to a Scalar Normal distribution as noted in Chapter 2.

Note that the mean of this marginal posterior distribution $\bar{B}$ can be written as

$$
\begin{aligned}
\bar{B} &= X'U(D^{-1}+U'U)^{-1}+B_0 D^{-1}(D^{-1}+U'U)^{-1}\\
&= \hat{B}[U'U(D^{-1}+U'U)^{-1}]+B_0[D^{-1}(D^{-1}+U'U)^{-1}], \quad (8.5.17)
\end{aligned}
$$

where we have defined the matrix $\hat{B} = X'U(U'U)^{-1}$. This posterior mean is a weighted combination of the prior mean B_0 from the prior distribution and the data mean $\hat{B}$ from the likelihood.

The above mean for the matrix of Regression coefficients can be written as

$$
\bar{B} = \hat{B}\frac{U'U}{n}\left[(nD)^{-1}+\frac{U'U}{n}\right]^{-1}+B_0(nD)^{-1}\left[(nD)^{-1}+\frac{U'U}{n}\right]^{-1},
$$

$$(8.5.18)$$

and as the sample size n increases, $(nD)^{-1}$ approaches the zero matrix, $\frac{U'U}{n}$ approaches a constant matrix, and the estimate of the matrix of Regression coefficients B is based only on the data mean $\hat{B}$ from the likelihood. Thus, the prior distribution has decreasing influence as the sample size increases. This is a feature of Bayesian estimators.

To integrate the joint posterior distribution in order to find the marginal posterior distribution of Σ, rearrange the terms in the exponent of the posterior distribution as in Chapter 6, and complete the square to find

$$
\begin{aligned}
p(B,\Sigma|X,U) &\propto |\Sigma|^{-\frac{(n+\nu)}{2}}e^{-\frac{1}{2}tr\Sigma^{-1}G}\\
&\times|\Sigma|^{-\frac{(q+1)}{2}}e^{-\frac{1}{2}tr\Sigma^{-1}(B-\bar{B})(U'U+D^{-1})(B-\bar{B})'}, \quad (8.5.19)
\end{aligned}
$$

where $\bar{B}$ and G are as previously defined. In the above equation, the last line with an additional multiplicative constant is a Matrix Normal distribution (Chapter 2). Using the definition of a Matrix Normal distribution for the purpose of integration as in Chapter 6, the marginal posterior distribution for Σ is

$$
p(\Sigma|X,U) \propto |\Sigma|^{-\frac{(n+\nu)}{2}}e^{-\frac{1}{2}tr\Sigma^{-1}G}, \quad (8.5.20)
$$

where the $p \times p$ matrix G is as defined above. It is easily seen that the mean of this marginal posterior distribution is

$$
\bar{\Sigma} = \frac{G}{n+\nu-2p-2}, \quad (8.5.21)
$$

while the mode is

$$\bar{\Sigma}_{mode} = \frac{G}{n+\nu}. \tag{8.5.22}$$

The mean and mode of an Inverted Wishart distribution are given in Chapter 2.

Since exact marginal estimates were found, a Gibbs sampling algorithm for computing exact sampling based marginal posterior means and variances does not need to be given.

8.5.2 Maximum a Posteriori

The joint posterior distribution may also be maximized with respect to B and Σ by direct differentiation as described in Chapter 6 to obtain maximum a posteriori estimates

$$\tilde{B} = \overset{\text{Arg Max}}{B}\, p(B|\tilde{\Sigma}, X, U) \tag{8.5.23}$$

$$= \tilde{B}(\tilde{\Sigma}, X, U) \tag{8.5.24}$$

$$\tilde{\Sigma} = \overset{\text{Arg Max}}{\Sigma}\, p(\Sigma|\tilde{B}, X, U) \tag{8.5.25}$$

$$= \tilde{\Sigma}(\tilde{B}, X, U). \tag{8.5.26}$$

The maximum a posteriori estimators are analogous to the maximum likelihood estimates found for the classical Regression model.

Upon performing some simplifying algebra, taking the natural logarithm, differentiating the joint posterior distribution in Equation 8.5.19 with respect to B, the result is

$$\frac{\partial}{\partial B}\log(p(B,\Sigma|X)) = \frac{\partial}{\partial B}tr(U'U + D^{-1})(B-\tilde{B})'\Sigma^{-1}(B-\tilde{B})$$

$$= 2\Sigma^{-1}(B-\tilde{B})(U'U + D^{-1}) \tag{8.5.27}$$

which upon evaluating this at $(B,\Sigma) = (\tilde{B},\tilde{\Sigma})$ and setting it equal to the null matrix yields a maxima of $\tilde{B} = \bar{B}$ as given in Equation 8.5.2. Note that a matrix derivative result as in Chapter 6 was used.

Upon differentiating the joint posterior distribution in Equation 8.5.19 with respect to Σ, the result is

$$\frac{\partial}{\partial \Sigma}\log(p(B,\Sigma|X)) = -\frac{(n+\nu+q+1)}{2}\frac{\partial}{\partial \Sigma}\log|\Sigma| + \frac{\partial}{\partial \Sigma}tr\Sigma^{-1}G$$

$$= -\frac{(n+\nu+q+1)}{2}[2\Sigma^{-1} - diag(\Sigma^{-1})]$$

$$+ \frac{1}{2}[2G^{-1} - diag(G^{-1})], \tag{8.5.28}$$

which upon setting this expression equal to the null matrix and evaluating at $(B, \Sigma) = (\tilde{B}, \tilde{\Sigma})$ yields a maxima of

$$\tilde{\Sigma} = \frac{G}{n + \nu + q + 1}. \tag{8.5.29}$$

Note that the estimator of $\tilde{B}$ is identical to that of the marginal mean $\bar{B}$, but the one of Σ, $\tilde{\Sigma}$ is different by a multiplicative factor. Conditional maximum a posteriori variance estimates can also be found. The conditional modal variance of the Regression coefficients is

$$var(B|\tilde{B}, \tilde{\Sigma}, X, U) = \tilde{\Sigma} \otimes (D^{-1} + U'U)^{-1} \tag{8.5.30}$$

or equivalently

$$var(\beta|\tilde{\beta}, \tilde{\Sigma}, X, U) = (D^{-1} + U'U)^{-1} \otimes \tilde{\Sigma},$$
$$= \tilde{\Delta},$$

where $\beta = vec(B)$.

Conditional modal intervals may be computed by using the conditional distribution for a particular parameter given the modal values of the others. The posterior conditional distribution of the matrix of Regression coefficients B given the modal values of the other parameters and the data is

$$p(B|\tilde{B}, \tilde{\Sigma}, U, X) \propto |(D^{-1} + U'U)|^{\frac{p}{2}} |\tilde{\Sigma}|^{-\frac{q+1}{2}}$$
$$\times e^{-\frac{1}{2} tr \tilde{\Sigma}^{-1}(B - \tilde{B})(D^{-1} + U'U)(B - \tilde{B})'}, \tag{8.5.31}$$

which may be also written in terms of vectors as

$$p(\beta|\tilde{\beta}, \tilde{\Sigma}, U, X) \propto |(D^{-1} + U'U)^{-1} \otimes \tilde{\Sigma}|^{-\frac{1}{2}}$$
$$\times e^{-\frac{1}{2}(\beta - \tilde{\beta})'[(D^{-1} + U'U)^{-1} \otimes \tilde{\Sigma}]^{-1}(\beta - \tilde{\beta})}. \tag{8.5.32}$$

It can be shown [17, 41] that the marginal distribution of any column of the matrix of B, B_k is Multivariate Normal

$$p(B_k|\tilde{B}_k, \tilde{\Sigma}, U, X) \propto |W_{kk}\tilde{\Sigma}|^{-\frac{1}{2}} e^{-\frac{1}{2}(B_k - \tilde{B}_k)'(W_{kk}\tilde{\Sigma})^{-1}(B_k - \tilde{B}_k)}, \tag{8.5.33}$$

where $W = (D^{-1} + U'U)^{-1}$ and W_{kk} is its kth diagonal element.

With the marginal distribution of a column of B, significance can be determined for the set of coefficients of an independent variable. Significance can be determined for a subset of coefficients by determining the marginal distribution of the subset within B_k which is also Multivariate Normal. With the subset being a singleton set, significance can be determined for a particular coefficient with the marginal distribution of the scalar coefficient which is

$$p(B_{kj}|\tilde{B}_{kj},\tilde{\Sigma}_{jj},U,X) \propto (W_{kk}\tilde{\Sigma}_{jj})^{-\frac{1}{2}}e^{-\frac{(B_{kj}-\tilde{B}_{kj})^2}{2W_{kk}\tilde{\Sigma}_{jj}}}, \tag{8.5.34}$$

where $\tilde{\Sigma}_{jj}$ is the jth diagonal element of $\tilde{\Sigma}$. Note that $\tilde{B}_{kj} = \tilde{\beta}_{jk}$ and that

$$z = \frac{(B_{kj}-\tilde{B}_{kj})}{\sqrt{W_{kk}\tilde{\Sigma}_{jj}}} \tag{8.5.35}$$

$$= \frac{(B_{kj}-\tilde{B}_{kj})}{\sqrt{W_{kk}G_{jj}/(n+\nu+q+1)}} \tag{8.5.36}$$

follows a Normal distribution with a mean of zero and variance of one.

8.6 Generalized Priors and Posterior

Using the generalized Conjugate procedure given in Chapter 4, the joint prior distribution $p(\beta,\Sigma)$ for the parameters $\beta = vec(B)$ and Σ, "vec" being the vectorization operator that stacks the columns of its matrix argument, is given by the product of the prior distribution $p(\beta)$ for the Regression coefficients and that for the error covariance matrix $p(\Sigma)$

$$p(\beta,\Sigma) = p(\beta)p(\Sigma). \tag{8.6.1}$$

These prior distributions are found from the generalized Conjugate procedure in Chapter 4 and given by

$$p(\beta) \propto |\Delta|^{-\frac{1}{2}}e^{-\frac{1}{2}(\beta-\beta_0)'\Delta^{-1}(\beta-\beta_0)} \tag{8.6.2}$$

and

$$p(\Sigma) \propto |\Sigma|^{-\frac{\nu}{2}}e^{-\frac{1}{2}tr\Sigma^{-1}Q}, \tag{8.6.3}$$

where the quantities Δ, β_0, ν, and Q are hyperparameters to be assessed. The matrices Δ, Σ, and Q are all positive definite. By specifying the hyperparameters, the entire joint prior distribution is determined.

By Bayes' rule, the joint posterior distribution for the unknown model parameters with specified generalized Conjugate priors is given by

$$p(\beta, \Sigma | X, U) \propto p(\Sigma)p(\beta)p(X | B, \Sigma, U), \qquad (8.6.4)$$

which becomes

$$p(\beta, \Sigma | X, U) \propto |\Delta|^{-\frac{1}{2}} e^{-\frac{1}{2}(\beta-\beta_0)'\Delta^{-1}(\beta-\beta_0)}$$
$$\times |\Sigma|^{-\frac{(n+\nu)}{2}} e^{-\frac{1}{2}tr\Sigma^{-1}[(X-UB')'(X-UB')+Q]} \qquad (8.6.5)$$

after inserting the generalized Conjugate prior distributions and the likelihood.

Now the joint posterior distribution is to be evaluated in order to obtain estimates of the parameters of the model.

8.7 Generalized Estimation and Inference

8.7.1 Marginalization

Marginal estimation of the parameters involves computing the marginal posterior distribution for each of the parameters then computing estimates from these marginal distributions.

To find the marginal posterior distribution of B or β, the joint posterior distribution Equation 8.6.5 must be integrated with respect to Σ. Integration of the joint posterior distribution is performed as described in Chapter 6 by first multiplying and dividing by

$$|(X - UB')'(X - UB') + Q|^{-\frac{(n+\nu-p-1)}{2}}$$

and then using the definition of the Inverted Wishart distribution given in Chapter 2 to get

$$p(\beta | X, U) \propto \frac{|\Delta|^{-\frac{1}{2}} e^{-\frac{1}{2}(\beta-\beta_0)'\Delta^{-1}(\beta-\beta_0)}}{|(X - UB')'(X - UB') + Q|^{-\frac{(n+\nu-p-1)}{2}}}. \qquad (8.7.1)$$

The above expression for the marginal distribution of the vector of Regression coefficients can be written as

$$p(\beta | X, U) \propto \frac{|\Delta|^{-\frac{1}{2}} e^{-\frac{1}{2}(\beta-\beta_0)'\Delta^{-1}(\beta-\beta_0)}}{|I_p + Q^{-\frac{1}{2}}(X - UB')'I_n(X - UB')Q^{-\frac{1}{2}}|^{-\frac{(n+\nu-p-1)}{2}}} \qquad (8.7.2)$$

by performing some algebra.

Sylvester's theorem [41]

$$|I_n + Y I_p Y'| = |I_p + Y' I_n Y| \tag{8.7.3}$$

is applied in the denominator of the marginal posterior distribution to obtain

$$p(\beta|X,U) \propto \frac{|\Delta|^{-\frac{1}{2}} e^{-\frac{1}{2}(\beta-\beta_0)'\Delta^{-1}(\beta-\beta_0)}}{|I_n + (X - U B')Q^{-1}(X - U B')'|^{-\frac{(n+\nu-p-1)}{2}}}. \tag{8.7.4}$$

Note the change that has occurred in the denominator. The large sample approximation that $|I + \Xi|^\alpha \cong e^{\alpha tr \Xi}$ is used [41] where α is a scalar and Ξ is an $n \times n$ positive definite matrix with eigenvalues which lie within the unit circle to obtain the approximate expression

$$p(\beta|X,U) \propto e^{-\frac{1}{2}(\beta-\beta_0)'\Delta^{-1}(\beta-\beta_0)} e^{-\frac{1}{2}tr[(X-UB')(\frac{Q}{n+\nu-p-1})^{-1}(X-UB')']} \tag{8.7.5}$$

for the marginal posterior distribution of the vector of Regression coefficients. After some algebra, the above is written as

$$p(\beta|X,U) \propto e^{-\frac{1}{2}(\beta-\breve{\beta})'\left[U'U \otimes \left(\frac{Q}{n+\nu-p-1}\right)^{-1} + \Delta^{-1}\right](\beta-\breve{\beta})}, \tag{8.7.6}$$

where the vector $\breve{\beta}$ is defined to be

$$\breve{\beta} = \left[\Delta^{-1} + U'U \otimes \left(\frac{Q}{n+\nu-p-1}\right)^{-1}\right]^{-1}$$
$$\times \left[\Delta^{-1}\beta_0 + U'U \otimes \left(\frac{Q}{n+\nu-p-1}\right)^{-1} \hat{\beta}\right] \tag{8.7.7}$$

and the vector $\hat{\beta}$ to be

$$\hat{\beta} = vec(\hat{B})$$
$$= vec(X'U(U'U)^{-1}). \tag{8.7.8}$$

The exact marginal mean and modal estimator for β from this approximate marginal posterior distribution is $\breve{\beta}$. Note that $\breve{\beta}$ is a weighted average of the prior mean from the prior distribution and the data mean from the likelihood.

To find the marginal posterior distribution of Σ, rearrange the terms in the exponent of the joint posterior distribution to arrive at

$$p(\beta, \Sigma | X, U) \propto |\Sigma|^{-\frac{(n+\nu)}{2}} e^{-\frac{1}{2} tr \Sigma^{-1} Q}$$
$$\times e^{-\frac{1}{2}[(\beta - \beta_0)' \Delta^{-1}(\beta - \beta_0) + (\beta - \hat{\beta})'(U'U \otimes \Sigma^{-1})(\beta - \hat{\beta})]}, \quad (8.7.9)$$

where the vector $\hat{\beta}$ is as previously defined.

Continuing on, complete the square of the terms in the exponent of the term involving β to find

$$p(\beta, \Sigma | X, U) \propto |\Sigma|^{-\frac{(n+\nu)}{2}} e^{-\frac{1}{2} tr \Sigma^{-1} [Q + (X - U\hat{B}')'(X - U\hat{B}')]}$$
$$\times e^{-\frac{1}{2}[(\beta - \tilde{\beta})'(\Delta^{-1} + U'U \otimes \Sigma^{-1})(\beta - \tilde{\beta})]}, \quad (8.7.10)$$

where again the vector $\tilde{\beta}$ has been defined as

$$\tilde{\beta} = [\Delta^{-1} + U'U \otimes \Sigma^{-1}]^{-1}[\Delta^{-1}\beta_0 + (U'U \otimes \Sigma^{-1})\hat{\beta}]. \quad (8.7.11)$$

Now, integration will be performed by recognition as in Chapter 6. Multiply and divide by the quantity

$$|\Delta^{-1} + U'U \otimes \Sigma^{-1}|^{-\frac{1}{2}} \quad (8.7.12)$$

and by integrating with the definition of a Multivariate Normal distribution as in Chapter 6, the marginal posterior distribution for Σ is given by

$$p(\Sigma | X, U) \propto \frac{|\Sigma|^{-\frac{(n+\nu)}{2}} e^{-\frac{1}{2} tr \Sigma^{-1} \{Q + (X - U\hat{B}')'(X - U\hat{B}')\}}}{|\Delta^{-1} + U'U \otimes \Sigma^{-1}|^{-\frac{1}{2}}}, \quad (8.7.13)$$

which by using a large sample result [41],

$$|\Delta^{-1} + U'U \otimes \Sigma^{-1}| = |\Sigma|^{-(q+1)} |U'U|^{-p} \quad (8.7.14)$$

is approximately for large samples

$$p(\Sigma | X, U) \propto |\Sigma|^{-\frac{(n+\nu-q-1)}{2}} e^{-\frac{1}{2} tr \Sigma^{-1} \{Q + (X - U\hat{B}')'(X - U\hat{B}')\}}. \quad (8.7.15)$$

From this approximate marginal posterior distribution which can easily be recognized as an Inverted Wishart distribution, the exact marginal posterior mean is

$$\breve{\Sigma} = \frac{Q + (X - U\hat{B}')'(X - U\hat{B}')}{n + \nu - q - 1 - 2p - 2}, \quad (8.7.16)$$

while the exact mode is

$$\breve{\Sigma}_{mode} = \frac{Q + (X - U\hat{B}')'(X - U\hat{B}')}{n + \nu - q - 1}. \quad (8.7.17)$$

8.7.2   Posterior Conditionals

Since approximations were made when finding the marginal posterior distributions for the vector of Regression coefficients β and the error covariance matrix Σ, a Gibbs sampling algorithm is given for computing exact sampling based quantities such as marginal mean and variance estimates of the parameters. The posterior conditional distributions of β and Σ are needed for the algorithm. The posterior conditional distribution for β is found by considering only those terms in the joint posterior distribution which involve β and is given by

$$p(\beta|\Sigma, X, U) \propto p(\beta)p(X|B, \Sigma, X, U)$$
$$\propto e^{-\frac{1}{2}(\beta-\tilde{\beta})'(\Delta^{-1}+U'U\otimes\Sigma^{-1})(\beta-\tilde{\beta})}. \qquad (8.7.18)$$

This is recognizable as a Multivariate Normal distribution for the vector of Regression coefficients for β whose mean and mode is

$$\tilde{\beta} = [\Delta^{-1}+U'U\otimes\Sigma^{-1}]^{-1}[\Delta^{-1}\beta_0 + (U'U\otimes\Sigma^{-1})\hat{\beta}], \qquad (8.7.19)$$

where the vector $\hat{\beta}$ is

$$\hat{\beta} = vec(X'U(U'U)^{-1}) \qquad (8.7.20)$$

which was found by completing the square in the exponent. Note that this is a weighted combination of the prior mean from the prior distribution and the data mean from the likelihood.

The posterior conditional distribution of the error covariance matrix Σ is similarly found by considering only those terms in the joint posterior distribution which involve Σ and is given by

$$p(\Sigma|\beta, X, U) \propto p(\Sigma)p(X|B, \Sigma, X, U)$$
$$\propto |\Sigma|^{-\frac{(n+\nu)}{2}}e^{-\frac{1}{2}tr\Sigma^{-1}[(X-UB')'(X-UB')+Q]}. \qquad (8.7.21)$$

This is easily recognized as being an Inverted Wishart distribution with mode

$$\tilde{\Sigma} = \frac{(X-UB')'(X-UB')+Q}{n+\nu} \qquad (8.7.22)$$

as described in Chapter 2.

8.7.3   Gibbs Sampling

To obtain marginal mean and variance estimates for the model parameters using the Gibbs sampling algorithm as described in Chapter 6, start with an initial value for the error covariance matrix Σ, say $\bar{\Sigma}_{(0)}$, and then cycle through

$$
\begin{aligned}
\bar{\beta}_{(l+1)} &= \text{a random variate from } p(\beta|\bar{\Sigma}_{(l)}, X, U) \\
&= A_\beta Y_\beta + M_\beta \quad\quad\quad\quad\quad\quad\quad (8.7.23) \\
\bar{\Sigma}_{(l+1)} &= \text{a random variate from } p(\Sigma|\bar{B}_{(l+1)}, X, U) \\
&= A_\Sigma (Y_\Sigma' Y_\Sigma)^{-1} A_\Sigma', \quad\quad\quad\quad\quad\quad (8.7.24)
\end{aligned}
$$

where

$$
\begin{aligned}
\bar{\beta}_{(l+1)} &= vec(\bar{B}_{(l+1)}) \\
A_\beta A_\beta' &= (\Delta^{-1} + U'U \otimes \bar{\Sigma}_{(l)}^{-1})^{-1} \\
M_\beta &= [\Delta^{-1} + U'U \otimes \bar{\Sigma}_{(l)}^{-1}]^{-1} [\Delta^{-1}\beta_0 + (U'U \otimes \bar{\Sigma}_{(l)}^{-1})\hat{\beta}] \\
A_\Sigma A_\Sigma' &= (X - U\bar{B}_{(l+1)}')'(X - U\bar{B}_{(l+1)}') + Q
\end{aligned}
$$

while Y_β is a $p(q+1) \times 1$ dimensional vector and Y_Σ is an $(n+\nu+p+1) \times p$, dimensional matrix whose respective elements are random variates from the standard Scalar Normal distribution. The formulas for the generation of random variates from the conditional posterior distributions are easily found from the methods in Chapter 6.

The first random variates called the "burn in" are discarded and after doing so, compute from the next L variates means of each of the parameters

$$
\bar{\beta} = \frac{1}{L} \sum_{l=1}^{L} \bar{\beta}_{(l)} \quad \text{and} \quad \bar{\Sigma} = \frac{1}{L} \sum_{l=1}^{L} \bar{\Sigma}_{(l)}
$$

which are the exact sampling based marginal mean parameters estimates from the posterior distribution. Exact sampling based marginal estimates of other quantities can also be found. Of interest is the estimate of the marginal posterior variance of the regression coefficients

$$
\begin{aligned}
\overline{var}(\beta|X, U) &= \frac{1}{L} \sum_{l=1}^{L} \bar{\beta}_{(l)} \bar{\beta}_{(l)}' - \bar{\beta}\bar{\beta}' \\
&= \bar{\Delta}.
\end{aligned}
$$

The covariance matrices of the other parameters follow similarly. With a specification of Normality for the marginal posterior distribution of the Regression coefficients, their distribution is

$$
p(\beta|X, U) \propto |\bar{\Delta}|^{-\frac{1}{2}} e^{-\frac{1}{2}(\beta - \bar{\beta})'\bar{\Delta}^{-1}(\beta - \bar{\beta})}, \quad\quad\quad (8.7.25)
$$

where $\bar{\beta}$ and $\bar{\Delta}$ are as previously defined.

To determine statistical significance with the Gibbs sampling approach, use the marginal distribution of the vector of Regression coefficients given above. General simultaneous hypotheses can be evaluated on the entire vector or a subset of it. Significance regarding the coefficient for a particular independent variable can be evaluated by computing marginal distributions. It can be shown that the marginal distribution of the kth column of the matrix of Regression coefficients B, $B_k = \beta'_k$ is Multivariate Normal

$$p(B_k|\bar{B}_k, X, U) \propto |\bar{\Delta}_k|^{-\frac{1}{2}} e^{-\frac{1}{2}(B_k - \bar{B}_k)' \bar{\Delta}_k^{-1}(B_k - \bar{B}_k)}, \qquad (8.7.26)$$

where $\bar{\Delta}_k$ is the covariance matrix of B_k found by taking the kth $p \times p$ submatrix along the diagonal of $\bar{\Delta}$.

Significance can be determined for a subset of coefficients of the kth column of B by determining the marginal distribution of the subset within B_k which is also Multivariate Normal. With the subset being a singleton set, significance can be determined for a particular Regression coefficient with the marginal distribution of the scalar coefficient which is

$$p(B_{kj}|\bar{B}_{kj}, X, U) \propto (\bar{\Delta}_{kj})^{-\frac{1}{2}} e^{-\frac{(B_{kj} - \bar{B}_{kj})^2}{2\bar{\Delta}_{kj}}}, \qquad (8.7.27)$$

where $\bar{\Delta}_{kj}$ is the jth diagonal element of $\bar{\Delta}_k$. Note that $\bar{B}_{kj} = \bar{\beta}_{jk}$ and that

$$z = \frac{(B_{kj} - \bar{B}_{kj})}{\sqrt{\bar{\Delta}_{kj}}} \qquad (8.7.28)$$

follows a Normal distribution with a mean of zero and variance of one.

8.7.4   Maximum a Posteriori

The joint posterior distribution may also be maximized with respect to the vector of Regression coefficients β and the error covariance matrix Σ to obtain maximum a posteriori estimates. For maximization of the posterior, the ICM algorithm is used. Using the posterior conditional distributions found for the Gibbs sampling algorithm, the ICM algorithm for determining maximum a posteriori estimates is to start with an initial value for the estimate of the matrix of Regression coefficients $\tilde{B}_{(0)}$ and cycle through

$$\tilde{\beta}_{(l+1)} = \overset{\text{Arg Max}}{\beta} \; p(\beta|\tilde{\Sigma}_{(l)}, X, U)$$

$$= [\Delta^{-1} + U'U \otimes \tilde{\Sigma}_{(l)}^{-1}]^{-1}[\Delta^{-1}\beta_0 + (U'U \otimes \tilde{\Sigma}_{(l)}^{-1})\hat{\beta}]$$

$$\tilde{\Sigma}_{(l+1)} = \overset{\text{Arg Max}}{\Sigma} \; p(\Sigma|\tilde{\beta}_{(l+1)}, X, U)$$

$$= \frac{(X - U\tilde{B}'_{(l+1)})'(X - U\tilde{B}'_{(l+1)}) + Q}{n + \nu}$$

until convergence is reached. The variables $\hat{B} = X'U(U'U)^{-1}$, $\hat{\beta} = vec(\hat{B})$, and $\tilde{\beta} = vec(\tilde{B})$ have been defined in the process. Conditional maximum a posteriori variance estimates can also be found. The conditional modal variance of the regression coefficients is

$$var(\beta|\tilde{\beta},\tilde{\Sigma},X,U) = (\Delta^{-1}+U'U \otimes \tilde{\Sigma}^{-1})^{-1}$$
$$= \tilde{\Delta},$$

where $\tilde{\beta}$ and $\tilde{\Sigma}$ are the converged value from the ICM algorithm.

Conditional modal intervals may be computed by using the conditional distribution for a particular parameter given the modal values of the others. The posterior conditional distribution of the vector of Regression coefficients β given the modal values of the other parameters and the data is

$$p(\beta|\tilde{\beta},\tilde{\Sigma},X,U) \propto |\tilde{\Delta}|^{-\frac{1}{2}}e^{-\frac{1}{2}(\beta-\tilde{\beta})'\tilde{\Delta}^{-1}(\beta-\tilde{\beta})}. \qquad (8.7.29)$$

To determine statistical significance with the ICM approach, use the marginal conditional distribution of the vector of Regression coefficients given above. General simultaneous hypotheses can be performed on the entire vector or a subset of it. Significance regarding the coefficient for a particular independent variable can be evaluated by computing marginal distributions. It can be shown that the marginal conditional distribution of the kth column of the matrix of Regression coefficients B, B_k is Multivariate Normal

$$p(B_k|\tilde{B}_k,\tilde{\Sigma},X,U) \propto |\tilde{\Delta}_k|^{-\frac{1}{2}}e^{-\frac{1}{2}(B_k-\tilde{B}_k)'\tilde{\Delta}_k^{-1}(B_k-\tilde{B}_k)}, \qquad (8.7.30)$$

where $\tilde{\Delta}_k$ is the covariance matrix of B_k found by taking the kth $p \times p$ submatrix along the diagonal of $\tilde{\Delta}$.

Significance can be evaluated for a subset of coefficients of the kth column of B by determining the marginal distribution of the subset within B_k which is also Multivariate Normal. With the subset being a singleton set, significance can be evaluated for a particular Regression coefficient with the marginal distribution of the scalar coefficient which is

$$p(B_{kj}|\bar{B}_{kj},X) \propto (\tilde{\Delta}_{kj})^{-\frac{1}{2}}e^{-\frac{(B_{kj}-\tilde{B}_{kj})^2}{2\tilde{\Delta}_{kj}}}, \qquad (8.7.31)$$

where $\tilde{\Delta}_{kj}$ is the jth diagonal element of $\tilde{\Delta}_k$. Note that $\tilde{B}_{kj} = \tilde{\beta}_{jk}$ and that

$$z = \frac{(B_{kj}-\tilde{B}_{kj})}{\sqrt{\tilde{\Delta}_{kj}}} \qquad (8.7.32)$$

follows a Normal distribution with a mean of zero and variance of one.

8.8   Interpretation

The main results of performing a Bayesian Regression are estimates of the matrix of Regression coefficients and the error covariance matrix. The results of a Bayesian Regression are described with the use of an example.

TABLE 8.1
Variables for Bayesian Regression example.

X Variables		U Variables	
X_1	Concentration at Time 0	U_1	Insulin Type
X_2	Concentration at Time 1	U_2	Dose Level
X_3	Concentration at Time 2	U_3	Insulin∗Dose Interaction
X_4	Concentration at Time 3		
X_5	Concentration at Time 4		
X_6	Concentration at Time 5		

As an illustrative example, consider data from a bioassay of insulin. The blood sugar concentration in 36 rabbits was measured in mg/100 ml every hour between 0 and 5 hours after administration of an insulin dose. There were two types of insulin preparation ("standard," $U_1 = -1$, and "test," $U_1 = 1$) each with two dose levels (0.75 units, $U_2 = -1$, and 1.50 units, $U_2 = 1$), from a study [67]. It is also believed that there is an interaction between preparation type and dose level so the interaction term $U_3 = U_1 * U_2$ is included.

The problem is to determine the relationship between the set of independent variables (the U's) and the dependent variables (the X's) described in Table 8.1. There are $n = 36$ observations of dimension $p = 5$ along with the $q = 3$ independent variables.

The data X and the design matrix U (including a column of ones for the intercept term) are given in Table 8.2. Hyperparameters for the prior distributions were assessed by performing a regression on a previous data set X_0 obtained from the same population on the previous day.

The estimate of the Regression coefficient matrix B defines a "fitted" line. The fitted line describes the linear relationship between the independent variables (the U's), and the dependent variables (the X's). The coefficient matrix has the interpretation that if all of the independent variables were held fixed except for one u_{ij}, which if increased to u_{ij}^*, the dependent variable x_{ij} increases to an amount x_{ij}^* given by

$$x_{ij}^* = \beta_{i0} + \cdots + \beta_{ij}u_{ij}^* + \cdots + \beta_{iq}u_{iq}. \qquad (8.8.1)$$

Regression coefficients are evaluated to determine whether they are statistically "large" meaning that the associated independent variable contributes

TABLE 8.2
Bayesian Regression data and design matrices.

X	1	2	3	4	5	6	U	e_n	1	2	3
1	96	37	31	33	35	41	1	1	1	1	1
2	90	47	48	55	68	89	2	1	1	1	1
3	99	49	55	64	74	97	3	1	1	1	1
4	95	33	37	43	63	92	4	1	1	1	1
5	107	62	62	85	110	117	5	1	1	1	1
6	81	40	43	45	49	55	6	1	1	1	1
7	95	49	56	63	68	88	7	1	1	1	1
8	105	53	57	69	103	106	8	1	1	1	1
9	97	50	53	59	82	96	9	1	1	1	1
10	97	54	57	66	80	89	10	1	1	-1	-1
11	105	66	83	95	97	100	11	1	1	-1	-1
12	105	49	54	56	70	90	12	1	1	-1	-1
13	106	79	92	95	99	100	13	1	1	-1	-1
14	92	46	51	57	73	91	14	1	1	-1	-1
15	91	61	64	71	80	90	15	1	1	-1	-1
16	101	51	63	91	95	96	16	1	1	-1	-1
17	87	53	55	57	78	89	17	1	1	-1	-1
18	94	57	70	81	94	96	18	1	1	-1	-1
19	98	48	55	71	91	96	19	1	-1	1	-1
20	98	41	43	61	89	101	20	1	-1	1	-1
21	103	60	56	61	76	97	21	1	-1	1	-1
22	99	36	43	57	89	102	22	1	-1	1	-1
23	97	44	51	58	85	105	23	1	-1	1	-1
24	95	41	45	49	59	78	24	1	-1	1	-1
25	109	65	62	72	93	104	25	1	-1	1	-1
26	91	57	60	61	67	83	26	1	-1	1	-1
27	99	43	48	52	61	86	27	1	-1	1	-1
28	102	51	56	81	97	103	28	1	-1	-1	1
29	96	57	55	72	85	89	29	1	-1	-1	1
30	111	84	83	91	101	102	30	1	-1	-1	1
31	105	57	67	83	100	103	31	1	-1	-1	1
32	105	57	61	70	90	98	32	1	-1	-1	1
33	98	55	67	88	94	95	33	1	-1	-1	1
34	98	69	72	89	98	98	34	1	-1	-1	1
35	90	53	61	78	94	95	35	1	-1	-1	1
36	100	60	63	67	77	104	36	1	-1	-1	1

to the dependent variable or statistically "small" meaning that the associated independent variable does not contribute to the dependent variable.

Table 8.4 contains the matrix of regression coefficients from an implementation of the aforementioned Conjugate prior model and used the data in Table 8.2. Exact analytic equations to compute marginal mean and maximum a posteriori estimates are used. The marginal mean and conditional modal

TABLE 8.3

Bayesian Regression prior data and design matrices.

X_0	1	2	3	4	5	6	U_0	e_n	1	2	3
1	96	54	61	63	93	103	1	1	1	1	1
2	98	57	63	75	99	104	2	1	1	1	1
3	104	77	88	91	113	110	3	1	1	1	1
4	109	63	60	67	85	109	4	1	1	1	1
5	98	59	65	72	95	103	5	1	1	1	1
6	104	59	62	74	89	97	6	1	1	1	1
7	97	63	70	72	101	102	7	1	1	1	1
8	101	54	64	77	97	100	8	1	1	1	1
9	107	59	67	61	69	99	9	1	1	1	1
10	96	63	81	97	101	97	10	1	1	-1	-1
11	99	48	70	94	108	104	11	1	1	-1	-1
12	102	61	78	81	99	104	12	1	1	-1	-1
13	112	67	76	100	112	112	13	1	1	-1	-1
14	92	49	59	83	104	103	14	1	1	-1	-1
15	101	53	63	86	104	102	15	1	1	-1	-1
16	105	63	77	94	111	107	16	1	1	-1	-1
17	99	61	74	76	89	92	17	1	1	-1	-1
18	99	51	63	77	99	103	18	1	1	-1	-1
19	98	53	62	71	81	101	19	1	-1	1	-1
20	103	62	65	96	101	105	20	1	-1	1	-1
21	102	54	60	57	64	69	21	1	-1	1	-1
22	108	83	67	80	106	108	22	1	-1	1	-1
23	92	56	60	61	73	79	23	1	-1	1	-1
24	102	61	59	71	91	101	24	1	-1	1	-1
25	94	51	53	55	86	83	25	1	-1	1	-1
26	95	55	58	59	71	85	26	1	-1	1	-1
27	103	47	59	64	92	100	27	1	-1	1	-1
28	120	46	44	58	118	108	28	1	-1	-1	1
29	95	65	75	85	96	95	29	1	-1	-1	1
30	99	59	73	82	109	109	30	1	-1	-1	1
31	105	50	58	84	107	107	31	1	-1	-1	1
32	97	67	89	104	118	118	32	1	-1	-1	1
33	97	46	50	59	78	91	33	1	-1	-1	1
34	102	63	67	74	83	98	34	1	-1	-1	1
35	104	69	81	98	104	105	35	1	-1	-1	1
36	101	65	69	72	93	95	36	1	-1	-1	1

posterior distributions are known and as described previously in this Chapter. With these posterior distributions, credibility intervals and hypotheses can be evaluated to determine whether the set or a subset of independent variables describe the observed relationship.

The prior mode, marginal mean, and maximum a posteriori values of the observation error variances and covariances are the elements of Table 8.5.

In terms of the Source Separation model, if a coefficient is statistically

TABLE 8.4
Prior, Gibbs, and ICM Bayesian Regression coefficients.

B_0	0	1	2	3
1	101.0000	0.0556	-0.3889	0.8889
2	58.6944	0.2500	0.5833	1.0278
3	66.3889	2.5556	-2.8889	0.6111
4	76.9444	3.0556	-6.6111	-0.9444
5	95.5278	2.6944	-6.3056	1.5278
6	100.2222	2.6111	-2.5556	2.7222
$\bar{B}$	0	1	2	3
1	99.6250	-0.6806	-0.5972	0.4861
2	55.9306	-0.4583	-2.5417	0.6806
3	62.0694	1.0417	-5.1806	-0.0417
4	72.4444	0.4722	-7.8889	-0.1389
5	88.9306	-0.4306	-6.4861	0.9306
6	96.7917	-0.3194	-2.5972	1.0139
$\tilde{B}$	0	1	2	3
1	99.6250	-0.6806	-0.5972	0.4861
2	55.9306	-0.4583	-2.5417	0.6806
3	62.0694	1.0417	-5.1806	-0.0417
4	72.4444	0.4722	-7.8889	-0.1389
5	88.9306	-0.4306	-6.4861	0.9306
6	96.7917	-0.3194	-2.5972	1.0139

"large," then the associated observed source contributes significantly to the observed mixture of sources.

Table 8.6 contains the matrix of individual marginal statistics for the coefficients. From this table, it is apparent which coefficients are statistically significant.

8.9   Discussion

Returning to the cocktail party problem, the matrix of Regression coefficients B where $B = (\mu, B_\star)$ contains the matrix of mixing coefficients $B_\star$ for the observed conversation (sources) U, and the population mean μ which is a vector of the overall background mean level at each microphone.

TABLE 8.5
Prior, Gibbs, and ICM Regression covariances.

Q/ν	1	2	3	4	5	6
1	30.3333	8.1327	-5.9198	2.8488	19.1790	18.5802
2		64.8086	55.9815	55.6574	29.8735	22.7747
3			80.3765	74.8858	29.8148	20.8056
4				128.1173	82.6636	57.0432
5					132.9506	75.2963
6						69.0247

$\bar{\Psi}$	1	2	3	4	5	6
1	45.6351	29.9234	25.0795	36.3755	56.8027	47.6303
2		115.7423	110.1676	104.0680	85.9636	57.8525
3			141.7941	139.9377	114.5536	76.4875
4				205.5766	188.4320	122.8870
5					277.5163	184.3142
6						171.0833

$\tilde{\Psi}$	1	2	3	4	5	6
1	34.8268	22.8363	19.1396	27.7602	43.3494	36.3494
2		88.3297	84.0753	79.4203	65.6038	44.1506
3			108.2113	106.7946	87.4225	58.3721
4				156.8874	143.8034	93.7822
5					211.7887	140.6608
6						130.5636

TABLE 8.6
Statistics for coefficients.

t_{65}	0	1	2	3
1	125.1369	-0.8548	-0.7502	0.6106
2	44.1133	-0.3615	-2.0047	0.5368
3	44.2298	0.7423	-3.6916	-0.0297
4	42.8731	0.2795	-4.6687	-0.0822
5	45.2974	-0.2193	-3.3037	0.4740
6	62.7914	-0.2072	-1.6849	0.6577
z	0	1	2	3
1	143.2445	-0.9785	-0.8587	0.6989
2	50.4966	-0.4138	-2.2947	0.6144
3	50.6300	0.8497	-4.2258	-0.0340
4	49.0769	0.3199	-5.3443	-0.0941
5	51.8520	-0.2510	-3.7818	0.5426
6	71.8775	-0.2372	-1.9287	0.7529

Exercises

1. Write the likelihood for all of the observations as

$$p(X|B,\Sigma,U) = \prod_{i=1}^{n} p(x_i|B,\Sigma,u_i)$$

and use the facts that $tr(\Upsilon\Xi) = tr(\Xi\Upsilon)$ to derive Equation 8.3.3. Note that the trace of a scalar is the scalar.

2. Given the posterior distribution, Equation 8.4.5, derive Gibbs sampling and ICM algorithms. Show that for the ICM algorithm, convergence is reached after one iteration when iterating in the order B, Σ.

3. Specify the prior distribution for the Regression coefficients B and the error covariance matrix Σ to be the vague priors

$$p(B) \propto (\text{a constant})$$

and

$$p(\Sigma) \propto |\Sigma|^{-\frac{(p+1)}{2}}.$$

Combine these prior distributions with the likelihood in Equation 8.3.3 to obtain a posterior distribution and derive equations for marginal parameter estimates.

4. Using the priors specified in Exercise 2, derive Gibbs sampling and ICM algorithms.

5. Specify the prior distribution for the Regression coefficients and the error covariance matrix to be the vague and Conjugate priors

$$p(B) \propto (\text{a constant})$$

and

$$p(\Sigma) \propto |\Sigma|^{-\frac{\nu}{2}} e^{-\frac{1}{2}\Sigma^{-1}Q}.$$

Combine these prior distributions with the likelihood in Equation 8.3.3 to obtain a posterior distribution and derive equations for marginal parameter estimates.

6. Using the priors specified in Exercise 4, derive Gibbs sampling and ICM algorithms.

7. Derive the ICM algorithm for maximizing the posterior distribution, Equation 8.4.5. Do this by taking the logarithm and differentiating with respect to B using the result [41] given in Chapter 6 that

$$\frac{\partial tr(\Xi B' \Upsilon B)}{\partial B} = 2\Upsilon B \Xi$$

and either differentiating with respect to Σ or using the mode or maximum value of an Inverted Wishart distribution.

8. In the Conjugate prior model, exact marginal mean estimates were computed. Derive the Gibbs sampling algorithm for the Conjugate prior model.

9

Bayesian Factor Analysis

9.1 Introduction

Now that the Bayesian Regression model has been discussed, the Bayesian Factor Analysis model is described. The Bayesian Factor Analysis model is similar to the Bayesian Source Separation model in that the factors, analogous to sources, are unobservable. However, there are differences that outline the Psychometric method.

The Factor Analysis model uses the correlations or covariances between a set of observed variables to describe them in terms of a smaller set of unobservable variables [50]. The unobserved variables called factors describe the underlying relationship among the original variables.

There are two main reasons why one would perform a Factor Analysis. The first is to explain the observed relationship among a set of observed variables in terms of a smaller number of unobserved variables or latent factors which underlie the observations. This smaller number of variables can be used to find a meaningful structure in the observed variables. This structure will aid in the interpretation and explanation of the process that has generated the observations.

The second reason one would carry out a Factor Analysis is for data reduction. Since the observed variables are represented in terms of a smaller number of unobserved or latent variables, the number of variables in the analysis is reduced, and so are the storage requirements. By having a smaller number of factors (vectors of smaller dimension) to work with that capture the essence of the observed variables, only this smaller number of factors is required to be stored. This smaller number of factors can also be used for further analysis to reduce computational requirements.

The structure of the Factor Analysis model is strikingly similar to the Source Separation model. However, its genesis comes from psychology and retains some of the specifics for the Psychometric model. In reading about the Factor Analysis model, the unobservable factors correspond to the unobservable sources.

9.2   The Bayesian Factor Analysis Model

The development of Bayesian Factor Analysis has been recent. Here is a description of the Factor Analysis model and a Bayesian approach which builds upon previous work [43, 44, 51, 54, 55, 65].

In the Bayesian Factor Analysis model, the p-dimensional observed values are denoted as x_i's just as in the Bayesian Regression model and the m-dimensional unobserved factor score vectors analogous to sources are denoted by f_i's. The f_i's are used to distinguish the unobservable factors from the observable regressors, the u_i's, and the unobservable sources, the s_i's.

Given a set of independent and identically distributed vector-valued observations x_i, $i = 1,\dots,n$, on p possibly correlated random variables, the Multivariate Factor Analysis model on the $m < p$, unobserved variable f_i, is

$$\begin{array}{ccccccc} (x_i|\mu,\Lambda,f_i) = & \mu & + & \Lambda & f_i & + & \epsilon_i, \\ (p \times 1) & (p \times 1) & & (p \times m) & (m \times 1) & & (p \times 1) \end{array} \tag{9.2.1}$$

where the p-dimensional overall population mean vector is

$$\mu = \begin{pmatrix} \mu_1 \\ \vdots \\ \mu_p \end{pmatrix}, \tag{9.2.2}$$

the matrix Λ of unobserved coefficients called "factor loadings" (analogous to the Regression coefficient matrix $B_\star$ in the Regression model and the mixing matrix Λ in the Source Separation model) describing the relationship between f_i and x_i,

$$\Lambda = \begin{pmatrix} \lambda_1' \\ \vdots \\ \lambda_p' \end{pmatrix}, \tag{9.2.3}$$

while ϵ_i is the error vector at time i is given by

$$\epsilon_i = \begin{pmatrix} \epsilon_{i1} \\ \vdots \\ \epsilon_{ip} \end{pmatrix}. \tag{9.2.4}$$

More specifically, a given element x_{ij} of observed vector x_i is represented by the model

$$\begin{array}{ccccccc} (x_{ij}|\mu_{ij},\lambda_j,f_i) = & \mu_{ij} & + & \lambda_j' & f_i & + & \epsilon_{ij}, \\ (1 \times 1) & (1 \times 1) & & (1 \times m) & (m \times 1) & & (1 \times 1) \end{array} \tag{9.2.5}$$

where $\lambda'_j = (\lambda_{j1}, \ldots, \lambda_{jm})$ is a vector of factor loadings that connects the m unobserved factors to the jth observation variable at observation number or time i, x_{ij}. This is also represented as

$$(x_{ij}|\mu_j, \lambda_j, f_i) = \mu_j + \sum_{k=1}^{m} \lambda_{jk} \; f_{ik} + \epsilon_{ij}. \tag{9.2.6}$$

Gathering all observed vectors into a matrix X, the Factor Analysis model may be written as

$$\begin{matrix} (X|\mu, \Lambda, F) = & e_n\mu' & + & F & \Lambda' & + & E, \\ (n \times p) & (n \times p) & & (n \times m) & (m \times p) & & (n \times p) \end{matrix} \tag{9.2.7}$$

where the matrix of observed vectors, the matrix of unobserved factors scores, and the matrix of error vectors are given by

$$X = \begin{pmatrix} x'_1 \\ \vdots \\ x'_n \end{pmatrix}, \qquad F = \begin{pmatrix} f'_1 \\ \vdots \\ f'_n \end{pmatrix}, \qquad \text{and} \qquad E = \begin{pmatrix} \epsilon'_1 \\ \vdots \\ \epsilon'_n \end{pmatrix}. \tag{9.2.8}$$

The ij^{th} element of the observed matrix X is given by the jth element of μ, μ_j plus the ith row of F multiplied by the jth column (row) of Λ' (Λ) plus the ij^{th} element of E.

9.3   Likelihood

In statistical models, there is inherent random variability or error which is characterized as having arisen from a distribution. As in the Regression model [1, 41] it is specified that the errors of observation are independent and Multivariate Normally distributed and represented by the Multivariate Normal distribution (Chapter 2)

$$p(\epsilon_i|\Sigma) \propto |\Sigma|^{-\frac{1}{2}} e^{-\frac{1}{2}\epsilon'_i \Sigma^{-1}\epsilon_i}, \tag{9.3.1}$$

where ϵ_i is the ith p-dimensional error vector, and Σ is the $p \times p$ error covariance matrix.

From this Multivariate Normal error specification, the distribution of the ith observation vector is the Multivariate Normal distribution

$$p(x_i|\mu, \Lambda, f_i, \Sigma) \propto |\Sigma|^{-\frac{1}{2}} e^{-\frac{1}{2}(x_i-\mu-\Lambda'f_i)'\Sigma^{-1}(x_i-\mu-\Lambda'f_i)}. \tag{9.3.2}$$

With the matrix representation of the model, the distribution of the matrix of observations is given by the Matrix Normal distribution

$$p(X|\mu,\Lambda,F,\Sigma) \propto |\Sigma|^{-\frac{n}{2}} e^{-\frac{1}{2}tr(X-e_n\mu'-F\Lambda')\Sigma^{-1}(X-e_n\mu'-F\Lambda')'}, \qquad (9.3.3)$$

where the observations are $X' = (x_1,\ldots,x_n)$ and the factor scores are $F' = (f_1,\ldots,f_n)$. The notation "tr" is the trace operator which yields the sum of the diagonal elements of its matrix argument.

The Factor Analysis model can be written in a similar form as the Bayesian Regression model. The overall mean vector μ and the factor loading matrix Λ are joined into a single matrix as

$$C = (\mu,\Lambda) = (C_1,\ldots,C_{m+1}) = \begin{pmatrix} c_1' \\ \vdots \\ c_p' \end{pmatrix}. \qquad (9.3.4)$$

An n-dimensional vector of ones e_n and the factor scores matrix F are also joined as $Z = (e_n, F)$.

Having joined these vectors and matrices, the Factor Analysis model is now written in the matrix formulation

$$\begin{array}{ccccc} (X|C,Z) = & Z & C' & + & E \\ n \times p & n \times (m+1) & (m+1) \times p & & (n \times p) \end{array} \qquad (9.3.5)$$

and the associated likelihood is the Matrix Normal distribution given by

$$p(X|C,Z,\Sigma) \propto |\Sigma|^{-\frac{n}{2}} e^{-\frac{1}{2}tr(X-ZC')\Sigma^{-1}(X-ZC')'}, \qquad (9.3.6)$$

where all variables are as defined above.

Both Conjugate and generalized Conjugate distributions are used to quantify our prior knowledge regarding various values of the model parameters.

9.4   Conjugate Priors and Posterior

In the model immediately described based on [60], which advances previous work [43, 44, 51, 54, 55, 65], Conjugate families of prior distributions for the model parameters are used. The joint prior distribution for the model parameters $C = (\mu,\Lambda)$ the matrix of coefficients, F the factor score matrix, and Σ the error covariance matrix is the product of the prior distribution for the factor score matrix multiplied by the prior distribution for the matrix of coefficients C given the error covariance matrix Σ multiplied by the prior distribution for the error covariance matrix Σ

$$p(F,C,\Sigma) = p(F)p(C|\Sigma)p(\Sigma), \qquad (9.4.1)$$

where the prior distribution for the model parameters from the Conjugate procedure outlined in Chapter 4 are the Matrix Normal distribution for the matrix of coefficients C, the Inverted Wishart distribution for the error covariance matrix Σ, and the Matrix Normal distribution for the matrix of factor scores F are given by

$$p(C|\Sigma) \propto |D|^{-\frac{p}{2}}|\Sigma|^{-\frac{m+1}{2}}e^{-\frac{1}{2}trD^{-1}(C-C_0)'\Sigma^{-1}(C-C_0)}, \qquad (9.4.2)$$

$$p(\Sigma) \propto |\Sigma|^{-\frac{\nu}{2}}e^{-\frac{1}{2}tr\Sigma^{-1}Q}, \qquad (9.4.3)$$

$$p(F) \propto e^{-\frac{1}{2}trF'F}, \qquad (9.4.4)$$

with Σ, D, and Q positive definite symmetric matrices. Thus, C conditional on Σ has elements which are jointly Normally distributed, and (C_0, D) are hyperparameters to be assessed; Σ follows an Inverted Wishart distribution, and (ν, Q) are hyperparameters to be assessed. The factor score vectors are independent and normally distributed random vectors with mean zero and identity covariance matrix which is consistent with the traditional orthogonal Factor Analysis model. The distributional specification for the factor scores is also present in non-Bayesian models as a model assumption [41, 50]. Note that Q and consequently $E(\Sigma)$ are diagonal, to represent traditional Psychometric views of the factor model containing "common" and "specific" factors.

If the vector of coefficients c is given by $c = vec(C)$, then from the prior specification, $var(c|\Sigma) = D \otimes \Sigma$. By Bayes' rule, the joint posterior distribution for the unknown model parameters F, C, and Σ is given by

$$p(F,C,\Sigma|X) \propto e^{-\frac{1}{2}trF'F}|\Sigma|^{-\frac{(n+\nu+m+1)}{2}}e^{-\frac{1}{2}tr\Sigma^{-1}G}, \qquad (9.4.5)$$

where the $p \times p$ matrix variable G has been defined to be

$$G = (X - ZC')'(X - ZC') + (C - C_0)D^{-1}(C - C_0)' + Q.$$

The joint posterior distribution must now be evaluated in order to obtain estimates of the matrix of factor scores F, the matrix containing the overall mean μ with the factor loadings Λ, and the error covariance matrix Σ. Marginal posterior mean and joint maximum a posteriori estimates of the parameters F, C, and Σ are found by the Gibbs sampling and iterated conditional modes (ICM) algorithms.

9.5   Conjugate Estimation and Inference

With the above joint posterior distribution from the Bayesian Factor Analysis model, it is not possible to obtain all or any of the marginal distributions and thus marginal estimates in closed form or explicit formulas for maximum a posteriori estimates from differentiation. For this reason, marginal mean estimates using the Gibbs sampling algorithm and maximum a posteriori estimates using the ICM algorithm are found.

9.5.1   Posterior Conditionals

Both the Gibbs sampling and ICM estimation procedures require the posterior conditional distributions. Gibbs sampling requires the posterior conditionals for the generation of random variates while ICM requires them for maximization by cycling through their modes or maxima.

The conditional posterior distribution of the matrix of factor scores F is found by considering only those terms in the joint posterior distribution which involve F and is given by

$$
\begin{aligned}
p(F|\mu,\Lambda,\Sigma,X) &\propto p(F)p(X|\mu,F,\Lambda,\Sigma) \\
&\propto e^{-\frac{1}{2}trF'F}|\Sigma|^{-\frac{n}{2}}e^{-\frac{1}{2}tr\Sigma^{-1}(X-e_n\mu-F\Lambda')'(X-e_n\mu-F\Lambda')} \\
&\propto e^{-\frac{1}{2}trF'F}e^{-\frac{1}{2}tr(X-e_n\mu'-F\Lambda')\Sigma^{-1}(X-e_n\mu'-F\Lambda')'}
\end{aligned}
$$

which after performing some algebra in the exponent can be written as

$$
p(F|\mu,\Lambda,\Sigma,X) \propto e^{-\frac{1}{2}tr(F-\tilde{F})(I_m+\Lambda'\Sigma^{-1}\Lambda)(F-\tilde{F})'}, \tag{9.5.1}
$$

where the matrix $\tilde{F}$ has been defined to be

$$
\tilde{F} = (X - e_n\mu')\Sigma^{-1}\Lambda(I_m + \Lambda'\Sigma^{-1}\Lambda)^{-1}. \tag{9.5.2}
$$

That is, the matrix of factor scores given the overall mean μ, the matrix of factor loadings Λ, the error covariance matrix Σ, and the data X is Matrix Normally distributed.

The conditional posterior distribution of the matrix C of factor loadings Λ and the overall mean μ is found by considering only those terms in the joint posterior distribution which involve C and is given by

$$
\begin{aligned}
p(C|F,\Sigma,X) &\propto p(C|\Sigma)p(X|F,C,\Sigma) \\
&\propto |\Sigma|^{-\frac{m+1}{2}}e^{-\frac{1}{2}tr\Sigma^{-1}(C-C_0)D^{-1}(C-C_0)'}
\end{aligned}
$$

$$\times |\Sigma|^{-\frac{n}{2}} e^{-\frac{1}{2} tr \Sigma^{-1} (X - ZC')'(X - ZC')}$$

$$\propto e^{-\frac{1}{2} tr \Sigma^{-1} [(C - C_0) D^{-1} (C - C_0)' + (X - ZC')'(X - ZC')]}$$

which after performing some algebra in the exponent becomes

$$p(C|F, \Sigma, X) \propto e^{-\frac{1}{2} tr \Sigma^{-1} (C - \tilde{C})(D^{-1} + Z'Z)(C - \tilde{C})'}, \qquad (9.5.3)$$

where the matrix $\tilde{C}$ has been defined to be

$$\tilde{C} = [C_0 D^{-1} + X'Z](D^{-1} + Z'Z)^{-1}.$$

Note that $\tilde{C}$ can be written as

$$\tilde{C} = C_0 [D^{-1} (D^{-1} + Z'Z)^{-1}] + \hat{C}[(Z'Z)(D^{-1} + Z'Z)^{-1}],$$

a weighted combination of the prior mean C_0 from the prior distribution and the data mean $\hat{C} = X'Z(Z'Z)^{-1}$ from the likelihood.

That is, the conditional posterior distribution of the matrix C (containing the mean vector μ and the factor loadings matrix Λ) given the factor scores F, the error covariance matrix Σ, and the data X is Matrix Normally distributed.

The conditional posterior distribution of the disturbance covariance matrix Σ is found by considering only those terms in the joint posterior distribution which involve Σ, and is given by

$$p(\Sigma|F, C, X) \propto p(\Sigma) p(C|\Sigma) p(X|F, C, \Sigma)$$

$$\propto |\Sigma|^{-\frac{\nu}{2}} e^{-\frac{1}{2} tr \Sigma^{-1} Q} |\Sigma|^{-\frac{m+1}{2}} e^{-\frac{1}{2} tr \Sigma^{-1} (C - C_0) D^{-1} (C - C_0)'}$$

$$\times |\Sigma|^{-\frac{n}{2}} e^{-\frac{1}{2} tr \Sigma^{-1} (X - ZC')'(X - ZC')}$$

$$\propto |\Sigma|^{-\frac{(n + \nu + m + 1)}{2}} e^{-\frac{1}{2} tr \Sigma^{-1} G}, \qquad (9.5.4)$$

where the $p \times p$ matrix G has been defined to be

$$G = (X - ZC')(X - ZC')' + (C - C_0) D^{-1} (C - C_0)' + Q. \qquad (9.5.5)$$

That is, the conditional distribution of the error covariance matrix Σ given the overall mean μ, the matrix of factor scores F, the matrix of factor loadings Λ, and the data X has an Inverted Wishart distribution.

The modes of these posterior conditional distributions are as described in Chapter 2 and are given by $\tilde{F}$, $\tilde{C}$, (both as defined above) and

$$\tilde{\Sigma} = \frac{G}{n + \nu + m + 1}, \qquad (9.5.6)$$

respectively.

9.5.2   Gibbs Sampling

To find marginal mean estimates of the model parameters from the joint posterior distribution using the Gibbs sampling algorithm, start with initial values for the matrix of factor scores F and the error covariance matrix Σ, say $\bar{F}_{(0)}$ and $\bar{\Sigma}_{(0)}$, and then cycle through

$$\bar{C}_{(l+1)} = \text{a random variate from } p(C|\bar{F}_{(l)}, \bar{\Sigma}_{(l)}, X)$$
$$= A_C Y_C B'_C + M_C, \tag{9.5.7}$$
$$\bar{\Sigma}_{(l+1)} = \text{a random variate from } p(\Sigma|\bar{F}_{(l)}, \bar{C}_{(l+1)}, X)$$
$$= A_\Sigma (Y'_\Sigma Y_\Sigma)^{-1} A'_\Sigma, \tag{9.5.8}$$
$$\bar{F}_{(l+1)} = \text{a random variate from } p(F|\bar{C}_{(l+1)}, \bar{\Sigma}_{(l+1)}, X)$$
$$= Y_F B'_F + M_F, \tag{9.5.9}$$

where

$$A_C A'_C = \bar{\Sigma}_{(l)},$$
$$B_C B'_C = (D^{-1} + \bar{Z}'_{(l)} \bar{Z}_{(l)})^{-1},$$
$$\bar{Z}_{(l)} = (e_n, \bar{F}_{(l)}),$$
$$M_C = (X' \bar{Z}_{(l)} + C_0 D^{-1})(D^{-1} + \bar{Z}'_{(l)} \bar{Z}_{(l)})^{-1},$$
$$A_\Sigma A'_\Sigma = (X - \bar{Z}_{(l)} \bar{C}'_{(l+1)})'(X - \bar{Z}_{(l)} \bar{C}'_{(l+1)})$$
$$+ (\bar{C}_{(l+1)} - C_0)D^{-1}(\bar{C}_{(l+1)} - C_0)' + Q,$$
$$B_F B'_F = (I_m + \bar{\Lambda}'_{(l+1)} \bar{\Sigma}^{-1}_{(l+1)} \bar{\Lambda}_{(l+1)})^{-1},$$
$$M_F = (X - e_n \bar{\mu}'_{(l+1)}) \bar{\Sigma}^{-1}_{(l+1)} \tilde{\Lambda}_{(l+1)} (I_m + \bar{\Lambda}'_{(l+1)} \bar{\Sigma}^{-1}_{(l+1)} \bar{\Lambda}_{(l+1)})^{-1}$$

while Y_C, Y_Σ, and Y_F are $p \times (m+1)$, $(n+\nu+m+1+p+1) \times p$, and $n \times m$ dimensional matrices respectively, whose elements are random variates from the standard Scalar Normal distribution. The formulas for the generation of random variates from the conditional posterior distributions are easily found from the methods in Chapter 6.

The first random variates called the "burn in" are discarded and after doing so, compute from the next L variates means of each of the parameters

$$\bar{F} = \frac{1}{L} \sum_{l=1}^{L} \bar{F}_{(l)} \qquad \bar{C} = \frac{1}{L} \sum_{l=1}^{L} \bar{C}_{(l)} \qquad \bar{\Sigma} = \frac{1}{L} \sum_{l=1}^{L} \bar{\Sigma}_{(l)}$$

which are the exact sampling-based marginal posterior mean estimates of the parameters. Exact sampling-based estimates of other quantities can also be found. Similar to Regression, there is interest in the estimate of the marginal posterior variance of the matrix containing the means and factor loadings

$$\overline{var}(c|X) = \frac{1}{L}\sum_{l=1}^{L} \bar{c}_{(l)}\bar{c}'_{(l)} - \bar{c}\bar{c}'$$
$$= \bar{\Delta},$$

where $c = vec(C)$ and $\bar{c} = vec(\bar{C})$.

The covariance matrices of the other parameters follow similarly. With a specification of Normality for the marginal posterior distribution of the vector containing the mean vector and factor loadings, their distribution is

$$p(c|X) \propto |\bar{\Delta}|^{-\frac{1}{2}} e^{-\frac{1}{2}(c-\bar{c})'\bar{\Delta}^{-1}(c-\bar{c})}, \qquad (9.5.10)$$

where $\bar{c}$ and $\bar{\Delta}$ are as previously defined.

To evaluate statistical significance with the Gibbs sampling approach, use the marginal distribution of the matrix containing the mean vector and factor loading matrix given above. General simultaneous hypotheses can be evaluated regarding the entire matrix containing the mean vector and the factor loading matrix, a submatrix, or the mean vector or a particular factor, or an element by computing marginal distributions. It can be shown that the marginal distribution of the kth column of the matrix containing the mean vector and factor loading matrix C, C_k is Multivariate Normal

$$p(C_k|\bar{C}_k, X) \propto |\bar{\Delta}_k|^{-\frac{1}{2}} e^{-\frac{1}{2}(C_k-\bar{C}_k)'\bar{\Delta}_k^{-1}(C_k-\bar{C}_k)}, \qquad (9.5.11)$$

where $\bar{\Delta}_k$ is the covariance matrix of C_k found by taking the kth $p \times p$ submatrix along the diagonal of $\bar{\Delta}$.

Significance can be determined for a subset of coefficients of the kth column of C by determining the marginal distribution of the subset within C_k which is also Multivariate Normal. With the subset being a singleton set, significance can be determined for a particular mean or loading with the marginal distribution of the scalar coefficient which is

$$p(C_{kj}|\bar{C}_{kj}, X) \propto (\bar{\Delta}_{kj})^{-\frac{1}{2}} e^{-\frac{(C_{kj}-\bar{C}_{kj})^2}{2\bar{\Delta}_{kj}}}, \qquad (9.5.12)$$

where $\bar{\Delta}_{kj}$ is the jth diagonal element of $\bar{\Delta}_k$. Note that $\bar{C}_{kj} = \bar{c}_{jk}$ and that

$$z = \frac{(C_{kj}-\bar{C}_{kj})}{\sqrt{\bar{\Delta}_{kj}}} \qquad (9.5.13)$$

follows a Normal distribution with a mean of zero and variance of one.

9.5.3 Maximum a Posteriori

The joint posterior distribution can also be maximized with respect to the matrix of coefficients C, the matrix of factor scores F, and the error covariance matrix Σ by using the ICM algorithm. To jointly maximize the joint posterior distribution using the ICM algorithm, start with an initial value for the matrix of factor scores F, say $\tilde{F}_{(0)}$, and then cycle through

$$\tilde{C}_{(l+1)} = \overset{\text{Arg Max}}{C}\; p(C|\tilde{F}_{(l)}, \tilde{\Sigma}_{(l)}, X)$$
$$= (X'\tilde{Z}_{(l)} + C_0 D^{-1})(D^{-1} + \tilde{Z}'_{(l)}\tilde{Z}_{(l)})^{-1},$$

$$\tilde{\Sigma}_{(l+1)} = \overset{\text{Arg Max}}{\Sigma}\; p(\Sigma|\tilde{C}_{(l+1)}, \tilde{F}_{(l)}, X)$$
$$= [(X - \tilde{Z}_{(l)}\tilde{C}'_{(l+1)})'(X - \tilde{Z}_{(l)}\tilde{C}'_{(l+1)})$$
$$+ (\tilde{C}_{(l+1)} - C_0)D^{-1}(\tilde{C}_{(l+1)} - C_0)' + Q]/(n + \nu + m + 1),$$

$$\tilde{F}_{(l+1)} = \overset{\text{Arg Max}}{F}\; p(F|\tilde{C}_{(l+1)}, \tilde{\Sigma}_{(l+1)}, X)$$
$$= (X - e_n\tilde{\mu}'_{(l+1)})\tilde{\Sigma}^{-1}_{(l+1)}\tilde{\Lambda}_{(l+1)}(I_m + \tilde{\Lambda}'_{(l+1)}\tilde{\Sigma}^{-1}_{(l+1)}\tilde{\Lambda}_{(l+1)})^{-1},$$

where the matrix $\tilde{Z}_{(l)} = (e_n, \tilde{F}_{(l)})$ has been defined and cycling continues until convergence is reached with the joint modal estimator for the unknown parameters $(\tilde{F}, \tilde{C}, \tilde{\Sigma})$. Conditional maximum a posteriori variance estimates can also be found. The conditional modal variance of the matrix containing the means and factor loadings is

$$var(C|\tilde{C}, \tilde{F}, \tilde{\Sigma}, X) = \tilde{\Sigma} \otimes (D^{-1} + \tilde{Z}'\tilde{Z})^{-1} \tag{9.5.14}$$

or equivalently

$$var(c|\tilde{c}, \tilde{F}, \tilde{\Sigma}, X) = (D^{-1} + \tilde{Z}'\tilde{Z})^{-1} \otimes \tilde{\Sigma} \tag{9.5.15}$$
$$= \tilde{\Delta}, \tag{9.5.16}$$

where $c = vec(C)$, while $\tilde{C}$, $\tilde{F}$, and $\tilde{\Sigma}$ are the converged value from the ICM algorithm.

To determine statistical significance with the ICM approach, use the conditional distribution of the matrix containing the mean vector and factor loading matrix which is

$$p(C|\tilde{C}, \tilde{F}, \tilde{\Sigma}, X) \propto |D^{-1} + \tilde{Z}'\tilde{Z}|^{\frac{1}{2}}|\tilde{\Sigma}|^{-\frac{1}{2}}e^{-\frac{1}{2}tr\tilde{\Sigma}^{-1}(C-\tilde{C})(D^{-1}+\tilde{Z}'\tilde{Z})(C-\tilde{C})'}. \tag{9.5.17}$$

That is,

$$C|\tilde{C},\tilde{F},\tilde{\Sigma},X \sim N\left(\tilde{C},\tilde{\Sigma}\otimes(D^{-1}+\tilde{Z}'\tilde{Z})^{-1}\right). \tag{9.5.18}$$

General simultaneous hypotheses can be evaluated regarding the entire matrix containing the mean vector and the factor loading matrix, a submatrix, or the mean vector or a particular factor, or an element by computing marginal conditional distributions.

It can be shown [17, 41] that the marginal conditional distribution of any column of the matrix containing the means and factor loadings C, C_k is Multivariate Normal

$$p(C_k|\tilde{C}_k,\tilde{F},\tilde{\Sigma},U,X) \propto |W_{kk}\tilde{\Sigma}|^{-\frac{1}{2}}e^{-\frac{1}{2}(C_k-\tilde{C}_k)'(W_{kk}\tilde{\Sigma})^{-1}(C_k-\tilde{C}_k)}, \tag{9.5.19}$$

where $W = (D^{-1}+U'U)^{-1}$ and W_{kk} is its kth diagonal element.

With the marginal distribution of a column of C, significance can be evaluated for the mean vector or a particular factor. Significance can be determined for a subset of coefficients by determining the marginal distribution of the subset within C_k which is also Multivariate Normal. With the subset being a singleton set, significance can be evaluated for a particular mean or loading with the marginal distribution of the scalar coefficient which is

$$p(C_{kj}|\tilde{C}_{kj},\tilde{F},\tilde{\Sigma}_{jj},U,X) \propto (W_{kk}\tilde{\Sigma}_{jj})^{-\frac{1}{2}}e^{-\frac{(C_{kj}-\tilde{C}_{kj})^2}{2W_{kk}\tilde{\Sigma}_{jj}}}, \tag{9.5.20}$$

where $\tilde{\Sigma}_{jj}$ is the jth diagonal element of $\tilde{\Sigma}$. Note that $\tilde{C}_{kj} = \tilde{c}_{jk}$ and that

$$z = \frac{(C_{kj}-\tilde{C}_{kj})}{\sqrt{W_{kk}\tilde{\Sigma}_{jj}}}$$

$$\tag{9.5.21}$$

follows a Normal distribution with a mean of zero and variance of one.

9.6   Generalized Priors and Posterior

The Conjugate prior distributions can be expanded to generalized Conjugate priors which permit greater freedom of assessment [56]. This extends previous work [51] in which available prior information regarding the parameter values was quantified using these generalized Conjugate prior distributions; however, independence was assumed between the overall mean and the factor loadings matrix.

The joint prior distribution $p(F, c, \Sigma)$ for the matrix of factor scores F, the vector containing the overall mean and factor loadings $c = vec(C)$, and the error covariance matrix Σ is given by the product of the prior distribution $p(F)$ for the factor loading matrix F with the prior distribution $p(c)$ for the vector of coefficients c and with the prior distribution $p(\Sigma)$ for the error covariance matrix Σ and is given by

$$p(F, c, \Sigma) = p(F)p(c)p(\Sigma). \tag{9.6.1}$$

These prior distributions are found from the generalized Conjugate procedure outlined in Chapter 4 and are given by

$$p(c) \propto |\Delta|^{-\frac{1}{2}} e^{-\frac{1}{2}(c-c_0)'\Delta^{-1}(c-c_0)}, \tag{9.6.2}$$

$$p(\Sigma) \propto |\Sigma|^{-\frac{\nu}{2}} e^{-\frac{1}{2}tr\Sigma^{-1}Q}, \tag{9.6.3}$$

$$p(F) \propto e^{-\frac{1}{2}trF'F}. \tag{9.6.4}$$

The hyperparameters Δ, c_0, ν, and Q are hyperparameters to be assessed. The matrices Δ, Σ, and Q are positive definite. By specifying these hyperparameters the joint prior distribution is determined.

By Bayes' rule, the joint posterior distribution for the unknown model parameters with specified generalized Conjugate prior distributions is given by

$$p(F, c, \Sigma | X) \propto p(F)p(c)p(\Sigma)p(X|F, C, \Sigma) \tag{9.6.5}$$

which is

$$p(F, c, \Sigma | X) \propto e^{-\frac{1}{2}trF'F} |\Delta|^{-\frac{1}{2}} e^{-\frac{1}{2}(c-c_0)'\Delta^{-1}(c-c_0)}$$
$$\times |\Sigma|^{-\frac{(n+\nu)}{2}} e^{-\frac{1}{2}tr\Sigma^{-1}[(X-ZC')'(X-ZC')+Q]} \tag{9.6.6}$$

after inserting the joint prior distribution and the likelihood.

The joint posterior distribution must now be evaluated in order to obtain estimates of the parameters.

9.7   Generalized Estimation and Inference

With the generalized Conjugate prior distributions, it is not possible to obtain all or any of the marginal distributions and thus marginal mean estimates in closed form. It is also not possible to obtain explicit formulas for maximum a posteriori estimates. For these reasons, marginal posterior mean and joint maximum a posteriori estimates are found using the Gibbs sampling and ICM algorithms.

9.7.1   Posterior Conditionals

Both the Gibbs sampling and ICM algorithms require the posterior conditionals. Gibbs sampling requires the conditionals for the generation of random variates while ICM requires them for maximization by cycling through their modes.

The conditional posterior distribution of the matrix of factor scores F is found by considering only those terms in the joint posterior distribution which only involve F and is given by

$$p(F|\mu, \Lambda, \Sigma, X) \propto p(F)p(X|\mu, F, \Lambda, \Sigma)$$
$$\propto e^{-\frac{1}{2}trF'F}|\Sigma|^{-\frac{n}{2}}e^{-\frac{1}{2}tr\Sigma^{-1}(X-F\Lambda')'(X-F\Lambda')}$$
$$\propto e^{-\frac{1}{2}trF'F}e^{-\frac{1}{2}tr(X-e_n\mu'-F\Lambda')\Sigma^{-1}(X-e_n\mu'-F\Lambda')'}$$

which after performing some algebra in the exponent can be written as

$$p(F|\mu, \Lambda, \Sigma, X) \propto e^{-\frac{1}{2}tr(F-\tilde{F})(I_m+\Lambda'\Sigma^{-1}\Lambda)(F-\tilde{F})'}, \qquad (9.7.1)$$

where the matrix $\tilde{F}$ has been defined to be

$$\tilde{F} = (X - e_n\mu')\Sigma^{-1}\Lambda(I_m + \Lambda'\Sigma^{-1}\Lambda)^{-1}. \qquad (9.7.2)$$

That is, the matrix of factor scores F given the overall mean μ, the factor loading matrix Λ, the error covariance matrix Σ, and the data X is Matrix Normally distributed.

The conditional posterior distribution of the vector of means and factor loadings is found by considering only those terms in the joint posterior distribution which involve c or C and is given by

$$p(c|F, \Sigma, X) \propto p(c)p(X|F, C, \Sigma)$$
$$\propto |\Delta|^{-\frac{1}{2}}e^{-\frac{1}{2}(c-c_0)'\Delta^{-1}(c-c_0)}$$
$$\times |\Sigma|^{-\frac{n}{2}}e^{-\frac{1}{2}tr\Sigma^{-1}(X-ZC')'(X-ZC')} \qquad (9.7.3)$$

which after performing some algebra in the exponent becomes

$$p(c|F, \Sigma, X) \propto e^{-\frac{1}{2}(c-\tilde{c})'[\Delta^{-1}+Z'Z\otimes\Sigma^{-1}](c-\tilde{c})}, \qquad (9.7.4)$$

where the vector $\tilde{c}$ has been defined to be

$$\tilde{c} = [\Delta^{-1} + Z'Z \otimes \Sigma^{-1}]^{-1}[\Delta^{-1}c_0 + (Z'Z \otimes \Sigma^{-1})\hat{c}] \qquad (9.7.5)$$

and the vector $\hat{c}$ has been defined to be

$$\hat{c} = vec[X'Z(Z'Z)^{-1}]. \tag{9.7.6}$$

Note that this is a weighted combination of the prior mean c_0 from the prior distribution and the data mean $\hat{c}$ from the likelihood.

That is, the conditional posterior distribution of the vector containing the overall mean vector μ and the factor loading vector $\lambda = vec(\Lambda)$ given the matrix of factor scores F, the error covariance matrix Σ, and the data X is Multivariate Normally distributed.

The conditional posterior distribution of the error covariance matrix Σ is found by considering only those terms in the joint posterior distribution which involve Σ and is given by

$$p(\Sigma|F,C,X) \propto p(\Sigma)p(X|F,C,\Sigma)$$
$$\propto |\Sigma|^{-\frac{(n+\nu)}{2}} e^{-\frac{1}{2}tr\Sigma^{-1}[(X-ZC')'(X-ZC')+Q]}. \tag{9.7.7}$$

That is, the posterior conditional distribution of the error covariance matrix Σ given the matrix of factor scores F, the overall mean μ, the matrix of factor loadings Λ, and the data X has an Inverted Wishart distribution.

The modes of these conditional posterior distributions are as described in Chapter 2 and given by $\tilde{F}$, $\tilde{c}$, (both as defined above) and

$$\tilde{\Sigma} = \frac{(X-ZC')'(X-ZC')+Q}{n+\nu}, \tag{9.7.8}$$

respectively.

9.7.2   Gibbs Sampling

To find marginal mean estimated of the parameters from the joint posterior distribution using the Gibbs sampling algorithm, start with initial values for the matrix of factor scores F and the error covariance matrix Σ, say $\bar{F}_{(0)}$ and $\bar{\Sigma}_{(0)}$, and then cycle through

$$\bar{c}_{(l+1)} = \text{a random variate from } p(c|\bar{F}_{(l)},\bar{\Sigma}_{(l)},X)$$
$$= A_c Y_c + M_c, \tag{9.7.9}$$
$$\bar{\Sigma}_{(l+1)} = \text{a random variate from } p(\Sigma|\bar{F}_{(l)},\bar{c}_{(l+1)},X)$$
$$= A_\Sigma (Y_\Sigma' Y_\Sigma)^{-1} A_\Sigma', \tag{9.7.10}$$
$$\bar{F}_{(l+1)} = \text{a random variate from } p(F|\bar{c}_{(l+1)},\bar{\Sigma}_{(l+1)},X)$$
$$= Y_F B_F' + M_F, \tag{9.7.11}$$

where

$$\hat{c}_{(l)} = vec[X'\bar{Z}_{(l)}(\bar{Z}'_{(l)}\bar{Z}_{(l)})^{-1}],$$

$$\bar{c}_{(l+1)} = [\Delta^{-1} + \bar{Z}'_{(l)}\bar{Z}_{(l)} \otimes \bar{\Sigma}^{-1}_{(l)}]^{-1}[\Delta^{-1}c_0 + (\bar{Z}'_{(l)}\bar{Z}_{(l)} \otimes \bar{\Sigma}^{-1}_{(l)})\hat{c}_{(l)}],$$

$$A_c A'_c = (\Delta^{-1} + \bar{Z}'_{(l)}\bar{Z}_{(l)} \otimes \bar{\Sigma}^{-1}_{(l)})^{-1},$$

$$M_c = [\Delta^{-1} + \bar{Z}'_{(l)}\bar{Z}_{(l)} \otimes \bar{\Sigma}^{-1}_{(l)}]^{-1}[\Delta^{-1}c_0 + (\bar{Z}'_{(l)}\bar{Z}_{(l)} \otimes \bar{\Sigma}^{-1}_{(l)})\hat{c}],$$

$$A_\Sigma A'_\Sigma = (X - \bar{Z}_{(l)}\bar{C}'_{(l+1)})'(X - \bar{Z}_{(l)}\bar{C}'_{(l+1)}) + Q,$$

$$B_F B'_F = (I_m + \bar{\Lambda}'_{(l+1)}\bar{\Sigma}^{-1}_{(l+1)}\bar{\Lambda}_{(l+1)})^{-1},$$

$$M_F = (X - e_n\bar{\mu}'_{(l+1)})\bar{\Sigma}^{-1}_{(l+1)}\bar{\Lambda}_{(l+1)}(I_m + \bar{\Lambda}'_{(l+1)}\bar{\Sigma}^{-1}_{(l+1)}\bar{\Lambda}_{(l+1)})^{-1}$$

while Y_c, Y_Σ, and Y_F are $p(m+1) \times 1$, $(n+\nu+p+1) \times p$, and $n \times m$ dimensional matrices whose respective elements are random variates from the standard Scalar Normal distribution. The formulas for the generation of random variates from the conditional posterior distributions are easily found from the methods in Chapter 6.

The first random variates called the "burn in" are discarded and after doing so, compute from the next L variates means of each of the parameters

$$\bar{F} = \frac{1}{L}\sum_{l=1}^{L}\bar{F}_{(l)} \qquad \bar{c} = \frac{1}{L}\sum_{l=1}^{L}\bar{c}_{(l)} \qquad \bar{\Sigma} = \frac{1}{L}\sum_{l=1}^{L}\bar{\Sigma}_{(l)}$$

which are the exact sampling-based marginal posterior mean estimates of the parameters. Exact sampling-based estimates of other quantities can also be found. Similar to Regression, there is interest in the estimate of the marginal posterior variance of the vector containing the means and factor loadings

$$\overline{var}(c|X) = \frac{1}{L}\sum_{l=1}^{L}\bar{c}_{(l)}\bar{c}'_{(l)} - \bar{c}\bar{c}'$$

$$= \bar{\Delta}.$$

The covariance matrices of the other parameters follow similarly. With a specification of Normality for the marginal posterior distribution of the vector containing the means and factor loadings, their distribution is

$$p(c|X) \propto |\bar{\Delta}|^{-\frac{1}{2}}e^{-\frac{1}{2}(c-\bar{c})'\bar{\Delta}^{-1}(c-\bar{c})}, \qquad (9.7.12)$$

where $\bar{c}$ and $\bar{\Delta}$ are as previously defined.

To evaluate statistical significance with the Gibbs sampling approach, use the marginal distribution of the vector c containing the means and factor loadings given above. General simultaneous hypotheses can be evaluated regarding the entire coefficient vector of means and loadings, a subset of it, or the coefficients for a particular factor by computing marginal distributions. It can be shown that the marginal distribution of the kth column of the matrix containing the means and factor loadings C, C_k is Multivariate Normal

$$p(C_k|\bar{C}_k, X, U) \propto |\bar{\Delta}_k|^{-\frac{1}{2}} e^{-\frac{1}{2}(C_k - \bar{C}_k)'\bar{\Delta}_k^{-1}(C_k - \bar{C}_k)}, \tag{9.7.13}$$

where $\bar{\Delta}_k$ is the covariance matrix of C_k found by taking the kth $p \times p$ submatrix along the diagonal of $\bar{\Delta}$.

Significance can be evaluated for a subset of means or coefficients of the kth column of C by determining the marginal distribution of the subset within C_k which is also Multivariate Normal. With the subset being a singleton set, significance can be determined for a particular mean or coefficient with the marginal distribution of the scalar coefficient which is

$$p(C_{kj}|\bar{C}_{kj}, X, U) \propto (\bar{\Delta}_{kj})^{-\frac{1}{2}} e^{-\frac{(C_{kj} - \bar{C}_{kj})^2}{2\bar{\Delta}_{kj}}}, \tag{9.7.14}$$

where $\bar{\Delta}_{kj}$ is the jth diagonal element of $\bar{\Delta}_k$. Note that $\bar{C}_{kj} = \bar{c}_{jk}$ and that

$$z = \frac{(C_{kj} - \bar{C}_{kj})}{\sqrt{\bar{\Delta}_{kj}}} \tag{9.7.15}$$

follows a Normal distribution with a mean of zero and variance of one.

9.7.3   Maximum a Posteriori

The joint posterior distribution can also be maximized with respect to the vector of coefficients c, the matrix of factor scores F, and the error covariance matrix Σ using the ICM algorithm. To maximize the joint posterior distribution using the ICM algorithm, start with initial values for the estimates of the matrix of factor score matrix $\tilde{F}$ and the error covariance matrix Σ, say $\tilde{F}_{(0)}$ and $\tilde{\Sigma}_{(0)}$, and then cycle through

$$\hat{c}_{(l)} = vec[X'\tilde{Z}_{(l)}(\tilde{Z}'_{(l)}\tilde{Z}_{(l)})^{-1}],$$

$$\tilde{c}_{(l+1)} = \overset{\text{Arg Max}}{c} \; p(c|\tilde{F}_{(l)}, \tilde{\Sigma}_{(l)}, X)$$

$$= [\Delta^{-1} + \tilde{Z}'_{(l)}\tilde{Z}_{(l)} \otimes \tilde{\Sigma}_{(l)}^{-1}]^{-1}[\Delta^{-1}c_0 + (\tilde{Z}'_{(l)}\tilde{Z}_{(l)} \otimes \tilde{\Sigma}_{(l)}^{-1})\hat{c}_{(l)}],$$

$$\tilde{\Sigma}_{(l+1)} = \overset{\text{Arg Max}}{\Sigma} \; p(\Sigma|\tilde{C}_{(l+1)}, \tilde{F}_{(l)}, X)$$

$$= \frac{(X - \tilde{Z}_{(l)}\tilde{C}'_{(l+1)})'(X - \tilde{Z}_{(l)}\tilde{C}'_{(l+1)}) + Q}{n + \nu},$$

$$\tilde{F}_{(l+1)} - \overset{\text{Arg Max}}{F} \; p(F|\tilde{C}_{(l+1)}, \tilde{\Sigma}_{(l+1)}, X)$$

$$= (X - e_n\tilde{\mu}'_{(l+1)})\tilde{\Sigma}^{-1}_{(l+1)}\tilde{\Lambda}_{(l+1)}(I_m + \tilde{\Lambda}'_{(l+1)}\tilde{\Sigma}^{-1}_{(l+1)}\tilde{\Lambda}_{(l+1)})^{-1}$$

where the matrix $\tilde{Z}_{(l)} = (e_n, \tilde{F}_{(l)})$ has been defined. Continue cycling until convergence is reached with the joint modal estimator for the unknown parameters $(\tilde{F}, \tilde{c}, \tilde{\Sigma})$. Conditional maximum a posteriori variance estimates can also be found. The conditional modal variance of the matrix containing the means and factor loadings is

$$var(c|\tilde{F}, \tilde{\Sigma}, X, U) = [\Delta^{-1} + \tilde{Z}'\tilde{Z} \otimes \tilde{\Sigma}^{-1}]^{-1} \qquad (9.7.16)$$
$$= \tilde{\Delta}, \qquad (9.7.17)$$

where $c = vec(C)$, while $\tilde{F}$ and $\tilde{\Sigma}$ are the converged value from the ICM algorithm.

Conditional modal intervals may be computed by using the conditional distribution for a particular parameter given the modal values of the others. The posterior conditional distribution of the matrix containing the means and factor loadings C given the modal values of the other parameters and the data is

$$p(c|\tilde{F}, \tilde{\Sigma}, X, U) \propto |\tilde{\Delta}|^{-\frac{1}{2}} e^{-\frac{1}{2}(c-\tilde{c})'\tilde{\Delta}^{-1}(c-\tilde{c})}. \qquad (9.7.18)$$

To determine statistical significance with the ICM approach, use the marginal conditional distribution of the matrix containing the means and factor loadings given above. General simultaneous hypotheses can be performed regarding the mean vector or the coefficient for a particular factor by computing marginal distributions. It can be shown that the marginal conditional distribution of the kth column C_k of the matrix C containing the overall mean vector and factor loading matrix is Multivariate Normal

$$p(C_k|\bar{C}_k, \tilde{\Sigma}, X, U) \propto |\tilde{\Delta}_k|^{-\frac{1}{2}} e^{-\frac{1}{2}(C_k-\bar{C}_k)'\tilde{\Delta}_k^{-1}(C_k-\bar{C}_k)}, \qquad (9.7.19)$$

where $\tilde{\Delta}_k$ is the covariance matrix of C_k found by taking the kth $p \times p$ submatrix along the diagonal of $\tilde{\Delta}$.

Significance can be determined for a subset of means or loadings of the kth column of C by determining the marginal distribution of the subset within C_k which is also Multivariate Normal. With the subset being a singleton set, significance can be determined for a particular mean or factor loading with the marginal distribution of the scalar coefficient which is

$$p(C_{kj}|\tilde{C}_{kj}, \tilde{F}, \tilde{\Sigma}_{jj}, X) \propto (\tilde{\Delta}_{kj})^{-\frac{1}{2}} e^{-\frac{(C_{kj}-\bar{C}_{kj})^2}{2\tilde{\Delta}_{kj}}}, \qquad (9.7.20)$$

where $\tilde{\Delta}_{kj}$ is the jth diagonal element of $\tilde{\Delta}_k$. Note that $\tilde{C}_{kj} = \tilde{c}_{jk}$ and that

$$z = \frac{(C_{kj} - \tilde{C}_{kj})}{\sqrt{\tilde{\Delta}_{kj}}}$$

follows a Normal distribution with a mean of zero and variance of one.

9.8 Interpretation

The main results of performing a Factor Analysis are estimates of the factor score matrix, the factor loading matrix, and the error covariance matrix. The results of a Factor Analysis are described with the use of an example.

Data are extracted from an example in [26] and have been used before in Bayesian Factor Analysis [43, 44]. Applicants for a particular position have been scored on fifteen variables which are listed in Table 9.1.

The aim of performing this Factor Analysis is to determine an underlying relationship between the original observed variables which is of lower dimension and to determine which applicants are candidates for being hired based on these factors.

TABLE 9.1
Variables for Bayesian Factor Analysis example.

X	Variables		
X_1	Form of letter application	X_2	Appearance
X_3	Academic ability	X_4	Likeabiliy
X_5	Self-confidence	X_6	Lucidity
X_7	Honesty	X_8	Salesmanship
X_9	Experience	X_{10}	Drive
X_{11}	Ambition	X_{12}	Grasp
X_{13}	Potential	X_{14}	Keenness to join
X_{15}	Suitability		

The applicants were scored on a ten-point scale on fifteen characteristics. There are $n = 48$ observations on applicants which consists of $p = 15$ observed variables. Table 9.2 contains the data for the applicant example. Note that there are only X's and not any U's.

Tables 9.3 and 9.4 contain the Gibbs sampling and ICM estimates of the factor scores. These are the new variables the applicants are rated on.

In a typical Factor Analysis, the sample size is usually large enough to estimate the variances of the x's as $\hat{\sigma}_1^2, \dots, \hat{\sigma}_p^2$, the maximum likelihood estimates, and scale the x's to have unit variance. Having done this helps with the interpretation of the factor loading matrix. The matrix of coefficients, Λ which was previously a matrix of covariances between x and f is now a matrix of correlations.

Table 9.5 contains prior along with estimated Gibbs sampling and ICM mean vectors with factor loading matrices. The analysis implemented the aforementioned Conjugate prior model. It has previously been determined to use a model with $m = 4$ factors [43]. The rows of the factor loading matrices have been rearranged for interpretation purposes. It is seen that factor 1 "loads heavily" for variables 5, 6, 8, 10, 11, 12, and 13; factor 2, heavily on variable 3; factor 3, heavily on variables 1, 9, and 15; while factor 4 loads heavily on variables 4 and 7.

Since the observed vectors were scaled by their standard deviations and the orthogonal factor model was used, the factor loading matrix is a matrix of correlations between the p observed variables and the m unobserved factors. For example, the correlation between observable variable 10 and unobservable factor 1 is 0.6898 when estimated by Gibbs sampling.

TABLE 9.2
Bayesian Factor Analysis data.

X	1	2	3	4	5	6	7	8	9	10	11	12	13	14	15
1	6	7	2	5	8	7	8	8	3	8	9	7	5	7	10
2	9	10	5	8	10	9	9	10	5	9	9	8	8	8	10
3	7	8	3	6	9	8	9	7	4	9	9	8	6	8	10
4	5	6	8	5	6	5	9	2	8	4	5	8	7	6	5
5	6	8	8	8	4	5	9	2	8	5	5	8	8	7	7
6	7	7	7	6	8	7	10	5	9	6	5	8	6	6	6
7	9	9	8	8	8	8	8	8	10	8	10	8	9	8	10
8	9	9	9	8	9	9	8	8	10	9	10	9	9	9	10
9	9	9	7	8	8	8	8	5	9	8	9	8	8	8	10
10	4	7	10	2	10	10	7	10	3	10	10	10	9	3	10
11	4	7	10	0	10	8	3	9	5	9	10	8	10	2	5
12	4	7	10	4	10	10	7	8	2	8	8	10	10	3	7
13	6	9	8	10	5	4	9	4	4	4	5	4	7	6	8
14	8	9	8	9	6	3	8	2	5	2	6	6	7	5	6
15	4	8	8	7	5	4	10	2	7	5	3	6	6	4	6
16	6	9	6	7	8	9	8	9	8	8	7	6	8	6	10
17	8	7	7	7	9	5	8	6	6	7	8	6	6	7	8
18	6	8	8	4	8	8	6	4	3	3	6	7	2	6	4
19	6	7	8	4	7	8	5	4	4	2	6	8	3	5	4
20	4	8	7	8	8	9	10	5	2	6	7	9	8	8	9
21	3	8	6	8	8	8	10	5	3	6	7	8	8	5	8
22	9	8	7	8	9	10	10	10	3	10	8	10	8	10	8
23	7	10	7	9	9	9	10	10	3	9	9	10	9	10	8
24	9	8	7	10	8	10	10	10	2	9	7	9	9	10	8
25	6	9	7	7	4	5	9	3	2	4	4	4	4	5	4
26	7	8	7	8	5	4	8	2	3	4	5	6	5	5	6
27	2	10	7	9	8	9	10	5	3	5	6	7	6	4	5
28	6	3	5	3	5	3	5	0	0	3	3	0	0	5	0
29	4	3	4	3	3	0	0	0	0	4	4	0	0	5	0
30	4	6	5	6	9	4	10	3	1	3	3	2	2	7	3
31	5	5	4	7	8	4	10	3	2	5	5	3	4	8	3
32	3	3	5	7	7	9	10	3	2	5	3	7	5	5	2
33	2	3	5	7	7	9	10	3	2	2	3	6	4	5	2
34	3	4	6	4	3	3	8	1	1	3	3	3	2	5	2
35	6	7	4	3	3	0	9	0	1	0	2	3	1	5	3
36	9	8	5	5	6	6	8	2	2	2	4	5	6	6	3
37	4	9	6	4	10	8	8	9	1	3	9	7	5	3	2
38	4	9	6	6	9	9	7	9	1	2	10	8	5	5	2
39	10	6	9	10	9	10	10	10	10	10	8	10	10	10	10
40	10	6	9	10	9	10	10	10	10	10	10	10	10	10	10
41	10	7	8	0	2	1	2	0	10	2	0	3	0	0	10
42	10	3	8	0	1	1	0	0	10	0	0	0	0	0	10
43	3	4	9	8	2	4	5	3	6	2	1	3	3	3	8
44	7	7	7	6	9	8	8	6	8	8	10	8	8	6	5
45	9	6	10	9	7	7	10	2	1	5	5	7	8	4	5
46	9	8	10	10	7	9	10	3	1	5	7	9	9	4	4
47	0	7	10	3	5	0	10	0	0	2	2	0	0	0	0
48	0	6	10	1	5	0	10	0	0	2	2	0	0	0	0

TABLE 9.3
Gibbs sampling estimates of factor scores.

$\bar{F}$	1	2	3	4
1	0.3077	-3.3603	-0.2125	-0.5730
2	0.7924	-1.6057	0.1965	0.2117
3	0.5037	-2.7629	-0.0012	-0.1239
4	-0.7681	0.4642	-0.2068	0.1312
5	-1.1131	0.1111	-0.1170	0.6952
6	-0.2053	-0.1476	0.3176	0.5089
7	0.3489	0.2003	0.7723	0.0877
8	0.5909	0.8982	0.7203	0.0415
9	0.1043	-0.4084	0.7848	0.1326
10	1.7144	2.1089	-0.1417	-1.3574
11	1.2927	1.9320	-0.7402	-2.8422
12	1.3329	2.0431	-0.6587	-0.7475
13	-1.2701	0.2614	-0.5932	1.0057
14	-1.1671	0.3324	-0.1722	0.6009
15	-1.1925	0.3136	-0.4554	0.8168
16	0.2348	-1.0655	0.1624	-0.1061
17	-0.0942	-0.0641	0.1559	0.0120
18	-0.5280	0.7515	-0.9473	-1.2914
19	-0.5392	0.6903	-0.6783	-1.3880
20	0.1663	-0.1587	-0.8480	0.7384
21	0.1888	-0.8514	-0.8602	0.8835
22	0.8396	-0.1468	-0.2744	0.6413
23	0.7156	-0.3827	-0.8944	0.6872
24	0.5293	-0.2395	-0.4618	1.1675
25	-1.4216	-0.2283	-1.1918	0.3591
26	-1.1812	-0.2195	-0.4864	0.4257
27	-0.1117	-0.2813	-1.6328	1.0079
28	-1.9465	-0.9311	-1.5140	-1.2838
29	-2.4544	-1.9705	-2.2364	-2.6975
30	-1.2091	-1.0386	-1.6438	0.3865
31	-0.9775	-1.8423	-1.4264	0.7280
32	-0.3479	-1.1174	-1.3412	1.1795
33	-0.6294	-1.0550	-1.5186	1.1185
34	-1.8538	-0.5416	-1.7054	-0.2570
35	-2.3996	-1.9346	-1.0801	-0.4663
36	-1.0747	-1.3773	-0.5516	-0.2966
37	0.5600	-0.5530	-1.7533	-0.7706
38	0.3733	-0.7114	-2.0248	-0.6473
39	0.8246	1.0423	1.0749	1.3483
40	0.9801	1.0611	1.0504	1.3164
41	-2.4371	0.5326	2.0433	-2.8728
42	-2.7783	0.7600	2.2499	-3.1261
43	-2.1129	1.1138	-0.3480	-0.0329
44	0.5688	-0.1781	-0.1073	-0.0994
45	-0.2459	2.1388	-0.0444	1.5419
46	0.1637	1.9385	-0.3786	1.6403
47	-2.0181	2.2786	-2.4526	-0.0478
48	-1.9193	2.4105	-2.2961	-0.4600

TABLE 9.4
ICM estimates of factor scores.

$\tilde{F}$	1	2	3	4
1	0.1606	-3.5584	-0.3575	-0.5349
2	0.7575	-1.6808	0.2195	0.3713
3	0.3921	-2.9206	-0.1091	-0.6416
4	-0.8869	0.5279	-0.4046	0.0312
5	-1.0895	0.2790	-0.1725	0.6625
6	-0.3319	-0.1462	0.1647	0.4682
7	0.3905	0.2531	0.8603	0.1704
8	0.6483	0.9351	0.8176	0.1400
9	0.1116	-0.3768	0.8034	0.2049
10	1.4590	2.0559	-0.4261	-1.3752
11	1.0815	1.9658	-0.9212	-2.9114
12	1.0971	2.0434	-0.9648	-0.8000
13	-1.2880	0.3323	-0.6155	1.0498
14	-1.2535	0.4241	-0.2638	0.6192
15	-1.2975	0.4498	-0.6034	0.7697
16	0.2113	-1.0312	0.1744	-0.0367
17	-0.2017	-0.1779	0.0809	0.0221
18	-0.7106	0.6368	-1.1290	-1.2534
19	-0.7226	0.6290	-0.9075	-1.4296
20	0.0829	-0.1970	-1.0102	0.8057
21	0.0443	-0.8392	-1.0544	0.9045
22	0.8245	-0.2741	-0.3225	0.7138
23	0.7634	-0.4100	-0.8217	0.8613
24	0.5545	-0.3338	-0.4949	1.2228
25	-1.5303	-0.2132	-1.3128	0.3743
26	-1.2976	-0.2034	-0.6532	0.3946
27	-0.2345	-0.2079	-1.7420	1.0966
28	-2.2280	-1.2265	-1.8600	-1.5166
29	-2.5958	-2.1770	-2.4250	-2.9581
30	-1.4420	-1.2948	-1.8597	0.3869
31	-1.1459	-2.0788	-1.6079	0.6640
32	-0.5964	-1.2671	-1.7646	0.9369
33	-0.8973	-1.2243	-1.9410	0.9053
34	-2.0689	-0.6754	-2.0180	-0.4497
35	-2.6429	-2.0414	-1.3609	-0.5105
36	-1.2520	-1.4339	-0.7874	-0.3159
37	0.2833	-0.6483	-1.9382	-0.6854
38	0.2040	-0.7895	-2.1215	-0.5485
39	0.8801	1.0039	1.0739	1.3098
40	1.0284	1.0084	1.0633	1.2868
41	-2.6775	0.6079	1.7383	-3.0415
42	-3.0521	0.7185	1.8513	-3.4691
43	-2.2026	1.1547	-0.5726	-0.2761
44	0.4625	-0.1958	-0.1921	-0.1354
45	-0.4851	2.0908	-0.4240	1.3945
46	-0.0210	1.9787	-0.6763	1.5681
47	-2.3946	2.2267	-2.7794	-0.1245
48	-2.3430	2.3154	-2.6884	-0.5719

TABLE 9.5
Prior, Gibbs, and ICM means and loadings.

C_0	0	1	2	3	4
5	7.5	.7	0	0	0
6	7.5	.7	0	0	0
8	7.5	.7	0	0	0
10	7.5	.7	0	0	0
11	7.5	.7	0	0	0
12	7.5	.7	0	0	0
13	7.5	.7	0	0	0
3	7.5	0	.7	0	0
1	7.5	0	0	.7	0
9	7.5	0	0	.7	0
15	7.5	0	0	.7	0
4	7.5	0	0	0	.7
7	7.5	0	0	0	.7
2	7.5	0	0	0	0
14	7.5	0	0	0	0
$\bar{C}$	0	1	2	3	4
5	7.5851	0.7867	-0.0494	-0.1259	0.0099
6	7.4334	0.7396	-0.0335	0.0362	0.1036
8	6.3612	0.7841	-0.0739	0.0834	-0.0483
10	6.6714	0.6898	-0.0519	0.1883	0.0413
11	7.0915	0.7839	-0.0725	0.0183	-0.0640
12	7.4275	0.6937	0.0126	0.1423	0.1298
13	7.0281	0.6432	0.0613	0.1839	0.2076
3	7.2963	0.0292	0.6941	0.0875	0.0335
1	7.1472	0.0460	-0.0934	0.7389	0.0590
9	5.9576	0.0276	0.0211	0.7905	0.0069
15	7.4903	0.2225	-0.0688	0.7138	0.0465
4	6.7395	0.0980	-0.0685	0.1495	0.6966
7	7.9370	0.1019	-0.0121	-0.1473	0.6985
2	7.4223	0.2957	-0.0158	0.0523	0.1103
14	6.4301	0.3017	-0.2900	0.1879	0.2995
$\tilde{C}$	0	1	2	3	4
5	7.6941	0.7828	-0.0472	-0.1676	-0.0111
6	7.6225	0.7315	-0.0212	-0.0087	0.0742
8	6.6418	0.7869	-0.0664	0.0645	-0.0738
10	6.9201	0.6839	-0.0439	0.1716	0.0120
11	7.3092	0.7885	-0.0630	0.0004	-0.0844
12	7.6515	0.6792	0.0356	0.1070	0.1103
13	7.2853	0.6247	0.0887	0.1609	0.1898
3	7.3538	0.0277	0.7191	0.0638	0.0270
1	7.3959	0.0049	-0.0942	0.7606	0.0622
9	6.3038	-0.0090	0.0423	0.8184	-0.0002
15	7.8412	0.1817	-0.0526	0.7233	0.0453
4	6.8584	0.0557	-0.0544	0.1834	0.7227
7	7.9147	0.0499	-0.0016	-0.1483	0.7339
2	7.5305	0.2822	0.0111	0.0836	0.1548
14	6.5979	0.2903	-0.2993	0.2279	0.3115

These factors which are based on Table 9.5 may be loosely interpreted as factor 1 being a measure of personality, factor 2 being a measure of academic ability, factor 3 being a measure of position match, and factor 4 being a measure of what can be described as charisma as presented in Table 9.6.

TABLE 9.6
Factors in terms of strong observation loadings.

Factor 1:	5 Self-confidence, 6 Lucidity, 8 Salesmanship, 10 Drive, 11 Ambition, 12 Grasp, 13 Potential
Factor 2:	3 Academic ability
Factor 3:	1 Form of letter application, 9 Experience, 15 Suitability
Factor 4:	4 Likeabiliy, 7 Honesty

The Gibbs sampling marginal values of the observation error variances and covariances are the elements of Table 9.7 while the ICM values of the observation error variances and covariances are the elements of Table 9.8. Note that the estimates from the two methods similar but with minor differences.

Table 9.9 contains the statistics for the individual posterior means and loadings. Inspection can reveal which are "large."

TABLE 9.7
Gibbs estimates of Factor Analysis covariances.

$\bar{\Psi}$	1	2	3	4	5	6	7	8	9	10	11	12	13	14	15
1	.2166	.0603	-.0014	.0681	.0116	.0063	-.0274	.0215	-.0359	.0148	.0527	.0151	.0227	.1258	-.0357
2		.4472	.0458	.0637	.0154	-.0270	.0589	.0462	.0268	-.0247	.0874	.0459	.0487	.0185	.0909
3			.0504	.0246	-.0138	-.0005	-.0107	.0126	.0315	.0137	.0144	.0165	.0303	.0084	.0143
4				.1451	-.0165	.0307	-.0595	.0513	.0419	.0325	.0537	.0258	.0534	.1035	.0452
5					.0876	-.0093	.0303	.0023	.0043	-.0123	.0138	-.0293	-.0320	.0181	-.0198
6						.0986	-.0317	.0112	.0003	-.0302	-.0240	.0439	-.0002	.0205	-.0051
7							.1206	-.0161	-.0115	-.0117	-.0189	-.0104	-.0303	-.0188	.0055
8								.1231	.0496	.0486	.0319	-.0083	.0067	.0773	.0538
9									.2205	.0511	.0424	.0194	.0398	.0710	.0220
10										.1771	.0297	-.0212	.0417	.1036	.0561
11											.1103	.0079	.0303	.0713	.0154
12												.1072	.0417	.0371	.0057
13													.1167	.0422	.0247
14														.2939	.0364
15															.1565

TABLE 9.8
ICM estimates of Factor Analysis covariances.

$\tilde{\Psi}$	1	2	3	4	5	6	7	8	9	10	11	12	13	14	15
1	.1731	.0287	.0009	.0359	.0097	.0028	-.0265	-.0017	-.0799	-.0100	.0315	.0043	.0040	.0808	-.0715
2		.4147	.0209	.0184	.0087	-.0425	.0318	.0275	-.0107	-.0481	.0718	.0214	.0166	-.0128	.0551
3			.0078	.0159	-.0093	-.0035	-.0101	.0085	.0123	.0091	.0086	.0029	.0098	.0185	.0054
4				.1006	-.0288	.0166	-.0830	.0271	.0104	.0100	.0306	.0042	.0263	.0643	.0090
5					.0817	-.0155	.0320	-.0088	.0002	-.0231	.0036	-.0352	-.0398	-.0033	-.0214
6						.0855	-.0250	-.0075	-.0081	-.0459	-.0404	.0312	-.0142	-.0035	-.0110
7							.0838	-.0087	-.0047	-.0029	-.0130	-.0086	-.0289	-.0300	.0048
8								.0882	.0186	.0162	.0047	-.0297	-.0198	.0384	.0271
9									.1531	.0186	.0147	.0000	.0099	.0300	-.0251
10										.1397	.0006	-.0419	.0140	.0634	.0277
11											.0827	-.0127	.0046	.0366	-.0070
12												.0867	.0189	.0085	-.0087
13													.0839	.0107	.0005
14														.2327	-.0080
15															.1136

TABLE 9.9

Statistics for means and loadings.

z_{Gibbs}	0	1	2	3	4
5	121.1806	16.5189	-1.1570	-2.4151	0.1896
6	113.8556	14.7298	-0.7452	0.6357	1.9246
8	85.5536	15.2162	-1.5884	1.3252	-0.7857
10	81.8379	10.5490	-0.9250	2.6814	0.5879
11	100.1574	16.0036	-1.4903	0.2825	-1.0762
12	110.8510	13.0914	0.2640	2.4543	2.3607
13	99.0115	11.2402	1.2095	3.1098	3.8215
3	98.4057	0.5196	16.2240	1.5184	0.5861
1	75.5597	0.5372	-1.3360	8.4302	0.6707
9	57.7264	0.3361	0.3013	10.2866	0.0884
15	84.6348	3.0311	-1.0764	10.4162	0.6551
4	82.3772	1.2793	-1.1176	2.0990	10.7559
7	105.5962	1.5913	-0.2195	-2.2840	11.2131
2	70.1764	2.9582	-0.1531	0.4033	0.9846
14	67.9739	3.5476	-4.3737	1.9789	3.4330
z_{ICM}	0	1	2	3	4
5	73.3077	22.3229	-1.6097	-4.6372	-0.3237
6	54.3596	20.3900	-0.7057	-0.2361	2.1094
8	42.5136	21.5955	-2.1822	1.7176	-2.0642
10	41.3721	14.9170	-1.1464	3.6325	0.2657
11	57.0133	22.3481	-2.1352	0.0103	-2.4375
12	56.3954	18.8087	1.1811	2.8743	3.1123
13	52.0442	17.5842	2.9896	4.3949	5.4445
3	276.2472	2.5567	79.5316	5.7244	2.5460
1	43.7899	0.0962	-2.2092	14.4615	1.2418
9	32.0644	-0.1868	1.0536	16.5418	-0.0045
15	46.4278	4.3946	-1.5214	16.9757	1.1163
4	50.7638	1.4320	-1.6729	4.5750	18.9345
7	71.0275	1.4054	-0.0556	-4.0504	21.0602
2	39.1745	3.5729	0.1679	1.0275	1.9980
14	33.8990	4.9054	-6.0528	3.7362	5.3657

9.9   Discussion

There has been some recent work related to the Bayesian Factor Analysis model. In a model which specifies independence between the overall mean μ vector and the factor loading matrix Λ with a vague prior distribution for the overall mean μ, robustness of assessed hyperparameters was investigated [34, 35] and the hyperparameters were found to be robust but most sensitive to the prior mean on the factor loadings Λ_0. This indicates that the most care should be taken when assessing this hyperparameter. For the same model, methods for assessing the hyperparameters were presented [20, 21]. A prior distribution was placed on the number of factors and then estimated a posteriori [45, 50]. For the same model, the process of estimating the overall mean by the sample mean was evaluated [55] and the parameters were estimated by Gibbs sampling and ICM [65]. The same Bayesian Factor Analysis model was extended to the case of the observation vectors and also the factor score vectors being correlated [50]. Bayesian Factor Analysis models which specified independence between the overall mean vector and the factor loadings matrix but took Conjugate and generalized Conjugate were introduced [51, 54].

Returning to the cocktail party problem, the matrix of factor loadings Λ is the mixing matrix which determines the contribution of each of the speakers to the mixed observations. The matrix of factor scores F is likened to the matrix of unobserved conversations S. The overall population mean μ is the overall background mean level at the microphones.

Exercises

1. Specify that the overall mean μ and the factor loading matrix Λ are independent with the prior distribution for the overall mean μ being the vague prior

$$p(\mu) \propto (\text{a constant}),$$

the distribution for the factor loading matrix being the Matrix Normal distribution

$$p(\Lambda|\Sigma) \propto |A|^{-\frac{p}{2}} |\Sigma|^{-\frac{m}{2}} e^{-\frac{1}{2} tr \Sigma^{-1}(\Lambda - \Lambda_0) A^{-1}(\Lambda - \Lambda_0)'},$$

and the others as in Equations 9.4.3–9.4.4.

Combine these prior distributions with the likelihood in Equation 9.3.3 to obtain a joint posterior distribution. Integrate the joint posterior distribution with respect to Σ then Λ to obtain a marginal posterior distribution for the matrix of factor scores. Use the large sample approximation $\frac{F'F}{n} = I_m$ to obtain an approximate marginal posterior distribution for the matrix of factor scores which is a Matrix Student T-distribution. Estimate the matrix of factor scores to be the mean of the approximate marginal posterior distribution, the matrix of factor loadings given the above mentioned factor scores, and then the error covariance matrix given the factor scores and loadings [43, 44].

2. Specify that the overall mean μ and the factor loading matrix Λ are independent with the prior distribution for the overall mean μ being the vague prior

$$p(\mu) \propto (\text{a constant}),$$

the prior distribution for the factor loading matrix being the Matrix Normal distribution

$$p(\Lambda|\Sigma) \propto |A|^{-\frac{p}{2}} |\Sigma|^{-\frac{m}{2}} e^{-\frac{1}{2} tr \Sigma^{-1}(\Lambda - \Lambda_0) A^{-1}(\Lambda - \Lambda_0)'},$$

and the others as in Equations 9.4.3-9.4.4.

Combine these prior distributions with the likelihood in Equation 9.3.3 to obtain a posterior distribution. Derive Gibbs sampling and ICM algorithms for marginal posterior mean and joint maximum a posteriori parameter estimates [65].

3. Specify that the overall mean μ and the factor loading matrix Λ are independent with the prior distribution for the overall mean μ being the Conjugate Normal prior

$$p(\mu|\Sigma) \propto |h\Sigma|^{-\frac{1}{2}} e^{-\frac{1}{2}tr\Sigma^{-1}(\mu-\mu_0)(h\Sigma)^{-1}(\mu-\mu_0)'},$$

the distribution for the factor loading matrix being

$$p(\Lambda|\Sigma) \propto |A|^{-\frac{p}{2}}|\Sigma|^{-\frac{m}{2}} e^{-\frac{1}{2}\Sigma^{-1}(\Lambda-\Lambda_0)A^{-1}(\Lambda-\Lambda_0)'},$$

and the others as in Equations 9.4.3-9.4.4.

Combine these prior distributions with the likelihood in Equation 9.3.3 to obtain a posterior distribution. Derive Gibbs sampling and ICM algorithms for marginal mean and joint maximum a posteriori parameter estimates [54].

4. Specify that μ and Λ are independent with the prior distribution for the overall mean μ to be the generalized Conjugate prior

$$p(\mu) \propto |\Gamma|^{-\frac{1}{2}} e^{-\frac{1}{2}(\mu-\mu_0)'\Gamma^{-1}(\mu-\mu_0)},$$

the distribution for the factor loading matrix being

$$p(\lambda) \propto |\Delta|^{-\frac{1}{2}} e^{-\frac{1}{2}(\lambda-\lambda_0)'\Delta^{-1}(\lambda-\lambda_0)},$$

where $\lambda = vec(\Lambda)$ and the other distributions are as in Equations 9.4.3–9.4.4.

Combine these prior distributions with the likelihood in Equation 9.3.3 to obtain a posterior distribution. Derive Gibbs sampling and ICM algorithms for marginal posterior mean and joint maximum a posteriori parameter estimates [51].

10

Bayesian Source Separation

10.1 Introduction

The Bayesian Source Separation model is different from the Bayesian Regression model in that the sources are unobserved and from the Bayesian Factor Analysis model in that there may be more or less sources than the observed dimension (the number of microphones). Further, in the Bayesian Factor Analysis model, the variance of the unobserved factor score vectors is a priori assumed to be unity and diagonal for Psychologic reasons. The Bayesian Source Separation model [52, 57, 58, 59, 60, 62, 63] allows the covariance matrix for the unobserved sources to have arbitrary variances. That is, the covariance matrix for the sources is not required to be diagonal and also the sources are allowed to have a mean other than zero.

With a general covariance matrix (one that is not constrained to be diagonal), the sources or speakers at the cocktail party are allowed to be dependent or correlated. There are other models which impose either the constraint of orthogonal sources [23] or the constraint of independent sources [6]. If the sources are truly orthogonal or independent, then such models would be appropriate (independent sources can be obtained here by imposing constraints). However, if the sources are not independent as in the "real-world" cocktail party problem, then an independence constraint would not be appropriate.

10.2 Source Separation Model

In the Bayesian approach to statistical inference, available prior information either from subjective expert experience, or prior experiments, is incorporated into the inferences. This prior information yields progressively less influence in the final results as the sample size increases, thus allowing the data to "speak the truth." The components of the source vectors are free to be correlated, as is frequently the case and not constrained to be statistically independent.

The constraint of independent sources models the situation where speakers at the cocktail party are talking without regard to the others at the party. The independent source model implies that the people are "babbling" incoherently. This is the case when we press play on several tape recorders with one speaker on each and record on others. This is not how conversations work. This does not model the true dynamics of a real cocktail party. Referring to Figure 1.1, focus on the two people in the left foreground. When they speak, they do not speak irrespective of each other. They speak interactively. For instance, the person on the left will speak and then fade out while the person on the right fades in to speak. They are not speaking at the same time. They are obviously negatively correlated or dependent in a negative fashion.

The linear synthesis Source Separation model which was motivated in Chapter 1 is given by

$$
\begin{array}{ccccc}
(x_i|\mu,\Lambda,s_i) = & \mu & + & \Lambda & s_i & + & \epsilon_i, \\
(p\times 1) & (p\times 1) & & (p\times m)\,(m\times 1) & & (p\times 1)
\end{array}
\tag{10.2.1}
$$

where for observations x_i at time increment i, $i = 1,\ldots,n$; $x_i = $ a p-dimensional observed vector, $x_i = (x_{i1},\ldots,x_{ip})'$; μ is an overall unobserved mean vector, $\mu = (\mu_1,\ldots,\mu_p)'$; $\Lambda = $ a $p\times m$ matrix of unobserved mixing constants, $\Lambda = (\lambda_1',\ldots,\lambda_p')'$; $s_i = $ the ith m-dimensional true unobservable source vector, $s_i = (s_{i1},\ldots,s_{1m})'$; and $\epsilon_i = $ the p-dimensional vector of errors or noise terms of the ith observed signal vector $\epsilon_i = (\epsilon_{i1},\ldots,\epsilon_{ip})'$.

Taking a closer look at the model, element (microphone) j in observed vector (at time) i is represented by the model

$$
\begin{array}{ccccc}
(x_{ij}|\mu_j,\lambda_j,s_i) = & \mu_j & + & \lambda_j' & s_i & + & \epsilon_{ij}, \\
(1\times 1) & (1\times 1) & & (1\times m)\,(m\times 1) & & (1\times 1)
\end{array}
\tag{10.2.2}
$$

in which the recorded or observed conversation x_{ij} for microphone j at time increment i is a linear mixture of the m true unobservable conversations at time increment i plus an overall (background) mean for microphone j and a random noise term ϵ_{ij}. This is also represented as

$$
(x_{ij}|\mu_j,\lambda_j,s_i) = \mu_j + \sum_{k=1}^{m}\lambda_{jk}\,s_{ik} + \epsilon_{ij}.
\tag{10.2.3}
$$

Analogous to the Regression and Factor Analysis models, the Source Separation model can be written in terms of matrices as

$$
\begin{array}{ccccc}
(X|\mu,\Lambda,S) = & e_n\mu' & + & S & \Lambda' & + & E, \\
(n\times p) & (n\times p) & & (n\times m)\,(m\times p) & & (n\times p)
\end{array}
\tag{10.2.4}
$$

where $X' = (x_1,\ldots,x_n)$, e_n is an n-dimensional vector of ones, $\mu = (\mu_1,\ldots,\mu_p)$, $S' = (s_1,\ldots,s_n)$, $\Lambda' = (\lambda_1,\ldots,\lambda_p)$, and $E' = (\epsilon_1,\ldots,\epsilon_n)$.

10.3   Source Separation Likelihood

Regarding the errors of the observations, it is specified that they are independent Normally distributed random vectors with mean zero and full positive definite covariance matrix Σ. From this error specification, it is seen that the observation vector x_i given the overall background mean μ, the mixing matrix Λ, the source vector s_i, and the error covariance matrix Σ is Multivariate Normally distributed with likelihood given by

$$p(x_i|\mu,\Lambda,s_i,\Sigma) \propto |\Sigma|^{-\frac{1}{2}} e^{-\frac{1}{2}(x_i-\mu-\Lambda s_i)'\Sigma^{-1}(x_i-\mu-\Lambda s_i)}. \qquad (10.3.1)$$

With the previously described matrix representation, the joint likelihood of all n observation vectors collected into the matrix X is given by

$$p(X|\mu,S,\Lambda,\Sigma) \propto |\Sigma|^{-\frac{n}{2}} e^{-\frac{1}{2}tr(X-e_n\mu'-S\Lambda')\Sigma^{-1}(X-e_n\mu'-S\Lambda')'}, \qquad (10.3.2)$$

where the variables are as previously defined.

The overall background mean vector μ and the mixing matrix Λ are joined into a single matrix as $C = (\mu, \Lambda)$. An n-dimensional vector of ones e_n and the source matrix S are also joined as $Z = (e_n, S)$. Having joined these vectors and matrices, the Source Separation model is now in a matrix representation given by

$$(X|C,Z) = \underset{n\times p}{} \quad \underset{n\times(m+1)}{Z} \quad \underset{(m+1)\times p}{C'} \quad + \quad \underset{(n\times p)}{E,} \qquad (10.3.3)$$

and its corresponding likelihood is given by the Matrix Normal distribution

$$p(X|C,Z,\Sigma) \propto |\Sigma|^{-\frac{n}{2}} e^{-\frac{1}{2}tr(X-ZC')\Sigma^{-1}(X-ZC')'}, \qquad (10.3.4)$$

where all variables are as previously defined and $tr(\cdot)$ denotes the trace operator.

Again, the objective is to unmix the sources by estimating the matrix containing them S, and to obtain knowledge about the mixing process by estimating the overall mean μ, the mixing matrix Λ, and the error covariance matrix Σ.

The advantage of the Bayesian statistical approach is that available prior information regarding parameters are quantified through probability distributions describing degrees of belief for various values. This prior knowledge is formally brought to bear in the problem through prior distributions and Bayes' rule. As stated earlier, the prior parameter values will have decreasing influence in the posterior estimates with increasing sample size, thus allowing the data to "speak the truth."

10.4   Conjugate Priors and Posterior

In the Bayesian Source Separation model [60] available information regarding values of the model parameters is quantified using Conjugate prior distributions. The joint prior distribution for the model parameters which are the matrix of coefficients C, the matrix of sources S, the covariance matrix for the sources R, and the error covariance matrix Σ is given by

$$p(S, R, C, \Sigma) = p(S|R)p(R)p(C|\Sigma)p(\Sigma), \qquad (10.4.1)$$

where the prior distribution for the parameters from the Conjugate procedure outlined in Chapter 4 are as follows

$$p(S|R) \propto |R|^{-\frac{n}{2}} e^{-\frac{1}{2}tr(S-S_0)R^{-1}(S-S_0)'}, \qquad (10.4.2)$$

$$p(R) \propto |R|^{-\frac{\eta}{2}} e^{-\frac{1}{2}trR^{-1}V}, \qquad (10.4.3)$$

$$p(\Sigma) \propto |\Sigma|^{-\frac{\nu}{2}} e^{-\frac{1}{2}tr\Sigma^{-1}Q}, \qquad (10.4.4)$$

$$p(C|\Sigma) \propto |D|^{-\frac{p}{2}} |\Sigma|^{-\frac{m}{2}} e^{-\frac{1}{2}tr\Sigma^{-1}(C-C_0)D^{-1}(C-C_0)'}, \qquad (10.4.5)$$

where Σ, R, V, Q, and D are positive definite matrices. The hyperparameters S_0, η, V, ν, Q, C_0, and D are to be assessed and having done so, completely determine the joint prior distribution.

The prior distributions for the combined matrix containing the overall mean μ with the mixing matrix Λ, the sources S, the source covariance matrix R, and the error covariance matrix Σ follow Normal, Normal, Inverted Wishart, and Inverted Wishart distributions respectively.

Note that both Σ and R are full positive definite symmetric covariance matrices which allow both the observed mixed signals (elements in the x_i's) and also the unobserved source components (elements in the s_i's) to be correlated. The prior mean of the sources is often taken to be constant for all observations and thus without loss of generality taken to be zero. Here an observation (time) varying source mean is adopted.

Upon using Bayes' rule, the joint posterior distribution for the unknown parameters is proportional to the product of the joint prior distribution and the likelihood and given by

$$p(S, R, C, \Sigma|X) \propto |\Sigma|^{-\frac{(n+\nu+m+1)}{2}} e^{-\frac{1}{2}tr\Sigma^{-1}G}$$
$$\times |R|^{-\frac{(n+\eta)}{2}} e^{-\frac{1}{2}trR^{-1}[(S-S_0)'(S-S_0)+V]}, \qquad (10.4.6)$$

where the $p \times p$ matrix variable G has been defined to be

$$G = (X - ZC')'(X - ZC') + (C - C_0)D^{-1}(C - C_0)' + Q. \quad (10.4.7)$$

This joint posterior distribution must now be evaluated in order to obtain parameter estimates of the sources S, the overall background mean/mixing matrix C, the source covariance matrix R, and the observation errors covariance matrix Σ. Marginal posterior mean and joint maximum a posteriori estimates of the parameters S, R, C, and Σ are found by the Gibbs sampling and ICM algorithms.

10.5 Conjugate Estimation and Inference

With the above joint posterior distribution for the Bayesian Source Separation model, it is not possible to obtain all or any of the marginal distributions and thus marginal estimates of the parameters in an analytic closed form. It is also not possible to obtain explicit formulas for maximum a posteriori estimates from differentiation. It is possible to use both Gibbs sampling, to obtain marginal parameter estimates and the ICM algorithm for maximum a posteriori estimates. For both estimation procedures, the posterior conditional distributions are required.

10.5.1 Posterior Conditionals

From the joint posterior distribution we can obtain the posterior conditional distributions for each of the model parameters.

The conditional posterior distribution for the overall mean/mixing matrix C is found by considering only the terms in the joint posterior distribution which involve C and is given by

$$
\begin{aligned}
p(C|S,R,\Sigma,X) &\propto p(C|\Sigma)p(X|C,S,\Sigma) \\
&\propto |\Sigma|^{-\frac{m+1}{2}} e^{-\frac{1}{2}tr\Sigma^{-1}(C-C_0)D^{-1}(C-C_0)'} \\
&\quad \times |\Sigma|^{-\frac{n}{2}} e^{-\frac{1}{2}tr\Sigma^{-1}(X-ZC')'(X-ZC')} \\
&\propto e^{-\frac{1}{2}tr\Sigma^{-1}[(C-C_0)D^{-1}(C-C_0)'+(X-ZC')'(X-ZC')]} \\
&\propto e^{-\frac{1}{2}tr\Sigma^{-1}(C-\tilde{C})(D^{-1}+Z'Z)(C-\tilde{C})'}, \quad (10.5.1)
\end{aligned}
$$

where the variable $\tilde{C}$, the posterior conditional mean and mode, has been defined and is given by

$$\tilde{C} = (C_0D^{-1} + X'Z)(D^{-1} + Z'Z)^{-1} \quad (10.5.2)$$

$$= C_0[D^{-1}(D^{-1}+Z'Z)^{-1}] + \hat{C}[(Z'Z)(D^{-1}+Z'Z)^{-1}]. \quad (10.5.3)$$

Note that the matrix $\tilde{C}$ can be written as a weighted combination of the prior mean C_0 from the prior distribution and the data mean $\hat{C} = X'Z(Z'Z)^{-1}$ from the likelihood.

The posterior conditional distribution for the matrix of coefficients C given the matrix of sources S, the source covariance matrix R, the error covariance matrix Σ, and the data matrix X is Matrix Normally distributed.

The conditional posterior distribution of the error covariance matrix Σ is found by considering only those terms in the joint posterior distribution which involve Σ and is given by

$$p(\Sigma|S,R,C,X) \propto p(\Sigma)p(C|\Sigma)p(X|S,C,\Sigma)$$
$$\propto |\Sigma|^{-\frac{\nu}{2}}e^{-\frac{1}{2}tr\Sigma^{-1}Q}|\Sigma|^{-\frac{m+1}{2}}e^{-\frac{1}{2}tr\Sigma^{-1}(C-C_0)D^{-1}(C-C_0)'}$$
$$\times |\Sigma|^{-\frac{n}{2}}e^{-\frac{1}{2}tr\Sigma^{-1}(X-ZC')'(X-ZC')}$$
$$\propto |\Sigma|^{-\frac{(n+\nu+m+1)}{2}}e^{-\frac{1}{2}tr\Sigma^{-1}G}, \quad (10.5.4)$$

where the $p \times p$ matrix G has been defined to be

$$G = (X - ZC')'(X - ZC') + (C - C_0)D^{-1}(C - C_0)' + Q \quad (10.5.5)$$

with a mode as described in Chapter 2 given by

$$\tilde{\Sigma} = \frac{G}{n+\nu+m+1}. \quad (10.5.6)$$

The posterior conditional distribution of the observation error covariance matrix Σ given the matrix of sources S, the source covariance matrix R, the matrix of coefficients C, and the data X is an Inverted Wishart.

The conditional posterior distribution for the sources S is found by considering only those terms in the joint posterior distribution which involve S and is given by

$$p(S|\mu,R,\Lambda,\Sigma,X) \propto p(S|R)p(X|\mu,\Lambda,S,\Sigma)$$
$$\propto |R|^{-\frac{n}{2}}e^{-\frac{1}{2}tr(S-S_0)R^{-1}(S-S_0)'}$$
$$\times |\Sigma|^{-\frac{n}{2}}e^{-\frac{1}{2}tr\Sigma^{-1}(X-e_n\mu'-S\Lambda')'(X-e_n\mu'-S\Lambda')}$$
$$\propto e^{-\frac{1}{2}tr(S-\tilde{S})(R^{-1}+\Lambda'\Sigma^{-1}\Lambda)(S-\tilde{S})'}, \quad (10.5.7)$$

where the matrix $\tilde{S}$ has been defined which is the posterior conditional mean and mode given by

$$\tilde{S} = [S_0R^{-1} + (X - e_n\mu')\Sigma^{-1}\Lambda](R^{-1}+\Lambda'\Sigma^{-1}\Lambda)^{-1}. \quad (10.5.8)$$

The conditional posterior distribution for the sources S given the overall mean vector μ, the source covariance matrix R, the matrix of mixing coefficients Λ, the error covariance matrix Σ, and the data matrix X is Matrix Normally distributed.

The conditional posterior distribution for the source covariance matrix R is found by considering only those terms in the joint posterior distribution which involve R and is given by

$$p(R|\mu,\Lambda,S,\Sigma,X) \propto p(R)p(S|R)p(X|\mu,\Lambda,S,\Sigma)$$
$$\propto |R|^{-\frac{\eta}{2}}e^{-\frac{1}{2}trR^{-1}V}|R|^{-\frac{n}{2}}e^{-\frac{1}{2}tr(S-S_0)R^{-1}(S-S_0)'}$$
$$\propto |R|^{-\frac{(n+\eta)}{2}}e^{-\frac{1}{2}trR^{-1}[(S-S_0)'(S-S_0)+V]}, \qquad (10.5.9)$$

with the posterior conditional mode as described in Chapter 2 given by

$$\tilde{R} = \frac{(S-S_0)'(S-S_0)+V}{n+\eta}. \qquad (10.5.10)$$

The conditional posterior distribution for the source covariance matrix R given the matrix of sources S, the error covariance matrix Σ, the matrix of means and mixing coefficients C, and the data matrix X is Inverted Wishart distributed.

10.5.2 Gibbs Sampling

To find marginal posterior mean estimates of the model parameters from the joint posterior distribution using the Gibbs sampling algorithm, start with initial values for the matrix of sources S and the error covariance matrix Σ, say $\bar{S}_{(0)}$ and $\bar{\Sigma}_{(0)}$, and then cycle through

$$\bar{C}_{(l+1)} = \text{a random variate from } p(C|\bar{S}_{(l)},\bar{R}_{(l)},\bar{\Sigma}_{(l)},X)$$
$$= A_C Y_C B_C' + M_C, \qquad (10.5.11)$$
$$\bar{\Sigma}_{(l+1)} = \text{a random variate from } p(\Sigma|\bar{S}_{(l)},\bar{R}_{(l)},\bar{C}_{(l+1)},X)$$
$$= A_\Sigma (Y_\Sigma' Y_\Sigma)^{-1} A_\Sigma', \qquad (10.5.12)$$
$$\bar{R}_{(l+1)} = \text{a random variate from } p(R|\bar{S}_{(l)},\bar{C}_{(l+1)},\bar{\Sigma}_{(l+1)},X)$$
$$= A_R (Y_R' Y_R)^{-1} A_R', \qquad (10.5.13)$$
$$\bar{S}_{(l+1)} = \text{a random variate from } p(S|\bar{R}_{(l+1)},\bar{C}_{(l+1)},\bar{\Sigma}_{(l+1)},X)$$
$$= Y_S B_S' + M_S, \qquad (10.5.14)$$

where

$$A_C A_C' = \bar{\Sigma}_{(l)},$$

$$B_C B_C' = (D^{-1} + \bar{Z}_{(l)}' \bar{Z}_{(l)})^{-1},$$
$$\bar{Z}_{(l)} = (e_n, \bar{S}_{(l)}),$$
$$M_C = (X'\bar{Z}_{(l)} + C_0 D^{-1})(D^{-1} + \bar{Z}_{(l)}' \bar{Z}_{(l)})^{-1}$$
$$A_\Sigma A_\Sigma' = (X - \bar{Z}_{(l)} \bar{C}_{(l+1)}')'(X - \bar{Z}_{(l)} \bar{C}_{(l+1)}')$$
$$\quad + (\bar{C}_{(l+1)} - C_0) D^{-1} (\bar{C}_{(l+1)} - C_0)' + Q,$$
$$A_R A_R' = (\bar{S}_{(l)} - S_0)'(\bar{S}_{(l)} - S_0) + V,$$
$$B_S B_S' = (\bar{R}_{(l+1)}^{-1} + \bar{\Lambda}_{(l+1)}' \bar{\Sigma}_{(l+1)}^{-1} \bar{\Lambda}_{(l+1)})^{-1},$$
$$M_S = [S_0 \bar{R}_{(l+1)}^{-1}(X - e_n \bar{\mu}_{(l+1)}')\bar{\Sigma}_{(l+1)}^{-1} \tilde{\Lambda}_{(l+1)}]$$
$$\quad \times (\bar{R}_{(l+1)}^{-1} + \bar{\Lambda}_{(l+1)}' \bar{\Sigma}_{(l+1)}^{-1} \bar{\Lambda}_{(l+1)})^{-1}$$

while Y_C, Y_Σ, Y_R, and Y_S are $p \times (m+1)$, $(n+\nu+m+1+p+1) \times p$, $(n+\eta+m+1) \times m$, and $n \times m$ dimensional matrices respectively, whose elements are random variates from the standard Scalar Normal distribution. The formulas for the generation of random variates from the conditional posterior distributions are easily found from the methods in Chapter 6.

The first random variates called the "burn in" are discarded and after doing so, compute from the next L variates means of each of the parameters

$$\bar{S} = \frac{1}{L}\sum_{l=1}^{L} \bar{S}_{(l)} \quad \bar{R} = \frac{1}{L}\sum_{l=1}^{L} \bar{R}_{(l)} \quad \bar{C} = \frac{1}{L}\sum_{l=1}^{L} \bar{C}_{(l)} \quad \bar{\Sigma} = \frac{1}{L}\sum_{l=1}^{L} \bar{\Sigma}_{(l)}$$

which are the exact sampling-based marginal posterior mean estimates of the parameters. Exact sampling-based estimates of other quantities can also be found. Similar to Bayesian Regression and Bayesian Factor Analysis, there is interest in the estimate of the marginal posterior variance of the vector containing the means and mixing coefficients

$$\overline{var}(c|X) = \frac{1}{L}\sum_{l=1}^{L} \bar{c}_{(l)} \bar{c}_{(l)}' - \bar{c}\bar{c}'$$
$$= \bar{\Delta},$$

where $c = vec(C)$.

The covariance matrices of the other parameters follow similarly. With a specification of Normality for the marginal posterior distribution of the matrix containing the mean vector and mixing matrix, their distribution is

$$p(c|X) \propto |\bar{\Delta}|^{-\frac{1}{2}} e^{-\frac{1}{2}(c-\bar{c})\bar{\Delta}^{-1}(c-\bar{c})'}, \qquad (10.5.15)$$

where $\bar{c}$ and $\bar{\Delta}$ are as previously defined.

To determine statistical significance with the Gibbs sampling approach, use the marginal distribution of the matrix containing the mean vector and mixing matrix given above. General simultaneous hypotheses can be evaluated regarding the entire matrix containing the mean vector and the mixing matrix, a submatrix, or the mean vector or a particular source, or an element by computing marginal distributions. It can be shown that the marginal distribution of the kth column of the matrix containing the mean vector and mixing matrix C, C_k is Multivariate Normal

$$p(C_k|\bar{C}_k, X) \propto |\bar{\Delta}_k|^{-\frac{1}{2}} e^{-\frac{1}{2}(C_k - \bar{C}_k)'\bar{\Delta}_k^{-1}(C_k - \bar{C}_k)}, \qquad (10.5.16)$$

where $\bar{\Delta}_k$ is the covariance matrix of C_k found by taking the kth $p \times p$ submatrix along the diagonal of $\bar{\Delta}$.

Significance can be evaluated for a subset of coefficients of the kth column of C by determining the marginal distribution of the subset within C_k which is also Multivariate Normal. With the subset being a singleton set, significance can be evaluated for a particular mean or mixing coefficient with the marginal distribution of the scalar coefficient which is

$$p(C_{kj}|\bar{C}_{kj}, X) \propto (\bar{\Delta}_{kj})^{-\frac{1}{2}} e^{-\frac{(C_{kj} - \bar{C}_{kj})^2}{2\bar{\Delta}_{kj}}}, \qquad (10.5.17)$$

where $\bar{\Delta}_{kj}$ is the jth diagonal element of $\bar{\Delta}_k$. Note that $\bar{C}_{kj} = \bar{c}_{jk}$ and that

$$z = \frac{(C_{kj} - \bar{C}_{kj})}{\sqrt{\bar{\Delta}_{kj}}} \qquad (10.5.18)$$

follows a Normal distribution with a mean of zero and variance of one.

10.5.3 Maximum a Posteriori

The joint posterior distribution can also be maximized with respect to the matrix of coefficients C, the matrix of sources S, the source covariance matrix R, and the error covariance matrix Σ by the ICM algorithm. To maximize the joint posterior distribution using the ICM algorithm, start with an initial value for the matrix of sources S, say $\tilde{S}_{(0)}$, and then cycle through

$$\tilde{C}_{(l+1)} = \overset{\text{Arg Max}}{C} \, p(C|\tilde{S}_{(l)}, \tilde{R}_{(l)}, \tilde{\Sigma}_{(l)}, X)$$
$$= [X'\tilde{Z}_{(l)} + C_0 D^{-1}](D^{-1} + \tilde{Z}'_{(l)}\tilde{Z}_{(l)})^{-1},$$

$$\tilde{\Sigma}_{(l+1)} = \overset{\text{Arg Max}}{\Sigma} \, p(\Sigma|\tilde{C}_{(l+1)}, \tilde{R}_{(l)}, \tilde{S}_{(l)}, X)$$
$$= [(X - \tilde{Z}_{(l)}\tilde{C}'_{(l+1)})'(X - \tilde{Z}_{(l)}\tilde{C}'_{(l+1)})$$
$$+ (\tilde{C}_{(l+1)} - C_0)D^{-1}(\tilde{C}_{(l+1)} - C_0)' + Q]/(n + m + \nu + 1),$$

$$\tilde{R}_{(l+1)} = \overset{\text{Arg Max}}{R}\ p(R|\tilde{S}_{(l)}, \tilde{C}_{(l+1)}, \tilde{\Sigma}_{(l+1)}, X)$$

$$= \frac{(\tilde{S}_{(l)} - S_0)'(\tilde{S}_{(l)} - S_0) + V}{n + \eta},$$

$$\tilde{S}_{(l+1)} = \overset{\text{Arg Max}}{S}\ p(S|\tilde{C}_{(l+1)}, \tilde{R}_{(l+1)}, \tilde{\Sigma}_{(l+1)}, X)$$

$$= [S_0 \tilde{R}_{(l+1)}^{-1} + (X - e_n \tilde{\mu}'_{(l+1)})\tilde{\Sigma}_{(l+1)}^{-1} \tilde{\Lambda}_{(l+1)}]$$

$$\times (\tilde{R}_{(l+1)}^{-1} + \tilde{\Lambda}'_{(l+1)} \tilde{\Sigma}_{(l+1)}^{-1} \tilde{\Lambda}_{(l+1)})^{-1},$$

where the matrix $\tilde{Z}_{(l)} = (e_n, \tilde{S}_{(l)})$ until convergence is reached. The converged values $(\tilde{S}, \tilde{R}, \tilde{C}, \tilde{\Sigma})$ are joint posterior modal (maximum a posteriori) estimators of the parameters. Conditional maximum a posteriori variance estimates can also be found. The conditional modal variance of the matrix containing the means and mixing coefficients is

$$var(C|\tilde{C}, \tilde{S}, \tilde{R}, \tilde{\Sigma}, X) = \tilde{\Sigma} \otimes (D^{-1} \otimes \tilde{Z}'\tilde{Z})^{-1}$$

or equivalently

$$var(c|\tilde{c}, \tilde{S}, \tilde{R}, \tilde{\Sigma}, X) = (D^{-1} \otimes \tilde{Z}'\tilde{Z})^{-1} \otimes \tilde{\Sigma}$$

$$= \bar{\Delta},$$

where $c = vec(C)$, while $\tilde{S}$, $\tilde{R}$, and $\tilde{\Sigma}$ are the converged value from the ICM algorithm.

To evaluate statistical significance with the ICM approach, use the conditional distribution of the matrix containing the mean vector and mixing matrix which is

$$p(C|\tilde{C}, \tilde{S}, \tilde{R}, \tilde{\Sigma}, X, \propto |D^{-1} + \tilde{Z}'\tilde{Z}|^{\frac{1}{2}}|\tilde{\Sigma}|^{-\frac{1}{2}} e^{-\frac{1}{2}tr\tilde{\Sigma}^{-1}(C-\tilde{C})(D^{-1}+\tilde{Z}'\tilde{Z})(C-\tilde{C})'}.$$

$$(10.5.19)$$

That is,

$$C|\tilde{C}, \tilde{S}, \tilde{R}, \tilde{\Sigma}, X, \sim N\left(\tilde{C}, \tilde{\Sigma} \otimes (D^{-1} + \tilde{Z}'\tilde{Z})^{-1}\right). \qquad (10.5.20)$$

General simultaneous hypotheses can be evaluated regarding the entire matrix containing the mean vector and the mixing matrix, a submatrix, or the mean vector or a particular source, or an element by computing marginal conditional distributions.

It can be shown [17, 41] that the marginal conditional distribution of any column of the matrix containing the means and mixing coefficients C, C_k is Multivariate Normal

$$p(C_k|\tilde{C}_k, \tilde{S}, \tilde{\Sigma}, U, X) \propto |W_{kk}\tilde{\Sigma}|^{-\frac{1}{2}} e^{-\frac{1}{2}(C_k - \tilde{C}_k)'(W_{kk}\tilde{\Sigma})^{-1}(C_k - \tilde{C}_k)}, \qquad (10.5.21)$$

where $W = (D^{-1} + U'U)^{-1}$ and W_{kk} is its kth diagonal element.

With the marginal distribution of a column of C, significance can be evaluated for the mean vector or a particular source. Significance can be evaluated for a subset of coefficients by determining the marginal distribution of the subset within C_k which is also Multivariate Normal. With the subset being a singleton set, significance can be determined for a particular mean or mixing coefficient with the marginal distribution of the scalar coefficient which is

$$p(C_{kj}|\tilde{C}_{kj}, \tilde{S}, \tilde{\Sigma}_{jj}, U, X) \propto (W_{kk}\tilde{\Sigma}_{jj})^{-\frac{1}{2}} e^{-\frac{(C_{kj}-\tilde{C}_{kj})^2}{2W_{kk}\tilde{\Sigma}_{jj}}}, \qquad (10.5.22)$$

where $\tilde{\Sigma}_{jj}$ is the jth diagonal element of $\tilde{\Sigma}$. Note that $\tilde{C}_{kj} = \tilde{c}_{jk}$ and that

$$z = \frac{(C_{kj} - \tilde{C}_{kj})}{\sqrt{W_{kk}\tilde{\Sigma}_{jj}}}$$

$$(10.5.23)$$

follows a Normal distribution with a mean of zero and variance of one.

10.6   Generalized Priors and Posterior

Generalized Conjugate prior distributions are assessed in order to quantify available prior information regarding values of the model parameters.

The joint prior distribution for the sources S, the source covariance matrix R, the error covariance matrix Σ, and the matrix of coefficients $c = vec(C)$ is given by

$$p(S, R, \Sigma, c) = p(S|R)p(R)p(\Sigma)p(c), \qquad (10.6.1)$$

where the prior distribution for the parameters from the generalized Conjugate procedure outlined in Chapter 4 are as follows

$$p(S|R) \propto |R|^{-\frac{n}{2}} e^{-\frac{1}{2}tr(S-S_0)R^{-1}(S-S_0)'}, \qquad (10.6.2)$$

$$p(R) \propto |R|^{-\frac{\eta}{2}} e^{-\frac{1}{2}trR^{-1}V}, \qquad (10.6.3)$$

$$p(\Sigma) \propto |\Sigma|^{-\frac{\nu}{2}} e^{-\frac{1}{2}tr\Sigma^{-1}Q}, \qquad (10.6.4)$$

$$p(c) \propto |\Delta|^{-\frac{1}{2}} e^{-\frac{1}{2}(c-c_0)'\Delta^{-1}(c-c_0)}, \qquad (10.6.5)$$

where Σ, R, V, Q, and Δ are positive definite matrices. The hyperparameters S_0, η, V, ν, Q, c_0, and Δ are to be assessed, and having done so, completely determine the joint prior distribution.

The prior distribution for the matrix of sources S is Matrix Normally distributed, the prior distribution for the source vector covariance matrix R is Inverted Wishart distributed, the vector $c = vec(C)$, $C = (\mu, \Lambda)$ containing the overall mean μ and the mixing matrix Λ is Multivariate Normal, and the prior distribution for the error covariance matrix Σ is Inverted Wishart distributed.

Note that both Σ and R are full positive definite covariance matrices allowing both the observed mixed signals (microphones) and also the unobserved source components (speakers) to be correlated. The mean of the sources is often taken to be constant for all observations and thus without loss of generality taken to be zero. An observation (time) varying source mean is adopted here.

Upon using Bayes' rule the joint posterior distribution for the unknown parameters with generalized Conjugate prior distributions for the model parameters is given by

$$p(S, R, C, \Sigma | X) = p(S|R)p(R)p(\Sigma)p(c)p(X|C, S, \Sigma), \qquad (10.6.6)$$

which is

$$
\begin{aligned}
p(S, R, C, \Sigma | X) \propto |\Sigma|^{-\frac{(n+\nu)}{2}} e^{-\frac{1}{2} tr\Sigma^{-1}[(X - ZC')'(X - ZC') \mid Q]} \\
\times |R|^{-\frac{(n+\eta)}{2}} e^{-\frac{1}{2} trR^{-1}[(S - S_0)'(S - S_0) + V]} \\
\times e^{-\frac{1}{2}(c - c_0)'\Delta^{-1}(c - c_0)}
\end{aligned}
\qquad (10.6.7)
$$

after inserting the joint prior distribution and the likelihood.

This joint posterior distribution must now be evaluated in order to obtain parameter estimates of the sources S, the overall mean/mixing matrix C, the covariance matrix for the sources R, and the observation error covariance matrix Σ.

10.7 Generalized Estimation and Inference

With the generalized Conjugate prior distributions for the parameters, it is not possible to obtain all or any of the marginal distributions or explicit maximum a posteriori estimates and thus marginal mean or maximum a posteriori estimates in closed form. For these reasons, marginal mean and maximum a posteriori estimates are found using the Gibbs sampling and ICM algorithms.

10.7.1 Posterior Conditionals

Both Gibbs sampling and ICM require the posterior conditionals. Gibbs sampling requires the conditionals for the generation of random variates while ICM requires them for maximization by cycling through their modes.

The conditional posterior distribution of the matrix of sources S is found by considering only those terms in the joint posterior distribution which involve S and is given by

$$
\begin{aligned}
p(S|\mu, R, \Lambda, \Sigma, X) &\propto p(S|R)p(X|\mu, S, \Lambda, \Sigma) \\
&\propto e^{-\frac{1}{2}tr(S-S_0)'R^{-1}(S-S_0)} \\
&\times e^{-\frac{1}{2}tr(X-e_n\mu'-S\Lambda')\Sigma^{-1}(X-e_n\mu'-S\Lambda')'}, \quad (10.7.1)
\end{aligned}
$$

which after performing some algebra in the exponent can be written as

$$
p(S|\mu, R, \Lambda, \Sigma, X) \propto e^{-\frac{1}{2}tr(S-\tilde{S})(R^{-1}+\Lambda'\Sigma^{-1}\Lambda)(S-\tilde{S})'}, \quad (10.7.2)
$$

where the matrix $\tilde{S}$ has been defined to be

$$
\tilde{S} = [S_0 R^{-1} + (X - e_n\mu')\Sigma^{-1}\Lambda](R^{-1} + \Lambda'\Sigma^{-1}\Lambda)^{-1}. \quad (10.7.3)
$$

That is, the matrix of sources S given the source covariance matrix R, the overall mean μ, the mixing matrix Λ, the error covariance matrix Σ, and the data matrix X is Matrix Normally distributed.

The conditional posterior distribution of the source covariance matrix R is found by considering only those terms in the joint posterior distribution which involve R and is given by

$$
\begin{aligned}
p(R|\mu, S, \Lambda, \Sigma, X) &\propto p(R)p(S|R) \\
&\propto |R|^{-\frac{\nu}{2}}e^{-\frac{1}{2}trR^{-1}V}|R|^{-\frac{n}{2}}e^{-\frac{1}{2}trR^{-1}(S-S_0)'(S-S_0)} \\
&\times |R|^{-\frac{(n+\nu)}{2}}e^{-\frac{1}{2}trR^{-1}[(S-S_0)'(S-S_0)+V]}. \quad (10.7.4)
\end{aligned}
$$

That is, the posterior conditional distribution of the source covariance matrix R given the overall mean μ, the matrix of sources S, the error covariance matrix Σ, the mixing matrix Λ, and the data matrix X has an Inverted Wishart distribution.

The conditional posterior distribution of the coefficient vector c containing the overall mean μ and mixing matrix Λ is found by considering only those terms in the joint posterior distribution which involve c or C and is given by

$$
\begin{aligned}
p(c|S, R, \Sigma, X) &\propto p(c)p(X|S, C, \Sigma) \\
&\propto |\Delta|^{-\frac{1}{2}}e^{-\frac{1}{2}(c-c_0)'\Delta^{-1}(c-c_0)} \\
&\times |\Sigma|^{-\frac{n}{2}}e^{-\frac{1}{2}tr\Sigma^{-1}(X-ZC')'(X-ZC')}, \quad (10.7.5)
\end{aligned}
$$

which after performing some algebra in the exponent becomes

$$p(c|S,R,\Sigma,X) \propto e^{-\frac{1}{2}(c-\tilde{c})'(\Delta^{-1}+Z'Z\otimes\Sigma^{-1})(c-\tilde{c})}, \qquad (10.7.6)$$

where the vector $\tilde{c}$ has been defined to be

$$\tilde{c} = (\Delta^{-1}+Z'Z\otimes\Sigma^{-1})^{-1}[\Delta^{-1}c_0+(Z'Z\otimes\Sigma^{-1})\hat{c}], \qquad (10.7.7)$$

and the vector $\hat{c}$ has been defined to be

$$\hat{c} = vec[X'Z(Z'Z)^{-1}]. \qquad (10.7.8)$$

Note that $\tilde{c}$ has been written as a weighted combination of the prior mean c_0 from the prior distribution and the data mean $\hat{c}$ from the likelihood.

The conditional posterior distribution of the vector c containing the overall mean μ and the mixing matrix Λ given the matrix of sources S, the source covariance matrix R, the error covariance matrix Σ, and the data matrix X is Multivariate Normally distributed.

The conditional posterior distribution of the error covariance matrix Σ is found by considering only those terms in the joint posterior distribution which involve Σ and is given by

$$p(\Sigma|S,R,C,X) \propto p(\Sigma)p(X|S,C,\Sigma)$$
$$\propto |\Sigma|^{-\frac{(n+\nu)}{2}}e^{-\frac{1}{2}tr\Sigma^{-1}[(X-ZC')'(X-ZC')+Q]}. \quad (10.7.9)$$

That is, the conditional distribution of the error covariance matrix Σ given the matrix of sources S, the source covariance matrix R, the overall mean μ, the mixing matrix Λ, and the data matrix X has an Inverted Wishart distribution.

The modes of these posterior conditional distributions are as described in Chapter 2 and given by $\tilde{S}$, $\tilde{c}$, (both as defined above)

$$\tilde{R} = \frac{(S-S_0)'(S-S_0)+V}{n+\eta}, \qquad (10.7.10)$$

and

$$\tilde{\Sigma} = \frac{(X-ZC')'(X-ZC')+Q}{n+\nu}, \qquad (10.7.11)$$

respectively.

10.7.2 Gibbs Sampling

To find marginal posterior mean estimates of the parameters from the joint posterior distribution using the Gibbs sampling algorithm, start with initial values for the matrix of sources S and the error covariance matrix Σ, say $\bar{S}_{(0)}$ and $\bar{\Sigma}_{(0)}$, and then cycle through

$$\bar{c}_{(l+1)} = \text{a random variate from } p(c|\bar{S}_{(l)}, \bar{\Sigma}_{(l)}, \bar{R}_{(l+1)}, X)$$
$$= A_c Y_c + M_c, \tag{10.7.12}$$
$$\bar{\Sigma}_{(l+1)} = \text{a random variate from } p(\Sigma|\bar{S}_{(l)}, \bar{R}_{(l+1)}, \bar{C}_{(l+1)}, X)$$
$$= A_\Sigma (Y'_\Sigma Y_\Sigma)^{-1} A'_\Sigma, \tag{10.7.13}$$
$$\bar{R}_{(l+1)} = \text{a random variate from } p(R|\bar{S}_{(l)}, \bar{C}_{(l+1)}, \bar{\Sigma}_{(l+1)}, X)$$
$$= A_R (Y'_R Y_R)^{-1} A'_R, \tag{10.7.14}$$
$$\bar{S}_{(l+1)} = \text{a random variate from } p(S|\bar{R}_{(l+1)}, \bar{C}_{(l+1)}, \bar{\Sigma}_{(l+1)}, X)$$
$$= Y_S B'_S + M_S, \tag{10.7.15}$$

where

$$\hat{c}_{(l)} = vec[X'\bar{Z}_{(l)}(\bar{Z}'_{(l)}\bar{Z}_{(l)})^{-1}],$$
$$\bar{c}_{(l+1)} = [\Delta^{-1} + \bar{Z}'_{(l)}\bar{Z}_{(l)} \otimes \bar{\Sigma}^{-1}_{(l)}]^{-1}[\Delta^{-1}c_0 + (\bar{Z}'_{(l)}\bar{Z}_{(l)} \otimes \bar{\Sigma}^{-1}_{(l)})\hat{c}_{(l)}],$$
$$A_c A'_c = (\Delta^{-1} + \bar{Z}'_{(l)}\bar{Z}_{(l)} \otimes \bar{\Sigma}^{-1}_{(l)})^{-1},$$
$$M_c = [\Delta^{-1} + \bar{Z}'_{(l)}\bar{Z}_{(l)} \otimes \bar{\Sigma}^{-1}_{(l)}]^{-1}[\Delta^{-1}c_0 + (\bar{Z}'_{(l)}\bar{Z}_{(l)} \otimes \bar{\Sigma}^{-1}_{(l)})\hat{c}],$$
$$A_\Sigma A'_\Sigma = (X - \bar{Z}_{(l)}\bar{C}'_{(l+1)})'(X - \bar{Z}_{(l)}\bar{C}'_{(l+1)}) + Q,$$
$$A_R A'_R = (\bar{S}_{(l)} - S_0)'(\bar{S}_{(l)} - S_0) + V,$$
$$B_S B'_S = (\bar{R}^{-1}_{(l+1)} + \bar{\Lambda}'_{(l+1)}\bar{\Sigma}^{-1}_{(l+1)}\bar{\Lambda}_{(l+1)})^{-1},$$
$$M_S = [S_0 R^{-1}_{(l+1)}(X - e_n \bar{\mu}'_{(l+1)})\bar{\Sigma}^{-1}_{(l+1)}\bar{\Lambda}_{(l+1)}]$$
$$\times (\bar{R}^{-1}_{(l+1)} + \bar{\Lambda}'_{(l+1)}\bar{\Sigma}^{-1}_{(l+1)}\bar{\Lambda}_{(l+1)})^{-1}$$

while Y_c, Y_Σ, Y_R, and Y_S are $p(m+1) \times 1$, $(n+\nu+p+1) \times p$, $(n+\eta+m+1) \times m$, and $n \times m$ dimensional matrices whose respective elements are random variates from the standard Scalar Normal distribution. The formulas for the generation of random variates from the conditional posterior distributions is easily found from the methods in Chapter 6.

The first random variates called the "burn in" are discarded and after doing so, compute from the next L variates means of each of the parameters

$$\bar{S} = \frac{1}{L}\sum_{l=1}^{L}\bar{S}_{(l)} \qquad \bar{R} = \frac{1}{L}\sum_{l=1}^{L}\bar{R}_{(l)} \qquad \bar{c} = \frac{1}{L}\sum_{l=1}^{L}\bar{c}_{(l)} \qquad \bar{\Sigma} = \frac{1}{L}\sum_{l=1}^{L}\bar{\Sigma}_{(l)}$$

which are the exact sampling-based marginal posterior mean estimates of the parameters. Exact sampling-based estimates of other quantities can also be found. Similar to Regression and Factor Analysis, there is interest in the estimate of the marginal posterior variance of the matrix containing the means and mixing coefficients

$$\overline{var}(c|X) = \frac{1}{L} \sum_{l=1}^{L} \bar{c}_{(l)} \bar{c}'_{(l)} - \bar{c}\bar{c}'$$

$$= \bar{\Delta}.$$

The covariance matrices of the other parameters follow similarly. With a specification of Normality for the marginal posterior distribution of the vector containing the means and mixing coefficients, their distribution is

$$p(c|X) \propto |\bar{\Delta}|^{-\frac{1}{2}} e^{-\frac{1}{2}(c-\bar{c})'\bar{\Delta}^{-1}(c-\bar{c})}, \tag{10.7.16}$$

where $\bar{c}$ and $\bar{\Delta}$ are as previously defined.

To evaluate statistical significance with the Gibbs sampling approach, use the marginal distribution of the vector c containing the means and mixing coefficients given above. General simultaneous hypotheses can be evaluated regarding the entire coefficient vector of means and mixing coefficients, a subset of it, or the mean vector or the coefficients for a particular source by computing marginal distributions. It can be shown that the marginal distribution of the kth column of the matrix containing the means and mixing coefficients C, C_k is Multivariate Normal

$$p(C_k|\bar{C}_k, X) \propto |\bar{\Delta}_k|^{-\frac{1}{2}} e^{-\frac{1}{2}(C_k-\bar{C}_k)'\bar{\Delta}_k^{-1}(C_k-\bar{C}_k)}, \tag{10.7.17}$$

where $\bar{\Delta}_k$ is the covariance matrix of C_k found by taking the kth $p \times p$ submatrix along the diagonal of $\bar{\Delta}$.

Significance can be determined for a subset of means or coefficients of the kth column of C by determining the marginal distribution of the subset within C_k which is also Multivariate Normal. With the subset being a singleton set, significance can be determined for a particular mean or coefficient with the marginal distribution of the scalar coefficient which is

$$p(C_{kj}|\bar{C}_{kj}, X) \propto (\bar{\Delta}_{kj})^{-\frac{1}{2}} e^{-\frac{(C_{kj}-\bar{C}_{kj})^2}{2\bar{\Delta}_{kj}}}, \tag{10.7.18}$$

where $\bar{\Delta}_{kj}$ is the jth diagonal element of $\bar{\Delta}_k$. Note that $\bar{C}_{kj} = \bar{c}_{jk}$ and that

$$z = \frac{(C_{kj}-\bar{C}_{kj})}{\sqrt{\bar{\Delta}_{kj}}} \tag{10.7.19}$$

follows a Normal distribution with a mean of zero and variance of one.

10.7.3   Maximum a Posteriori

The joint posterior distribution can also be maximized with respect to the vector of coefficients c, the matrix of sources S, the source covariance matrix R, and the error covariance matrix Σ using the ICM algorithm. To maximize the joint posterior distribution using the ICM algorithm, start with initial values for the matrix of sources $\tilde{S}$ and the error covariance matrix Σ, say $\tilde{S}_{(0)}$ and $\tilde{\Sigma}_{(0)}$, and then cycle through

$$\hat{c}_{(l)} = vec[X'\tilde{Z}_{(l)}(\tilde{Z}'_{(l)}\tilde{Z}_{(l)})^{-1}],$$

$$\tilde{c}_{(l+1)} = \underset{c}{\text{Arg Max}}\ p(c|\tilde{S}_{(l)},\tilde{R}_{(l+1)},\tilde{\Sigma}_{(l)},X)$$
$$= [\Delta^{-1} + \tilde{Z}'_{(l)}\tilde{Z}_{(l)} \otimes \tilde{\Sigma}_{(l)}^{-1}]^{-1}[\Delta^{-1}c_0 + (\tilde{Z}'_{(l)}\tilde{Z}_{(l)} \otimes \tilde{\Sigma}_{(l)}^{-1})\hat{c}_{(l)}],$$

$$\tilde{\Sigma}_{(l+1)} = \underset{\Sigma}{\text{Arg Max}}\ p(\Sigma|\tilde{C}_{(l+1)},\tilde{R}_{(l)},\tilde{S}_{(l)},X)$$
$$= \frac{(X - \tilde{Z}_{(l)}\tilde{C}'_{(l+1)})'(X - \tilde{Z}_{(l)}\tilde{C}'_{(l+1)}) + Q}{n+\nu},$$

$$\tilde{R}_{(l+1)} = \underset{R}{\text{Arg Max}}\ p(R|\tilde{S}_{(l)},\tilde{C}_{(l+1)},\tilde{\Sigma}_{(l+1)},X)$$
$$= \frac{(\tilde{S}_{(l)} - S_0)'(\tilde{S}_{(l)} - S_0) + V}{n+\eta}$$

$$\tilde{S}_{(l+1)} = \underset{S}{\text{Arg Max}}\ p(S|\tilde{C}_{(l+1)},\tilde{R}_{(l+1)},\tilde{\Sigma}_{(l+1)},X)$$
$$= [S_0\tilde{R}_{(l+1)}^{-1} + (X - e_n\bar{\mu}'_{(l+1)})\tilde{\Sigma}_{(l+1)}^{-1}\tilde{\Lambda}_{(l+1)}]$$
$$\times (\tilde{R}_{(l+1)}^{-1} + \tilde{\Lambda}'_{(l+1)}\tilde{\Sigma}_{(l+1)}^{-1}\tilde{\Lambda}_{(l+1)})^{-1},$$

where the matrix $\tilde{Z}_{(l)} = (e_n, \tilde{S}_{(l)})$ until convergence is reached with the joint modal (maximum a posteriori) estimator for the unknown model parameters $(\tilde{R}, \tilde{S}, \tilde{c}, \tilde{\Sigma})$. Conditional maximum a posteriori variance estimates can also be found. The conditional modal variance of the matrix containing the means and mixing coefficients is

$$var(c|\tilde{S}, \tilde{R}, \tilde{\Sigma}, X, U) = [\Delta^{-1} + \tilde{Z}'\tilde{Z} \otimes \tilde{\Sigma}^{-1}]^{-1}$$
$$= \tilde{\Delta},$$

where $c = vec(C)$, while $\tilde{S}$, $\tilde{R}$, and $\tilde{\Sigma}$ are the converged values from the ICM algorithm.

Conditional modal intervals may be computed by using the conditional distribution for a particular parameter given the modal values of the others. The posterior conditional distribution of the matrix containing the means and mixing coefficients C given the modal values of the other parameters and the data is

$$p(c|\tilde{S},\tilde{\Sigma},X,U) \propto |\tilde{\Delta}|^{-\frac{1}{2}} e^{-\frac{1}{2}(c-\tilde{c})'\tilde{\Delta}^{-1}(c-\tilde{c})}. \tag{10.7.20}$$

To evaluate statistical significance with the ICM approach, use the marginal conditional distribution of the matrix containing the means and mixing coefficients given above. General simultaneous hypotheses can be evaluated regarding the mean vector or the coefficient for a particular source by computing marginal distributions. It can be shown that the marginal conditional distribution of the kth column C_k of the matrix C containing the overall mean vector and mixing matrix is Multivariate Normal

$$p(C_k|\tilde{C}_k,\tilde{\Sigma},X,U) \propto |\tilde{\Delta}_k|^{-\frac{1}{2}} e^{-\frac{1}{2}(C_k-\tilde{C}_k)'\tilde{\Delta}_k^{-1}(C_k-\tilde{C}_k)}, \tag{10.7.21}$$

where $\tilde{\Delta}_k$ is the covariance matrix of C_k found by taking the kth $p \times p$ submatrix along the diagonal of $\tilde{\Delta}$.

Significance can be determined for a subset of means or mixing coefficients of the kth column of C by determining the marginal distribution of the subset within C_k which is also Multivariate Normal. With the subset being a singleton set, significance can be determined for a particular mean or mixing coefficient with the marginal distribution of the scalar coefficient which is

$$p(C_{kj}|\tilde{C}_{kj},\tilde{S},\tilde{\Sigma}_{jj},X) \propto (\tilde{\Delta}_{kj})^{-\frac{1}{2}} e^{-\frac{(C_{kj}-\tilde{C}_{kj})^2}{2\tilde{\Delta}_{kj}}}, \tag{10.7.22}$$

where $\tilde{\Delta}_{kj}$ is the jth diagonal element of $\tilde{\Delta}_k$. Note that $\tilde{C}_{kj} = \tilde{c}_{jk}$ and that

$$z = \frac{(C_{kj}-\tilde{C}_{kj})}{\sqrt{\tilde{\Delta}_{kj}}}$$

follows a Normal distribution with a mean of zero and variance of one.

10.8 Interpretation

Although the main focus after having performed a Bayesian Source Separation is the separated sources, there are others. The mixing coefficients are the amplitudes which determine the relative contribution of the sources. A "small" coefficient indicates that the particular source does not significantly contribute to the observed mixed signal. A "large" coefficient indicates that the particular source significantly contributes to the observed mixed signal. Whether a mixing coefficient is large or small depends on its associated statistic.

TABLE 10.1
Variables for Bayesian Source Separation example.

X	Variables		
X_1	Species 1	X_2	Species 2
X_3	Species 3	X_4	Species 4
X_5	Species 5	X_6	Species 6
X_7	Species 7	X_8	Species 8
X_9	Species 9		

Consider the following data which consist of a core sample taken from the ocean floor [24]. In the core sample, plankton content is measured for $p = 9$ species listed as species 1-9 in Table 10.1 at one hundred and ten depths. (The seventh species has been dropped from the original data because nearly all values were zero.)

FIGURE 10.1
Mixed and unmixed signals.

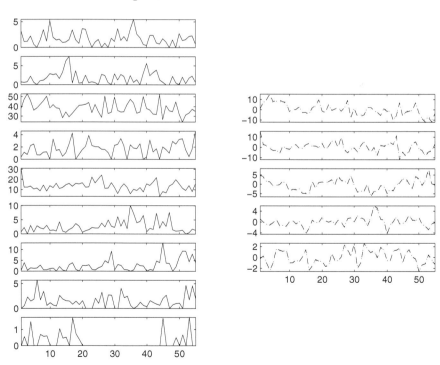

The plankton content is used to infer approximate climatological conditions which existed on Earth. The many species which coexist at different times (core depths) consist of contributions from several different sources. The Source Separation model will tell us what the sources are, and how they contribute to each of the plankton types.

TABLE 10.2
Covariance hyperparameter for coefficients.

D	1	2	3	4	5	6
1	0.0182	-0.0000	0.0000	-0.0000	0.0000	0.0000
2		0.0004	-0.0000	0.0000	-0.0000	-0.0000
3			0.0007	-0.0000	-0.0000	-0.0000
4				0.0018	-0.0000	0.0000
5					0.0054	-0.0000
6						0.0089

The $n_0 = 55$ odd observations which are in Table 10.9 were used as prior data to assess the model hyperparameters for the analysis of the $n = 55$ even observations in Table 10.8. The plankton species in the core samples are believed to be made up from $m = 5$ different sources. The first five normalized scores from a principal component analysis were used for the prior mean of the sources. The remaining hyperparameters were assessed from the results of a Regression of the prior data on the prior source mean.

TABLE 10.3
Prior mode, Gibbs, and ICM source covariance.

V/η	1	2	3	4	5
1	42.1132	0.0000	-0.0000	0.0000	0.0000
2		25.0674	0.0000	0.0000	0.0000
3			10.2049	0.0000	-0.0000
4				3.3642	0.0000
5					2.0436
$\bar{R}$	1	2	3	4	5
1	36.8064	-1.0408	0.1357	-0.0161	-0.1414
2		21.1973	-0.5267	-0.1098	-0.0124
3			8.4600	-0.0989	0.0238
4				3.0521	0.0015
5					1.9396
$\tilde{R}$	1	2	3	4	5
1	25.5538	-0.0001	0.1350	-0.0783	0.2615
2		15.9102	-0.0426	0.0012	-0.0506
3			6.3911	0.0188	-0.0199
4				2.1210	0.0312
5					1.1658

The assessed prior hyperparameter C_0 for the mean vector and mixing matrix was assessed from $C_0 = XZ_0(Z_0'Z_0)^{-1}$ where $Z_0 = (e_n, S_0)$. The assessed prior covariance matrix D for the coefficient matrix was assessed from $D = (Z_0'Z_0)^{-1}$ and presented in Table 10.2. The assessed values for η was $\eta = n_0 = 55$. The scale matrix Q for the error covariance matrix was assessed from $Q = (X - UC_0')'(X - UC_0')$ in Table 10.5 and $\nu = n_0 = 55$.

The observed mixed signals in Tables 10.8 and 10.9 along with the estimated sources in Tables 10.11 and 10.12 are displayed in Figure 10.1.

Refer to Table 10.3 which contains the (top) prior mode, (middle) Gibbs, and (bottom) ICM source covariance matrices. We can see the variances of the sources are far from unity as is specified in the Factor Analysis model.

Table 10.4 has the matrix containing the prior mean vector and mixing matrix along with the Gibbs sampling and ICM estimated ones using Conjugate prior distributions. From the estimated mixing coefficient values, it is difficult to discern which are "large" and which are "small."

This motivates the need for relative mixing coefficient statistics which are a measure of relative size. In order to compute the statistics for the mixing coefficients, the error covariance matrices are needed.

The (top) prior mode, (middle) Gibbs, and (bottom) ICM error covariance matrices are displayed in Table 10.5. The covariance values have been rounded to two decimal places for presentation purposes. If these covariance matrices were converted to correlation matrices, then it can be seen that several of the values are above one half.

The statistics for the mean vector and mixing coefficients are displayed in Table 10.6. The rows of the mean/coefficient matrix have been rearranged for increased interpretability. It is seen that Species 3 primarily is made up of Source 1, Species 7 contains a positive mixture of Source 3 and negative mixtures of Sources 1 and 2, Species 5 consists primarily of Source 3, Species 2 consists of negative mixtures of Sources 3 and 4, Species 6 consists of Source 4 and possibly 5, Species 4 consists of a negative mixture of Source 5 and possibly a negative one of 4, and Species 8 consists of Source 5. Table 10.7 lists the species which correspond to the sources.

TABLE 10.4
Prior, Gibbs, and ICM mixing coefficients.

C_0	0	1	2	3	4	5
1	1.6991	-0.0248	0.0361	-0.0718	0.1107	0.0962
2	1.5746	-0.0455	0.0609	-0.2164	-0.4228	-0.0158
3	38.9093	0.9288	-0.0066	0.3409	0.0027	0.0184
4	1.6265	0.0116	-0.0689	0.0112	0.2694	-0.6050
5	13.1707	-0.1431	0.8980	0.3874	0.0870	0.0557
6	2.2433	-0.0111	-0.0807	-0.1280	0.8041	0.4115
7	3.3637	-0.3376	-0.4087	0.8122	0.0039	-0.0174
8	1.4693	-0.0025	-0.1012	0.0732	-0.2828	0.6719
9	0.2135	-0.0017	0.0030	-0.0170	0.0467	-0.0002
$\bar{C}$	0	1	2	3	4	5
1	1.6804	-0.0241	0.0105	-0.0530	0.1297	-0.0360
2	1.6143	-0.0363	0.0500	-0.2098	-0.3291	-0.0128
3	38.6167	0.6343	-0.0391	0.1755	-0.0121	0.4518
4	1.6694	-0.0040	-0.0163	-0.0064	0.1581	-0.3663
5	13.2820	-0.0676	0.7008	0.3610	0.0802	-0.0502
6	2.4086	-0.0004	-0.0447	-0.1226	0.5575	0.3451
7	3.1186	-0.2359	-0.2633	0.5060	0.0289	-0.1600
8	1.4778	-0.0119	-0.0576	0.0405	-0.1631	0.2755
9	0.2297	0.0033	-0.0037	0.0089	-0.0089	-0.0222
$\tilde{C}$	0	1	2	3	4	5
1	1.6746	-0.0308	0.0152	-0.0695	0.1736	-0.0490
2	1.6040	-0.0469	0.0713	-0.2691	-0.4442	-0.0171
3	38.6004	0.8555	-0.0619	0.1944	0.0150	0.5738
4	1.6701	-0.0045	-0.0231	-0.0081	0.2130	-0.4976
5	13.2669	-0.1088	0.8972	0.4302	0.0958	-0.0596
6	2.3959	0.0015	-0.0580	-0.1567	0.7543	0.4597
7	3.1709	-0.3219	-0.3504	0.6878	0.0248	-0.1853
8	1.4857	-0.0153	-0.0751	0.0578	-0.2199	0.3809
9	0.2306	0.0042	-0.0049	0.0109	-0.0104	-0.0303

TABLE 10.5
Prior, Gibbs, and ICM covariances.

$\frac{Q}{\nu}$	1	2	3	4	5	6	7	8	9
1	1.53	-0.01	0.52	-0.34	0.21	0.05	-0.68	-0.03	0.12
2		2.03	-1.27	-0.45	1.23	-0.43	-0.74	-0.22	-0.02
3			54.94	0.46	-12.81	-0.60	-8.33	-2.27	-0.96
4				2.12	-1.67	0.31	0.60	-0.89	0.01
5					18.71	-2.00	-3.36	-0.84	-0.07
6						4.80	0.74	-0.32	0.25
7							12.95	1.81	-0.03
8								3.23	-0.09
9									0.27
$\bar{\Psi}$	1	2	3	4	5	6	7	8	9
1	1.43	-0.01	0.80	-0.28	0.08	-0.00	-0.61	-0.06	0.11
2		1.76	-0.86	-0.37	0.97	-0.26	-0.55	-0.24	-0.01
3			47.42	0.20	-10.86	-0.43	-6.97	-2.15	-0.85
4				1.78	-1.35	0.24	0.53	-0.66	0.02
5					15.53	-1.71	-2.75	-0.65	-0.09
6						4.15	0.74	-0.30	0.22
7							10.63	1.44	0.02
8								2.76	-0.07
9									0.24
$\tilde{\Psi}$	1	2	3	4	5	6	7	8	9
1	1.33	-0.01	1.05	-0.30	-0.02	-0.11	-0.52	-0.01	0.11
2		1.38	-0.31	-0.28	0.68	-0.04	-0.16	-0.24	-0.00
3			38.67	0.34	-9.08	-0.49	-4.76	-2.17	-0.84
4				1.56	-1.18	0.19	0.41	-0.51	0.01
5					10.97	-1.28	-1.99	-0.34	-0.07
6						3.38	0.83	-0.25	0.22
7							7.72	1.17	0.01
8								2.50	-0.07
9									0.23

TABLE 10.6
Statistics for means and loadings.

z_{Gibbs}	0	1	2	3	4	5
3	54.8773	5.6098	-0.2840	0.7928	-0.0318	0.9182
7	9.1505	-4.4542	-3.8932	4.6551	0.1601	-0.6934
5	31.8876	-1.0324	8.2695	2.7367	0.3548	-0.1722
2	11.7241	-1.6765	1.8071	-4.8271	-4.3483	-0.1311
6	11.6744	-0.0118	-1.0876	-1.8774	4.8424	2.3436
4	12.6162	-0.1925	-0.6144	-0.1538	2.1872	-3.8858
8	9.0991	-0.4708	-1.7873	0.7872	-1.8362	2.3665
1	14.3622	-1.2867	0.4391	-1.4076	1.9977	-0.4199
9	4.8283	0.4346	-0.3866	0.5967	-0.3387	-0.6567
z_{ICM}	0	1	2	3	4	5
3	65.0052	8.6843	-0.4979	0.9815	0.0443	1.3159
7	11.9498	-7.3128	-6.3027	7.7711	0.1638	-0.9512
5	41.9351	-2.0735	13.5378	4.0770	0.5316	-0.2563
2	14.2853	-2.5192	3.0324	-7.1856	-6.9418	-0.2071
6	13.6452	0.0506	-1.5773	-2.6756	7.5374	3.5650
4	13.9991	-0.2264	-0.9248	-0.2048	3.1326	-5.6794
8	9.8461	-0.6123	-2.3762	1.1479	-2.5572	3.4372
1	15.1846	-1.6856	0.6574	-1.8882	2.7613	-0.6056
9	5.0139	0.5510	-0.5104	0.7078	-0.3982	-0.8961

TABLE 10.7
Sources in terms of strong mixing coefficients.

Source 1:	Species 3, 7
Source 2:	Species 5, 7
Source 3:	Species 2, 7, 5
Source 4:	Species 2, 6, 4
Source 5:	Species 6, 8

TABLE 10.8
Plankton data.

X	1	2	3	4	5	6	7	8	9
1	3.203	0.712	37.722	0.356	30.961	0.712	0.356	0.000	0.000
2	1.124	0.562	47.191	1.124	12.360	2.247	3.933	0.562	0.562
3	1.149	0.766	52.874	0.766	12.261	0.000	0.383	2.299	0.000
4	2.222	2.222	45.926	2.222	13.333	2.963	1.481	1.481	1.481
5	0.621	0.621	36.025	2.484	10.519	0.621	1.242	1.863	0.000
6	0.000	0.000	38.298	0.709	11.348	2.837	1.418	5.674	0.000
7	1.379	1.034	42.069	0.690	8.621	2.069	2.759	1.724	0.690
8	3.429	1.143	45.714	1.143	14.286	1.714	0.571	3.429	0.571
9	1.198	1.796	50.299	1.198	8.383	2.994	0.599	0.599	0.599
10	5.143	2.857	38.286	0.000	13.714	1.143	1.143	1.143	0.000
11	1.961	2.614	41.830	3.268	11.765	1.307	1.307	0.654	0.000
12	1.422	2.844	38.389	1.422	16.114	0.948	0.000	0.474	0.000
13	1.571	1.571	37.696	1.571	10.995	4.188	2.094	2.618	1.047
14	0.926	3.241	28.241	0.463	12.037	0.926	0.463	1.852	0.463
15	1.036	6.218	34.197	1.036	14.508	0.518	0.000	1.554	0.518
16	1.485	7.426	29.208	2.475	15.842	1.485	2.970	1.485	0.000
17	3.404	0.426	32.766	4.255	13.191	2.128	3.830	0.851	1.700
18	1.449	3.623	36.957	0.000	15.942	3.623	0.725	1.449	0.720
19	0.772	0.386	40.927	0.772	15.444	2.703	0.000	0.772	0.380
20	3.627	0.518	41.451	1.554	16.580	0.518	2.591	1.554	0.000
21	3.509	2.456	42.105	2.105	12.281	1.053	2.456	0.000	0.000
22	1.449	0.483	43.961	3.865	12.560	1.449	2.899	0.000	0.000
23	0.000	0.741	33.333	2.222	22.222	2.222	0.741	0.000	0.000
24	1.026	0.513	42.051	2.051	16.410	2.051	0.513	2.051	0.000
25	1.523	0.000	34.518	2.030	20.305	2.030	1.523	1.015	0.000
26	0.000	2.703	28.649	1.622	24.324	3.784	2.162	3.243	0.000
27	0.800	2.400	50.400	1.600	11.200	2.400	4.800	0.000	0.000
28	0.000	0.543	32.609	1.087	11.413	4.891	3.804	2.717	0.000
29	1.762	0.000	33.921	0.000	16.740	2.643	9.251	2.643	0.000
30	1.136	2.841	49.432	2.273	11.932	2.273	0.568	0.000	0.000
31	3.636	1.212	35.758	2.424	6.061	6.061	3.030	0.000	0.000
32	1.342	2.685	34.228	3.356	12.081	2.685	2.685	4.027	0.000
33	2.158	2.158	34.532	2.158	15.826	5.036	0.719	2.158	0.000
34	1.235	0.000	41.975	0.000	12.346	1.852	0.617	2.469	0.000
35	3.550	2.367	47.337	2.367	5.917	10.059	0.000	0.592	0.000
36	5.455	0.606	43.636	1.818	10.303	7.273	0.605	0.000	0.000
37	2.609	1.304	33.043	1.739	9.130	3.913	3.478	0.435	0.000
38	1.899	0.000	34.177	2.532	12.025	4.430	2.532	1.266	0.000
39	0.595	2.976	50.000	0.000	7.738	6.548	2.381	0.595	0.000
40	0.372	5.576	37.918	0.372	15.613	0.743	0.000	0.372	0.000
41	2.362	2.362	36.220	3.150	14.173	1.969	0.787	1.575	0.000
42	2.381	3.175	32.143	1.190	17.460	1.587	0.397	1.190	0.000
43	0.858	3.863	31.760	1.717	21.888	7.296	4.721	0.858	0.000
44	0.658	1.316	52.632	0.000	3.289	1.974	3.947	0.658	0.000
45	1.689	0.676	26.689	2.027	8.108	4.392	13.176	2.027	1.689

X	1	2	3	4	5	6	7	8	9
46	1.064	0.000	40.957	1.596	6.915	2.660	3.723	2.660	0.000
47	0.000	0.000	35.533	1.015	13.706	7.614	3.553	0.000	0.000
48	1.471	2.206	34.559	2.941	15.441	1.471	0.000	0.735	0.000
49	0.000	0.498	44.776	2.488	19.900	0.995	1.990	0.995	0.498
50	2.717	0.000	32.065	3.261	15.761	1.087	6.522	1.087	0.000
51	1.342	2.013	24.161	3.356	11.409	1.342	9.396	0.000	0.671
52	1.548	0.310	31.269	1.548	9.288	0.000	9.288	4.644	0.000
53	2.183	1.747	33.188	0.437	13.974	0.437	4.367	1.747	1.747
54	2.286	2.286	37.143	1.714	8.000	1.714	8.000	4.571	0.000
55	0.658	0.658	34.868	4.605	15.789	1.316	3.947	1.974	0.000

TABLE 10.9

Plankton prior data.

X_0	1	2	3	4	5	6	7	8	9
1	1.792	0.489	43.485	0.814	25.570	0.651	0.163	0.000	0.163
2	2.364	1.709	47.009	0.855	20.513	1.709	1.282	0.427	0.000
3	0.671	1.007	43.624	3.020	15.436	1.007	0.336	0.671	0.336
4	1.990	0.498	53.234	3.980	6.965	0.000	0.498	0.995	0.000
5	1.786	1.190	49.405	1.786	10.714	1.786	0.595	0.595	0.000
6	1.418	0.000	46.099	2.837	9.220	4.255	0.709	2.836	0.000
7	0.498	0.498	48.756	0.000	5.970	1.990	0.498	2.985	0.000
8	0.662	0.000	46.358	0.000	11.921	0.000	1.987	3.311	0.000
9	2.899	2.899	42.995	0.000	14.010	1.449	2.415	2.415	0.483
10	1.887	2.516	38.994	3.145	7.547	2.516	1.258	1.258	0.000
11	3.067	0.613	37.423	1.227	13.497	2.761	1.227	0.000	0.307
12	1.515	2.020	37.374	1.010	12.626	2.020	0.000	0.505	0.000
13	1.630	1.630	36.957	2.174	10.870	2.174	0.000	0.000	0.000
14	1.826	3.196	36.073	0.913	12.329	2.283	0.457	0.913	0.457
15	1.379	2.414	35.517	0.345	11.679	0.345	0.000	4.828	0.000
16	0.649	3.896	39.610	3.896	13.636	1.299	0.543	0.649	0.000
17	1.087	0.000	42.391	1.630	15.761	1.630	2.174	1.087	0.000
18	1.429	0.476	42.381	2.857	10.952	1.905	0.476	0.952	1.900
19	1.685	1.685	48.315	2.809	10.674	1.124	1.124	1.124	0.000
10	1.266	1.266	37.975	2.532	18.143	3.376	2.110	0.422	0.000
21	1.869	1.402	37.850	2.804	12.617	2.336	9.813	0.467	0.930
22	0.904	0.904	44.578	1.205	14.759	0.602	1.506	0.602	0.000
23	1.299	0.649	38.961	0.325	17.208	1.945	4.545	1.948	0.000
24	2.513	4.523	35.176	1.005	20.603	0.000	0.000	0.000	0.000
25	0.565	0.565	44.068	3.955	10.169	1.695	9.605	3.390	0.000
26	0.508	0.000	40.609	0.508	21.827	0.508	3.046	0.000	0.000
27	0.629	4.403	39.623	0.629	10.063	3.145	5.660	5.031	0.000
28	1.630	0.543	54.348	2.174	7.609	3.804	1.630	2.717	0.000
29	1.622	1.081	32.973	2.162	11.892	3.784	9.780	0.541	0.000
30	1.418	0.000	36.879	0.709	11.348	4.255	4.965	4.965	0.709
31	0.893	3.561	33.036	5.357	13.393	2.679	4.464	0.893	0.893
32	3.448	1.478	29.064	3.448	14.778	4.433	2.955	0.000	0.000
33	4.435	2.419	33.468	0.806	17.742	3.226	0.000	4.032	0.000
34	0.000	4.545	38.636	0.000	15.152	1.515	2.273	2.273	0.758
35	1.508	1.508	38.191	0.503	3.518	1.508	1.508	2.010	0.503
36	5.344	0.000	39.695	1.527	13.740	6.870	0.763	0.000	0.000
37	0.000	0.000	38.095	3.571	4.762	9.524	3.571	0.000	1.190
38	1.604	1.604	33.690	0.000	19.251	2.139	3.209	3.209	0.535
39	2.041	0.816	36.327	2.041	20.000	2.449	2.449	1.224	0.408
40	0.000	6.130	35.249	0.000	10.728	0.000	0.383	0.383	0.000
41	3.582	5.373	38.209	0.896	17.015	0.896	0.000	0.896	0.299
42	2.105	4.211	26.842	1.053	13.684	4.737	5.263	2.105	0.000
43	0.455	0.909	37.273	0.455	24.091	3.182	0.455	0.455	0.909
44	2.769	1.231	43.385	1.231	2.769	4.000	6.462	3.077	0.000
45	3.448	0.575	35.632	1.149	14.368	0.000	4.598	0.575	0.000

Multivariate Bayesian Statistics

X_0	1	2	3	4	5	6	7	8	9
46	1.533	0.000	35.249	0.383	9.195	2.682	13.793	1.533	0.000
47	1.394	0.348	36.585	1.045	8.014	3.833	6.969	1.394	0.000
48	1.970	2.463	39.901	0.493	15.764	3.941	0.985	0.493	0.493
49	1.613	0.403	42.742	1.210	16.129	2.823	2.823	0.403	0.000
50	0.448	0.448	40.359	4.484	12.556	2.242	6.278	0.897	0.000
51	1.887	1.887	34.906	1.415	12.264	1.415	3.302	1.415	0.472
52	1.633	0.816	24.898	2.449	6.531	0.408	12.245	2.041	0.000
53	1.093	0.546	31.694	1.639	14.208	0.000	19.672	4.372	0.000
54	1.878	0.469	24.883	1.878	14.085	1.408	9.390	0.939	0.000
55	3.911	2.793	32.961	1.117	14.525	1.117	2.793	0.559	0.000

TABLE 10.10

Prior sources.

S_0	1	2	3	4	5
1	3.6149	12.6825	4.0799	0.4612	-0.2944
2	7.1521	7.6238	3.8249	0.3161	0.2023
3	5.1598	3.2538	0.3356	-0.0632	-1.7127
4	15.2554	-4.4849	0.6459	-1.0708	-2.6534
5	11.0592	-1.0505	0.4543	-0.1117	-0.7537
6	8.2063	-3.0009	-1.0149	1.8473	0.9696
7	11.2025	-5.4905	-1.3124	-1.3693	1.6291
8	7.6609	-0.6361	1.7594	-2.3160	1.3145
9	3.8920	1.3196	0.7227	-1.6788	1.5253
10	1.5624	-4.2317	-4.1069	-0.1962	-1.2202
11	-0.7028	1.3016	-2.1855	1.3025	-0.3576
12	-0.2300	1.0737	-3.5947	-0.3564	-0.3915
13	-0.3380	-0.5615	-4.3846	0.2477	-1.4599
14	-1.6171	0.6572	-4.0716	-0.7554	0.0124
15	-1.8337	-0.0017	-4.1467	-3.3739	2.1201
16	1.4962	1.6719	-2.2090	-0.9924	-2.3625
17	3.3596	2.7586	1.6629	0.4339	-0.3165
18	4.5899	-0.9119	-1.7738	0.5243	-1.2499
19	9.8718	-1.3394	0.5319	-0.7428	-1.3385
20	-1.1304	4.9008	0.4796	1.9486	-0.5406
21	-3.0642	-3.1156	4.6031	0.7750	-1.4913
22	5.7309	2.3271	1.3865	-0.8565	-0.8441
23	-0.8891	3.1361	2.8324	-0.0328	1.1677
24	-3.5279	8.6545	-1.6456	-2.0993	-1.0984
25	3.2165	-5.6941	6.2049	-0.2878	-0.6181
26	0.5586	8.1180	4.2066	-0.0011	-0.5954
27	0.2002	-3.9653	0.5069	-2.1345	3.0200
28	15.7550	-4.7438	1.8297	1.0191	1.1627
29	-7.4803	-3.8323	2.5490	1.7341	-0.6052
30	-2.1294	-2.8353	0.2431	0.8822	3.5626
31	-5.8898	-0.3510	-1.4622	0.6291	-2.6885
32	-9.2742	1.5786	-3.5336	3.0033	-1.1024
33	-4.7051	5.3833	-3.1385	0.1500	3.0859
34	-0.2758	2.4375	-0.6078	-2.5024	1.1381
35	1.3419	-7.8329	-5.3471	-1.8751	0.2044
36	1.4589	1.3861	-2.2417	5.2097	1.4438
37	0.4345	-8.3575	-3.9381	6.5848	0.2063
38	-5.6870	5.5015	0.5682	-0.5077	2.3459
39	-3.0381	6.4700	1.1192	1.2976	0.0628
40	-2.1995	-0.3325	-5.2858	-4.2932	-1.0557
41	-0.2766	5.3482	-2.3177	-2.1927	-0.1166
42	-12.0879	-0.2870	-3.2471	0.6110	1.5717
43	-2.0622	11.0301	1.3244	1.8362	0.9329
44	4.5609	-10.8968	-0.0979	0.2247	1.5995
45	-3.6077	0.8983	0.6598	-0.9732	-1.0658

S_0	1	2	3	4	5
46	-6.2946	-7.8674	5.9732	0.3227	0.5144
47	-2.5983	-6.2559	0.2136	1.1774	0.5527
48	1.2758	3.3990	-1.1071	1.2221	0.9443
49	3.3661	2.8668	2.1198	1.3968	0.0288
50	0.5678	-2.0059	2.9503	1.2206	-2.2747
51	-3.5799	-0.6490	-1.7519	-0.8967	-0.3592
52	-14.9961	-9.5153	0.3231	-1.6923	-1.6445
53	-12.2752	-5.8850	11.9566	-2.1282	0.6192
54	-15.1326	-1.5071	0.7682	0.0891	-1.1288
55	-5.6267	1.8602	-2.3135	-0.9663	-0.5969

TABLE 10.11
Gibbs estimated sources.

$\bar{S}$	1	2	3	4	5
1	0.9807	15.6149	4.9057	1.1081	-0.1951
2	7.6055	2.8646	3.9698	-0.0716	0.2639
3	9.6302	1.6182	1.4932	-0.8031	-1.0248
4	15.2314	-1.0222	1.2876	-1.1253	-2.3789
5	8.0468	-3.0071	-0.7450	-0.2822	-1.0089
6	8.6775	-3.0436	-0.6280	1.0225	1.6290
7	9.4814	-6.0544	-0.2174	-1.3684	1.3358
8	7.7334	1.2350	1.6243	-1.5465	1.1107
9	7.2569	-1.0552	0.2612	-1.5892	1.3960
10	-1.9127	-1.9787	-3.9417	0.0928	-0.9757
11	0.0741	1.2153	-2.6727	0.6031	-0.9816
12	0.1553	2.7110	-2.9621	-0.7201	-0.4769
13	2.4942	-0.6772	-2.8096	0.1257	-0.8011
14	-2.8384	-0.9523	-4.7943	-1.6783	0.1043
15	-2.6866	2.4285	-4.9798	-4.3689	1.4187
16	-4.2748	3.8976	-5.1482	-2.3088	-2.3416
17	1.9142	1.6006	1.5756	0.6781	-1.5672
18	3.9248	1.3067	-2.0418	-0.2850	-0.3889
19	9.8123	-0.1879	0.9218	-0.1547	-0.6778
20	-1.1903	4.1789	1.1927	1.6698	-0.7015
21	-3.1564	-2.1541	1.5573	0.4897	-1.6518
22	5.2444	1.2583	1.3740	-0.0676	-1.4785
23	-0.6582	6.0503	2.4677	0.4769	0.7760
24	0.6486	7.1883	-0.3399	-1.0240	-0.6335
25	1.9693	-0.2651	4.5535	0.7289	-0.5858
26	-1.4475	10.4732	2.0957	0.0082	0.1105
27	2.0043	-2.3173	1.8399	-1.9020	2.3047
28	9.2866	-5.1963	0.3418	1.2520	1.4795
29	-7.7186	-2.3700	4.6203	1.5911	0.2233
30	1.4263	-0.4446	-0.3194	0.0516	2.5013
31	-6.4891	-4.1211	-4.0640	2.2281	-1.9756
32	-6.3127	1.8598	-3.9654	1.6800	-0.9687
33	-4.9825	5.7280	-3.6544	0.9995	2.6256
34	2.1692	0.2236	0.0881	-1.5464	1.5986
35	2.3613	-5.0694	-6.6207	1.2189	0.9052
36	-0.5200	0.0357	-3.7793	5.9551	1.2934
37	-2.4959	-7.6663	-4.1741	5.1509	-0.0393
38	-5.5737	2.0416	-0.7126	0.9923	1.8707
39	0.9322	1.2373	-0.9587	0.7867	1.3443
40	-1.5557	2.3431	-4.8692	-4.4719	-0.6164
41	-1.5259	4.6410	-3.1214	-1.1761	-0.5867
42	-10.5555	2.8372	-3.4738	0.1699	1.1763
43	-5.6385	11.3828	-0.1632	1.9829	1.3539
44	6.9574	-11.8097	0.5968	-0.6437	1.6987
45	-6.4213	-4.3600	2.7961	-0.6655	-0.9720

$\bar{S}$	1	2	3	4	5
46	-2.3506	-8.5414	3.2495	0.3405	0.7791
47	-1.4567	-3.9941	0.3060	2.5545	1.4352
48	-0.2274	3.5452	-2.3818	0.7657	0.1226
49	6.9381	5.5400	4.1900	0.5332	-0.2802
50	-3.0303	-1.1452	2.8620	1.6998	-2.5665
51	-8.5665	-2.8726	-0.5945	-0.8719	-1.3653
52	-12.1401	-8.6555	1.9367	-1.5648	-1.1537
53	-7.6439	-4.2261	8.3341	-2.8372	0.4667
54	-11.8889	-3.4816	0.3105	-0.6313	-0.6106
55	-3.6070	3.3189	-0.4605	-0.2526	-1.2679

TABLE 10.12

ICM estimated sources.

$\tilde{S}$	1	2	3	4	5
1	0.4899	14.5145	4.5071	1.0927	-0.2315
2	6.9053	1.8323	3.5958	-0.1215	0.2405
3	9.7792	1.2122	1.4778	-0.8028	-1.0118
4	14.4503	-0.5136	1.2786	-0.9314	-2.4179
5	7.0907	-3.0951	-0.8586	-0.2755	-1.0750
6	8.2539	-2.8721	-0.5496	0.8661	1.4973
7	8.3046	-5.4934	0.0080	-1.2320	1.2450
8	7.0406	1.4542	1.4681	-1.3204	1.1038
9	6.8416	-1.1500	0.1584	-1.4614	1.4064
10	-1.8991	-1.3645	-3.5401	0.1384	-1.0253
11	0.1033	1.0660	-2.4731	0.4460	-0.9366
12	0.1227	2.7583	-2.5841	-0.6775	-0.4761
13	2.9057	-0.6468	-2.3662	0.2029	-0.8006
14	-2.7731	-0.9990	-4.4835	-1.6639	0.0846
15	-2.9552	2.7171	-4.5537	-4.1590	1.4409
16	-4.2508	3.7226	-5.0439	-2.2657	-2.3691
17	1.2587	1.2114	1.3578	0.6423	-1.4816
18	3.9268	1.5191	-1.9320	-0.2931	-0.4898
19	9.3672	-0.0612	0.8758	-0.0257	-0.7625
20	-1.0664	3.6590	1.1118	1.4867	-0.6761
21	-2.4913	-1.7258	0.8635	0.3771	-1.5980
22	4.6917	0.8820	1.3025	0.0330	-1.4372
23	-0.8445	5.9142	2.1302	0.4665	0.7865
24	1.1053	6.3389	-0.1244	-0.7830	-0.5295
25	1.9905	0.5701	3.8052	0.7934	-0.6001
26	-1.3647	9.6680	1.5065	0.0127	0.0810
27	1.6113	-1.6881	1.9844	-1.7399	2.3240
28	7.8789	-4.9566	0.1180	1.1929	1.2565
29	-6.8021	-1.9096	4.3907	1.4590	0.1934
30	1.2311	0.1059	-0.3359	-0.1058	2.5217
31	-5.6701	-4.3043	-4.1307	2.2532	-1.9239
32	-5.2401	1.6640	-3.7472	1.3927	-0.9051
33	-5.1508	5.3068	-3.3875	0.9692	2.6413
34	2.2469	0.0121	0.1681	-1.3107	1.6087
35	2.5091	-4.1341	-6.0621	1.5724	0.8760
36	-0.8734	-0.2395	-3.7032	5.5596	1.2169
37	-2.5916	-6.9029	-3.8893	4.5502	-0.1520
38	-5.6731	1.4270	-0.8668	1.0290	1.9274
39	1.5083	0.3557	-1.2108	0.6666	1.3105
40	-1.2330	2.6784	-4.2421	-4.0828	-0.6014
41	-1.6943	4.1259	-2.9150	-0.9705	-0.5250
42	-9.7477	3.2048	-3.2242	0.0645	1.2387
43	-5.7349	10.1939	-0.3842	1.8287	1.2970
44	6.7774	-10.7151	0.7029	-0.7140	1.6126
45	-6.3299	-4.7704	2.8196	-0.5370	-0.9357

$\tilde{S}$	1	2	3	4	5
46	-1.4643	-7.7089	2.4053	0.2631	0.8422
47	-1.0564	-3.2776	0.2526	2.5518	1.3459
48	-0.6876	3.1744	-2.3780	0.6098	0.1191
49	6.7672	5.3878	4.0594	0.4279	-0.2528
50	-2.9901	-0.9787	2.5267	1.6197	-2.5275
51	-8.8425	-2.9188	-0.3269	-0.7978	-1.3129
52	-10.4148	-7.5808	1.9425	-1.4108	-1.0169
53	-6.2772	-3.3309	6.7783	-2.7691	0.6239
54	-10.2317	-3.3483	0.1529	-0.6803	-0.4715
55	-3.2689	3.2464	-0.1322	-0.1348	-1.1372

10.9 Discussion

After having estimated the model parameters, the estimates of the sources as well as the mixing matrix are now available. The estimated matrix of sources corresponds to the unobservable signals or conversations emitted from the mouths of the speakers at the cocktail party. Row i of the estimated source matrix is the estimate of the unobserved source vector at time i and column j of the estimated source matrix is the estimate of the unobserved conversation of speaker j at the party for all n time increments.

Exercises

1. Specify that μ and Λ are independent with the prior distribution for the overall mean μ being the vague prior

$$p(\mu) \propto (\text{a constant}),$$

the distribution for the factor loading matrix being

$$p(\Lambda|\Sigma) \propto |A|^{-\frac{p}{2}} |\Sigma|^{-\frac{m}{2}} e^{-\frac{1}{2} tr \Sigma^{-1}(\Lambda-\Lambda_0)A^{-1}(\Lambda-\Lambda_0)'},$$

and the others as in Equations 10.4.2-10.4.4.

Combine these prior distributions with the likelihood in Equation 10.3.2 to obtain a posterior distribution. Derive Gibbs sampling and ICM algorithms for marginal mean and joint maximum a posteriori parameter estimates [52, 59].

2. Specify that μ and Λ are independent with the prior distribution for the overall mean μ being the Conjugate prior

$$p(\mu|\Sigma) \propto |h\Sigma|^{-\frac{1}{2}} e^{-\frac{1}{2}(\mu-\mu_0)(h\Sigma)^{-1}(\mu-\mu_0)'},$$

the distribution for the factor loading matrix being

$$p(\Lambda|\Sigma) \propto |A|^{-\frac{p}{2}} |\Sigma|^{-\frac{m}{2}} e^{-\frac{1}{2} tr \Sigma^{-1}(\Lambda-\Lambda_0)A^{-1}(\Lambda-\Lambda_0)'},$$

and the others as in Equations 10.4.2-10.4.2.

Combine these prior distributions with the likelihood in Equation 10.3.2 to obtain a posterior distribution. Derive Gibbs sampling and ICM algorithms for marginal mean and joint maximum a posteriori parameter estimates [61].

3. Specify that μ and Λ are independent with the prior distribution for the overall mean μ being the generalized Conjugate prior

$$p(\mu) \propto |\Gamma|^{-\frac{1}{2}} e^{-\frac{1}{2}(\mu-\mu_0)'\Gamma^{-1}(\mu-\mu_0)},$$

the distribution for the factor loading matrix being

$$p(\lambda) \propto |\Delta|^{-\frac{1}{2}} e^{-\frac{1}{2}(\lambda-\lambda_0)'\Delta^{-1}(\lambda-\lambda_0)},$$

and the others as in Equations 10.4.2-10.4.4.

Combine these prior distributions with the likelihood in Equation 10.3.2 to obtain a posterior distribution. Derive Gibbs sampling and ICM algorithms for marginal mean and joint maximum a posteriori parameter estimates [61].

4. Specify that the prior distribution for the matrix of sources S is

$$p(S) = \begin{cases} 1 & S = S_0 \\ 0 & S \neq S_0 \end{cases}, \tag{10.9.1}$$

and as a result there is no variability or covariance matrix R and thus no $p(R)$. Show that by taking the $(e_n, S_0) = U$, $(\mu, \Lambda) = B$, and the Conjugate prior distributions for the unknown model parameters, the resulting model is the Bayesian Regression model given in Chapter 8.

11

Unobservable and Observable Source Separation

11.1 Introduction

There may be instances where some sources may be specified to be observable while others are not. An example of such a situation is when we recognize that a stereo at a cocktail party is tuned to a particular radio station. We record the radio station in isolation; thus the source is said to be observable, but the associated mixing coefficients for this source are still unknown. The following model allows for such situations. The following model is a combination of the Bayesian Regression model for observable sources introduced in Chapter 8 and the Bayesian Source Separation model for unobservable sources described in Chapter 10. Either model may be obtained as a special case by setting either the number unobservable or observable sources to be zero.

11.2 Model

Consider the model at time i, in which p-dimensional vector-valued observations x_i, on p possibly correlated random variables are observed as well as $(q+1)$-dimensional vector-valued sources (including a vector of ones for the overall mean) on u_i, but m sources, at each time increment s_i, $i = 1, \ldots, n$, are unobservable. The $(q+1)$-dimensional observable sources are denoted by u_i as in Regression with coefficients denoted by B and the m-dimensional unobservable sources by s_i with coefficients given by Λ. The mixing coefficients for the observable sources denoted by B will be referred to as Regression coefficients or the matrix of Regression coefficients. The unobservable and observable Source Separation model is given by

$$\begin{array}{cccccccc} (x_i|B,u_i,\Lambda,s_i) = & B & u_i & + & \Lambda & s_i & + & \epsilon_i, \\ (p \times 1) & [p \times (q+1)] & [(q+1) \times 1] & & (p \times m) & (m \times 1) & & (p \times 1) \end{array}$$

$$(11.2.1)$$

where the variables are as previously defined.

That is, the observed signal x_{ij} for microphone j at time (observation) i contains a linear combination of the $(q+1)$ observable sources $1, u_{i1}, \ldots, u_{iq}$ (where the first element of the u_i's is a 1 for the overall mean μ_j) in addition to a linear combination of the m unobserved source components $s_{i1}, \ldots, s_{im}$ with amplitudes or mixing coefficients $\lambda_{j1}, \ldots, \lambda_{jm}$. This combined model can be written in terms of vectors to describe the observed signal at microphone j at time i as

$$x_{ij} = \beta_j' u_i + \lambda_j' s_i + \epsilon_{ij}, \tag{11.2.2}$$

where $u_i = (1, u_{i1}, \ldots, u_{iq})'$, $\beta_j = (\beta_{j0}, \ldots, \beta_{jq})'$, $\lambda_j = (\lambda_{j1}, \ldots, \lambda_{jm})'$, and $s_i = (s_{i1}, \ldots, s_{im})'$. If any or all of the unobservable sources were specified to be observable, they could be grouped into the u's and their Regression or observed mixing coefficients computed. This is also represented as

$$(x_{ij}|\mu_j, \lambda_j, s_i, u_i) = \mu_j + \sum_{t=1}^{q} \beta_{jt} \, u_{it} + \sum_{k=1}^{m} \lambda_{jk} \, s_{ik} + \epsilon_{ij}. \tag{11.2.3}$$

The recorded or observed conversation for microphone j at time increment i is a linear mixture of the $(q+1)$ observable sources and the m unobservable sources at time increment i and a random noise term. The observed sources contains a 1 for an overall mean for microphone j.

The unobservable and observable Source Separation model that describes all observations for all microphones can be written in terms of matrices as

$$(X|B, U, \Lambda, S) = \underset{[n \times (q+1)]}{U} \underset{[(q+1) \times p]}{B'} + \underset{(n \times m)}{S} \underset{(m \times p)}{\Lambda'} + \underset{(n \times p)}{E},$$
$$\underset{(n \times p)}{} \tag{11.2.4}$$

where $X' = (x_1, \ldots, x_n)$, $U' = (u_1, \ldots, u_n)$, $\Lambda' = (\lambda_1, \ldots, \lambda_p)$ $S' = (s_1, \ldots, s_n)$, and $E' = (\epsilon_1, \ldots, \epsilon_n)$.

11.3   Likelihood

Regarding the errors of the observations, it is specified that they are independent Multivariate Normally distributed random vectors with mean zero and full positive definite symmetric covariance matrix Σ. From this error specification, it is seen that the observation vector x_i given the source vector s_i, the Regression coefficient matrix B, the observable source vector u_i, the mixing matrix Λ, and the error covariance matrix Σ is Multivariate Normally distributed with likelihood given by

$$p(x_i|B, u_i, \Lambda, s_i, \Sigma) \propto |\Sigma|^{-\frac{1}{2}} e^{-\frac{1}{2}(x_i - Bu_i - \Lambda s_i)'\Sigma^{-1}(x_i - Bu_i - \Lambda s_i)}. \quad (11.3.1)$$

With the previously described matrix representation, the joint likelihood of all n observation vectors collected into the observations matrix X is given by the Matrix Normal Distribution

$$p(X|U, B, S, \Lambda, \Sigma) \propto |\Sigma|^{-\frac{n}{2}} e^{-\frac{1}{2}tr(X - UB' - S\Lambda')\Sigma^{-1}(X - UB' - S\Lambda')'}, \quad (11.3.2)$$

where the variables are as previously defined.

The Regression and the mixing coefficient matrices B and Λ are joined into a single coefficient matrix as $C = (B, \Lambda)$. The observable and unobservable source matrices U and S are also joined as $Z = (U, S)$. Having joined these matrices, the unobservable and observable Source Separation model is now in a matrix representation given by

$$\begin{array}{cccc} (X|C, Z) = & Z & C' & + & E, \\ n \times p & n \times (m+q+1) & (m+q+1) \times p & (n \times p) \end{array} \quad (11.3.3)$$

and its corresponding likelihood is given by the Matrix Normal distribution

$$p(X|C, Z, \Sigma) \propto |\Sigma|^{-\frac{n}{2}} e^{-\frac{1}{2}tr(X - ZC')\Sigma^{-1}(X - ZC')'}, \quad (11.3.4)$$

where all variables are as defined above and $tr(\cdot)$ denotes the trace operator.

Again, the objective is to unmix the unobservable sources by estimating the matrix containing them S and to obtain knowledge about the mixing process by estimating the Regression coefficients (coefficients for the observable sources) B, the matrix of mixing coefficients (coefficients for the unobservable sources) Λ, and the error covariance matrix Σ.

Both Conjugate and generalized Conjugate distributions are utilized in order to quantify our prior knowledge regarding value of the parameters.

11.4   Conjugate Priors and Posterior

The unobservable and observable Bayesian Source Separation model that was just described, is based on previous work [52, 57, 59].

When quantifying available prior information regarding the parameters of interest, Conjugate prior distributions are specified as described in Chapter 4. The joint prior distribution for the model parameters which are the matrix of (Regression/mixing) coefficients C, the matrix of sources S, the source covariance matrix R, and the error covariance matrix Σ is given by

$$p(S, R, C, \Sigma) = p(S|R)p(R)p(C|\Sigma)p(\Sigma), \qquad (11.4.1)$$

where the prior distributions for the model parameters from the Conjugate procedure outlined in Chapter 4 are given by

$$p(S|R) \propto |R|^{-\frac{n}{2}} e^{-\frac{1}{2} tr R^{-1}(S-S_0)'(S-S_0)}, \qquad (11.4.2)$$

$$p(R) \propto |R|^{-\frac{\eta}{2}} e^{-\frac{1}{2} tr R^{-1} V}, \qquad (11.4.3)$$

$$p(\Sigma) \propto |\Sigma|^{-\frac{\nu}{2}} e^{-\frac{1}{2} tr \Sigma^{-1} Q}, \qquad (11.4.4)$$

$$p(C|\Sigma) \propto |D|^{-\frac{p}{2}} |\Sigma|^{-\frac{m+q+1}{2}} e^{-\frac{1}{2} tr \Sigma^{-1}(C-C_0)D^{-1}(C-C_0)'}, \qquad (11.4.5)$$

where the matrices Σ, R, V, D, and Q are positive definite. The hyperparameters S_0, η, V, C_0, D, ν, and Q are to be assessed and having done so completely determines the joint prior distribution. The prior distributions for the parameters are Matrix Normal for the matrix of sources where the source components are free to be correlated, Matrix Normal for the matrix of Regression/mixing coefficients, while the observation error and source covariance matrices are taken to be Inverted Wishart distributed.

Note that both Σ and R are full positive definite symmetric covariance matrices allowing both the observed mixed signals (the elements in the x_i's) and also the unobserved source components (the elements in the s_i's) to be correlated. The mean of the sources is free to be general but often taken to be constant for all observations and thus without loss of generality taken to be zero. Here, an observation (time) varying source mean is adopted.

Upon using Bayes' rule, the joint posterior distribution for the unknown parameters is proportional to the product of the joint prior distribution and the likelihood and is given by

$$p(S, R, C, \Sigma|U, X) \propto |\Sigma|^{-\frac{(n+\nu+m+q+1)}{2}} e^{-\frac{1}{2} tr \Sigma^{-1} G}$$
$$\times |R|^{-\frac{(n+\eta)}{2}} e^{-\frac{1}{2} tr R^{-1}[(S-S_0)'(S-S_0)+V]}, \qquad (11.4.6)$$

where the $p \times p$ matrix G has been defined to be

$$G = (X - ZC')'(X - ZC') + (C - C_0)D^{-1}(C - C_0)' + Q. \qquad (11.4.7)$$

This joint posterior distribution must now be evaluated in order to obtain parameter estimates of the sources S, the Regression/mixing matrix C, the errors of the sources R, and the errors of observation Σ. Marginal posterior mean and joint maximum a posteriori estimates of the parameters S, R, C, Σ are found by the Gibbs sampling and ICM algorithms.

11.5   Conjugate Estimation and Inference

With the above posterior distribution, it is not possible to obtain marginal distributions and thus marginal estimates for any of the parameters in an analytic closed form. It is also not possible to obtain explicit formulas for maximum a posteriori estimates. It is possible to use both Gibbs sampling, as described in Chapter 6 to obtain marginal parameter estimates and the ICM algorithm for finding maximum a posteriori estimates. For both estimation procedures, the posterior conditional distributions are needed.

11.5.1   Posterior Conditionals

From the joint posterior distribution we can obtain the posterior conditional distribution for each of the model parameters.

The conditional posterior distribution for the Regression/mixing matrix C is found by considering only the terms in the joint posterior distribution which involve C and is given by

$$
\begin{aligned}
p(C|S,R,\Sigma,U,X) &\propto p(C|\Sigma)p(X|C,S,\Sigma,U)\\
&\propto |\Sigma|^{-\frac{m+q+1}{2}}e^{-\frac{1}{2}tr\Sigma^{-1}(C-C_0)D^{-1}(C-C_0)'}\\
&\quad\times|\Sigma|^{-\frac{n}{2}}e^{-\frac{1}{2}tr\Sigma^{-1}(X-ZC')'(X-ZC')}\\
&\propto e^{-\frac{1}{2}tr\Sigma^{-1}[(C-C_0)D^{-1}(C-C_0)'+(X-ZC')'(X-ZC')]}\\
&\propto e^{-\frac{1}{2}tr\Sigma^{-1}(C-\tilde{C})(D^{-1}+Z'Z)(C-\tilde{C})'},
\end{aligned}
\tag{11.5.1}
$$

where the variable $\tilde{C}$, the posterior conditional mean and mode, has been defined and is given by

$$
\tilde{C} = [C_0D^{-1}+X'Z](D^{-1}+Z'Z)^{-1}
\tag{11.5.2}
$$

$$
= C_0[D^{-1}(D^{-1}+Z'Z)^{-1}]+\hat{C}[(Z'Z)(D^{-1}+Z'Z)^{-1}].
\tag{11.5.3}
$$

Note that the matrix $\tilde{C}$ can be written as a weighted combination of the prior mean C_0 from the prior distribution and the data mean $\hat{C}=X'Z(Z'Z)^{-1}$ from the likelihood.

The conditional distribution for the matrix of Regression and mixing coefficients C given the matrix of unobservable sources S, the source covariance matrix R, the error covariance matrix Σ the matrix of observable sources U, and the data matrix X is Matrix Normally distributed.

The conditional posterior distribution of the observation error covariance matrix Σ is found by considering only the terms in the joint posterior distribution which involve Σ and is given by

$$p(\Sigma|S,R,C,U,X) \propto p(\Sigma)p(\Lambda|\Sigma)p(X|C,S,\Sigma,U)$$

$$\propto |\Sigma|^{-\frac{\nu}{2}}e^{-\frac{1}{2}tr\Sigma^{-1}Q}|\Sigma|^{-\frac{m+q+1}{2}}e^{-\frac{1}{2}tr\Sigma^{-1}(C-C_0)D^{-1}(C-C_0)'}$$

$$\times |\Sigma|^{-\frac{n}{2}}e^{-\frac{1}{2}tr\Sigma^{-1}(X-ZC')'(X-ZC')}$$

$$\propto |\Sigma|^{-\frac{(n+\nu+m+q+1)}{2}}e^{-\frac{1}{2}tr\Sigma^{-1}G}, \tag{11.5.4}$$

where the $p \times p$ matrix G has been defined to be

$$G = (X - ZC')'(X - ZC') + (C - C_0)D^{-1}(C - C_0)' + Q, \tag{11.5.5}$$

with a mode as discussed in Chapter 2 given by

$$\tilde{\Sigma} = \frac{G}{n + \nu + m + q + 1}. \tag{11.5.6}$$

The posterior conditional distribution of the observation error covariance matrix Σ given the matrix of unobservable sources S, the source covariance matrix R, the matrix of Regression/mixing coefficients C, the observable sources U, and the data X is an Inverted Wishart.

The conditional posterior distribution for the sources S is found by considering only those terms in the joint posterior distribution which involve S and is given by

$$p(S|B,\Lambda,R,\Sigma,U,X) \propto p(S|R)p(X|B,\Lambda,S,\Sigma,U)$$

$$\propto |R|^{-\frac{n}{2}}e^{-\frac{1}{2}trR^{-1}(S-S_0)'(S-S_0)}$$

$$\times |\Sigma|^{-\frac{n}{2}}e^{-\frac{1}{2}tr\Sigma^{-1}(X-UB'-S\Lambda')'(X-UB'-S\Lambda')}$$

$$\propto e^{-\frac{1}{2}tr(S-\tilde{S})(R^{-1}+\Lambda'\Sigma^{-1}\Lambda)(S-\tilde{S})'}, \tag{11.5.7}$$

where the matrix $\tilde{S}$ has been defined which is the posterior conditional mean and mode and is given by

$$\tilde{S} = [S_0 R^{-1} + (X - UB')\Sigma^{-1}\Lambda](R^{-1} + \Lambda'\Sigma^{-1}\Lambda)^{-1}. \tag{11.5.8}$$

The conditional posterior distribution for the sources S given the matrix of Regression coefficients B, the matrix of mixing coefficients Λ, the source covariance matrix R, the error covariance matrix Σ, the matrix of observable sources U, and the matrix of data X is Matrix Normally distributed.

The conditional posterior distribution for the source covariance matrix R is found by considering only the terms in the joint posterior distribution which involve R and is given by

$$p(R|C,S,\Sigma,U,X) \propto p(R)p(S|R)p(X|C,S,\Sigma,U)$$

$$\propto |R|^{-\frac{\eta}{2}} e^{-\frac{1}{2}trR^{-1}V} |R|^{-\frac{n}{2}} e^{-\frac{1}{2}tr(S-S_0)R^{-1}(S-S_0)'}$$

$$\propto |R|^{-\frac{(n+\eta)}{2}} e^{-\frac{1}{2}trR^{-1}[(S-S_0)'(S-S_0)+V]}, \qquad (11.5.9)$$

with the posterior conditional mode as described in Chapter 2 given by

$$\tilde{R} = \frac{(S-S_0)'(S-S_0)+V}{n+\eta}. \qquad (11.5.10)$$

The conditional posterior distribution for the source covariance matrix R given the matrix of Regression/mixing coefficients C, the matrix of sources S, the error covariance matrix Σ, the matrix of observable sources U, and the matrix of data X is Inverted Wishart distributed.

11.5.2 Gibbs Sampling

To find marginal posterior mean estimates of the parameters from the joint posterior distribution using the Gibbs sampling algorithm, start with initial values for the matrix of sources S and the error covariance matrix Σ, say $\bar{S}_{(0)}$ and $\bar{\Sigma}_{(0)}$, and then cycle through

$$\bar{C}_{(l+1)} = \text{a random variate from } p(C|\bar{S}_{(l)}, \bar{R}_{(l)}, \bar{\Sigma}_{(l)}, U, X)$$
$$= A_C Y_C B_C' + M_C, \qquad (11.5.11)$$
$$\bar{\Sigma}_{(l+1)} = \text{a random variate from } p(\Sigma|\bar{S}_{(l)}, \bar{R}_{(l)}, \bar{C}_{(l+1)}, U, X)$$
$$= A_\Sigma (Y_\Sigma' Y_\Sigma)^{-1} A_\Sigma', \qquad (11.5.12)$$
$$\bar{R}_{(l+1)} = \text{a random variate from } p(R|\bar{S}_{(l)}, \bar{C}_{(l+1)}, \bar{\Sigma}_{(l+1)}, U, X)$$
$$= A_R (Y_R' Y_R)^{-1} A_R', \qquad (11.5.13)$$
$$\bar{S}_{(l+1)} = \text{a random variate from } p(S|\bar{R}_{(l+1)}, \bar{C}_{(l+1)}, \bar{\Sigma}_{(l+1)}, U, X)$$
$$= Y_S B_S' + M_S, \qquad (11.5.14)$$

where

$$A_C A_C' = \bar{\Sigma}_{(l)},$$
$$B_C B_C' = (D^{-1} + \bar{Z}_{(l)}' \bar{Z}_{(l)})^{-1},$$
$$\bar{Z}_{(l)} = (U, \bar{S}_{(l)}),$$
$$M_C = (X' \bar{Z}_{(l)} + C_0 D^{-1})(D^{-1} + \bar{Z}_{(l)}' \bar{Z}_{(l)})^{-1}$$
$$A_\Sigma A_\Sigma' = (X - \bar{Z}_{(l)} \bar{C}_{(l+1)}')'(X - \bar{Z}_{(l)} \bar{C}_{(l+1)}') +$$
$$(\bar{C}_{(l+1)} - C_0) D^{-1}(\bar{C}_{(l+1)} - C_0)' + Q,$$

$$A_R A_R' = (\bar{S}_{(l)} - S_0)'(\bar{S}_{(l)} - S_0) + V,$$
$$B_S B_S' = (\bar{R}_{(l+1)}^{-1} + \bar{\Lambda}_{(l+1)}' \bar{\Sigma}_{(l+1)}^{-1} \bar{\Lambda}_{(l+1)})^{-1},$$
$$M_S = [S_0 \bar{R}_{(l+1)}^{-1} + (X - U\bar{B}_{(l+1)}')\bar{\Sigma}_{(l+1)}^{-1} \bar{\Lambda}_{(l+1)}]$$
$$\times (\bar{R}_{(l+1)}^{-1} + \bar{\Lambda}_{(l+1)}' \bar{\Sigma}_{(l+1)}^{-1} \bar{\Lambda}_{(l+1)})^{-1}$$

while Y_C, Y_Σ, Y_R, and Y_S are $p \times (m+q+1)$, $(n+\nu+m \mid 1+p+1) \times p$, $(n+\eta+m+1) \times m$, and $n \times m$ dimensional matrices respectively, whose elements are random variates from the standard Scalar Normal distribution. The formulas for the generation of random variates from the conditional posterior distributions are easily found from the methods in Chapter 6.

The first random variates called the "burn in" are discarded and after doing so, compute from the next L variates means of the parameters

$$\bar{S} = \frac{1}{L}\sum_{l=1}^{L} \bar{S}_{(l)} \qquad \bar{R} = \frac{1}{L}\sum_{l=1}^{L} \bar{R}_{(l)} \qquad \bar{C} = \frac{1}{L}\sum_{l=1}^{L} \bar{C}_{(l)} \qquad \bar{\Sigma} = \frac{1}{L}\sum_{l=1}^{L} \bar{\Sigma}_{(l)}$$

which are the exact sampling-based marginal posterior mean estimates of the parameters. Exact sampling-based estimates of other quantities can also be found. Similar to Bayesian Regression, Bayesian Factor Analysis, and Bayesian Source Separation, there is interest in the estimate of the marginal posterior variance of the matrix containing the Regression and mixing coefficients

$$\overline{var}(c|\bar{c}, X, U) = \frac{1}{L}\sum_{l=1}^{L} \bar{c}_{(l)}\bar{c}_{(l)}' - \bar{c}\bar{c}'$$
$$= \bar{\Delta},$$

where $c = vec(C)$.

The covariance matrices of the other parameters follow similarly. With a specification of Normality for the marginal posterior distribution of the vector containing the Regression and mixing coefficients, their distribution is

$$p(c|\bar{c}, X, U) \propto |\bar{\Delta}|^{-\frac{1}{2}} e^{-\frac{1}{2}(c-\bar{c})\bar{\Delta}^{-1}(c-\bar{c})'}, \tag{11.5.15}$$

where $\bar{c}$ and $\bar{\Delta}$ are as previously defined.

To evaluate statistical significance with the Gibbs sampling approach, use the marginal distribution of the matrix containing the Regression and mixing coefficients given above. General simultaneous hypotheses can be evaluated regarding the entire matrix containing the Regression and mixing coefficients, a submatrix, or a particular independent variable or source, or an element by computing marginal distributions. It can be shown that the marginal distribution of the kth column of the matrix containing the Regression and mixing coefficients C, C_k is Multivariate Normal

$$p(C_k|\bar{C}_k,X,U) \propto |\bar{\Delta}_k|^{-\frac{1}{2}} e^{-\frac{1}{2}(C_k-\bar{C}_k)'\bar{\Delta}_k^{-1}(C_k-\bar{C}_k)}, \tag{11.5.16}$$

where $\bar{\Delta}_k$ is the covariance matrix of C_k found by taking the kth $p \times p$ submatrix along the diagonal of $\bar{\Delta}$.

Significance can be evaluated for a subset of coefficients of the kth column of C by determining the marginal distribution of the subset within C_k which is also Multivariate Normal. With the subset being a singleton set, significance can be evaluated for a particular coefficient with the marginal distribution of the scalar coefficient which is

$$p(C_{kj}|\bar{C}_{kj},X,U) \propto (\bar{\Delta}_{kj})^{-\frac{1}{2}} e^{-\frac{(C_{kj}-\bar{C}_{kj})^2}{2\bar{\Delta}_{kj}}}, \tag{11.5.17}$$

where $\bar{\Delta}_{kj}$ is the jth diagonal element of $\bar{\Delta}_k$. Note that $\bar{C}_{kj} = \bar{c}_{jk}$ and that

$$z = \frac{(C_{kj}-\bar{C}_{kj})}{\sqrt{\bar{\Delta}_{kj}}} \tag{11.5.18}$$

follows a Normal distribution with a mean of zero and variance of one.

11.5.3 Maximum a Posteriori

The joint posterior distribution can also be maximized with respect to the matrix of coefficients C, the error covariance matrix Σ, the matrix of sources S, and the source covariance matrix R by using the ICM algorithm. To jointly maximize the joint posterior distribution using the ICM algorithm, start with an initial value for the matrix of sources S, say $\tilde{S}_{(0)}$, and then cycle through

$$\tilde{C}_{(l+1)} = \overset{\text{Arg Max}}{C} \; p(C|\tilde{S}_{(l)}, \tilde{R}_{(l)}, \tilde{\Sigma}_{(l)}, X)$$
$$= (X'\tilde{Z}_{(l)} + C_0 D^{-1})(D^{-1} + \tilde{Z}'_{(l)}\tilde{Z}_{(l)})^{-1},$$

$$\tilde{\Sigma}_{(l+1)} = \overset{\text{Arg Max}}{\Sigma} \; p(\Sigma|\tilde{C}_{(l+1)}, \tilde{R}_{(l)}, \tilde{S}_{(l)}, X)$$
$$= [(X - \tilde{Z}_{(l)}\tilde{C}'_{(l+1)})'(X - \tilde{Z}_{(l)}\tilde{C}'_{(l+1)})$$
$$+ (\tilde{C}_{(l+1)} - C_0)D^{-1}(\tilde{C}_{(l+1)} - C_0)' + Q]/(n+\nu+m+q+1),$$

$$\tilde{S}_{(l+1)} = \overset{\text{Arg Max}}{S} \; p(S|\tilde{C}_{(l+1)}, \tilde{R}_{(l)}, \tilde{\Sigma}_{(l+1)}, X)$$
$$= [S_0\tilde{R}_{(l)}^{-1} + (X - U\tilde{B}'_{(l+1)})\tilde{\Sigma}_{(l+1)}^{-1}\tilde{\Lambda}_{(l+1)}]$$
$$\times (\tilde{R}_{(l)}^{-1} + \tilde{\Lambda}'_{(l+1)}\tilde{\Sigma}_{(l+1)}^{-1}\tilde{\Lambda}_{(l+1)})^{-1},$$

$$\tilde{R}_{(l+1)} = \overset{\text{Arg Max}}{R} \; p(R|\tilde{S}_{(l+1)}, \tilde{C}_{(l+1)}, \tilde{\Sigma}_{(l+1)}, X)$$

$$= \frac{(\tilde{S}_{(l+1)} - S_0)'(\tilde{S}_{(l+1)} - S_0) + V}{n + \eta},$$

where the matrix $\tilde{Z}_{(l)} = (U, \tilde{S}_{(l)})$ until convergence is reached. The converged values $(\tilde{S}, \tilde{R}, \tilde{C}, \tilde{\Sigma})$ are joint posterior modal (maximum a posteriori) estimates of the parameters. Conditional maximum a posteriori variance estimates can also be found. The conditional modal variance of the matrix containing the Regression and mixing coefficients is

$$var(C | \tilde{C}, \tilde{S}, \tilde{R}, \tilde{\Sigma}, X, U) = \tilde{\Sigma} \otimes (D^{-1} \otimes \tilde{Z}'\tilde{Z})^{-1}$$

or equivalently

$$var(c | \tilde{c}, \tilde{S}, \tilde{R}, \tilde{\Sigma}, X, U) = (D^{-1} \otimes \tilde{Z}'\tilde{Z})^{-1} \otimes \tilde{\Sigma}$$
$$= \tilde{\Delta},$$

where $c = vec(C)$, $\tilde{S}$, $\tilde{R}$, and $\tilde{\Sigma}$ are the converged value from the ICM algorithm.

To evaluate statistical significance with the ICM approach, use the conditional distribution of the matrix containing the Regression and mixing coefficients which is

$$p(C | \tilde{C}, \tilde{S}, \tilde{R}, \tilde{\Sigma}, X, U) \propto |D^{-1} + \tilde{Z}'\tilde{Z}|^{\frac{1}{2}} |\tilde{\Sigma}|^{-\frac{1}{2}} e^{-\frac{1}{2} tr \tilde{\Sigma}^{-1}(C - \tilde{C})(D^{-1} + \tilde{Z}'\tilde{Z})(C - \tilde{C})'},$$
(11.5.19)

That is,

$$C | \tilde{C}, \tilde{S}, \tilde{R}, \tilde{\Sigma}, X, U \sim N\left(\tilde{C}, \tilde{\Sigma} \otimes (D^{-1} + \tilde{Z}'\tilde{Z})^{-1}\right). \quad (11.5.20)$$

General simultaneous hypotheses can be evaluated regarding the entire matrix containing the Regression and mixing coefficients, a submatrix, or the coefficients of a particular independent variable or source, or an element by computing marginal conditional distributions.

It can be shown [17, 41] that the marginal conditional distribution of any column of the matrix containing the Regression and mixing coefficients C, C_k is Multivariate Normal

$$p(C_k | \tilde{C}_k, \tilde{S}, \tilde{\Sigma}, U, X) \propto |W_{kk}\tilde{\Sigma}|^{-\frac{1}{2}} e^{-\frac{1}{2}(C_k - \tilde{C}_k)'(W_{kk}\tilde{\Sigma})^{-1}(C_k - \tilde{C}_k)}, \quad (11.5.21)$$

where $W = (D^{-1} + \tilde{Z}'\tilde{Z})^{-1}$ and W_{kk} is its kth diagonal element.

With the marginal distribution of a column of C, significance can be determined for a particular independent variable or source. Significance can be determined for a subset of coefficients by determining the marginal distribution of the subset within C_k which is also Multivariate Normal. With the subset being a singleton set, significance can be determined for a particular coefficient with the marginal distribution of the scalar coefficient which is

$$p(C_{kj}|\tilde{C}_{kj}, \tilde{S}, \tilde{\Sigma}_{jj}, U, X) \propto (W_{kk}\tilde{\Sigma}_{jj})^{-\frac{1}{2}} e^{-\frac{(C_{kj} - \tilde{C}_{kj})^2}{2W_{kk}\tilde{\Sigma}_{jj}}}, \qquad (11.5.22)$$

where $\tilde{\Sigma}_{jj}$ is the jth diagonal element of $\tilde{\Sigma}$. Note that $\tilde{C}_{kj} = \tilde{c}_{jk}$ and that

$$z = \frac{(C_{kj} - \tilde{C}_{kj})}{\sqrt{W_{kk}\tilde{\Sigma}_{jj}}}$$

$$(11.5.23)$$

follows a Normal distribution with a mean of zero and variance of one.

11.6   Generalized Priors and Posterior

Generalized Conjugate prior distributions are assessed in order to quantify available prior information regarding values of the model parameters. The joint prior distribution for the sources S, the source covariance matrix R, the vector of coefficients $c = vec(C)$, and the error covariance matrix Σ is given by

$$p(S, R, \Sigma, c) = p(S|R)p(R)p(\Sigma)p(c), \qquad (11.6.1)$$

where the prior distribution for the parameters from the generalized Conjugate procedure outlined in Chapter 4 are as follows

$$p(S|R) \propto |R|^{-\frac{n}{2}} e^{-\frac{1}{2}tr(S - S_0)R^{-1}(S - S_0)'}, \qquad (11.6.2)$$

$$p(R) \propto |R|^{-\frac{\eta}{2}} e^{-\frac{1}{2}tr R^{-1}V}, \qquad (11.6.3)$$

$$p(\Sigma) \propto |\Sigma|^{-\frac{\nu}{2}} e^{-\frac{1}{2}tr\Sigma^{-1}Q}, \qquad (11.6.4)$$

$$p(c) \propto |\Delta|^{-\frac{1}{2}} e^{-\frac{1}{2}(c - c_0)'\Delta^{-1}(c - c_0)}, \qquad (11.6.5)$$

where Σ, R, V, Q, and Δ are positive definite matrices. The hyperparameters S_0, η, V, ν, Q, c_0, and Δ are to be assessed. Upon assessing the hyperparameters, the joint prior distribution is completely determined.

The prior distribution for the matrix of sources S is Matrix Normally distributed, the prior distribution for the source vector covariance matrix R is

Inverted Wishart distributed, the vector of combined Regression/mixing coefficients $c = vec(C)$, $C = (B, \Lambda)$ is Multivariate Normally distributed, the prior distribution for the error covariance matrix Σ is Inverted Wishart distributed.

Note that both Σ and R are full covariance matrices allowing both the observed mixed signals (microphones) and the unobserved source components (speakers) to be correlated. The mean of the sources is often taken to be constant for all observations and thus without loss of generality taken to be zero. An observation (time) varying source mean is adopted here.

Upon using Bayes' rule the joint posterior distribution for the unknown parameters with generalized Conjugate prior distributions for the model parameters is given by

$$p(S, R, c, \Sigma | U, X) \propto p(S|R)p(R)p(\Sigma)p(c)p(X|C, Z, \Sigma), \qquad (11.6.6)$$

which is

$$p(S, R, c, \Sigma | U, X) \propto |\Sigma|^{-\frac{(n+\nu)}{2}} e^{-\frac{1}{2} tr \Sigma^{-1}[(X-ZC')'(X-ZC')+Q]}$$
$$\times |R|^{-\frac{(n+\eta)}{2}} e^{-\frac{1}{2} tr R^{-1}[(S-S_0)'(S-S_0)+V]}$$
$$\times |\Delta|^{-\frac{1}{2}} e^{-\frac{1}{2}(c-c_0)'\Delta^{-1}(c-c_0)}, \qquad (11.6.7)$$

after inserting the joint prior distribution and the likelihood.

This joint posterior distribution must now be evaluated in order to obtain parameter estimates of the matrix of sources S, the vector of Regression/mixing coefficients c, the sources covariance matrix R, and the error covariance matrix Σ.

11.7   Generalized Estimation and Inference

With the generalized Conjugate prior distributions, it is not possible to obtain all or any of the marginal distributions or explicit expressions for maxima and thus marginal mean and joint maximum a posteriori estimates in closed form. For these reasons, marginal mean and joint maximum a posteriori estimates are found using the Gibbs sampling and ICM algorithms.

11.7.1   Posterior Conditionals

Both the Gibbs sampling and ICM require the posterior conditionals. Gibbs sampling requires the conditionals for the generation of random variates while ICM requires them for maximization by cycling through their modes or maxima.

The conditional posterior distribution of the matrix of sources S is found by considering only the terms in the joint posterior distribution which involve S and is given by

$$p(S|B,R,\Lambda,\Sigma,U,X) \propto p(S|R)p(X|B,S,\Lambda,\Sigma,U)$$
$$\propto e^{-\frac{1}{2}tr(S-S_0)'R^{-1}(S-S_0)}$$
$$\times e^{-\frac{1}{2}tr(X-UB'-S\Lambda')\Sigma^{-1}(X-UB'-S\Lambda')'},$$

which after performing some algebra in the exponent can be written as

$$p(S|B,R,\Lambda,\Sigma,U,X) \propto e^{-\frac{1}{2}tr(S-\tilde{S})(R^{-1}+\Lambda'\Sigma^{-1}\Lambda)(S-\tilde{S})'}, \qquad (11.7.1)$$

where the matrix $\tilde{S}$ has been defined to be

$$\tilde{S} = [S_0 R^{-1} + (X-UB')\Sigma^{-1}\Lambda](R^{-1}+\Lambda'\Sigma^{-1}\Lambda)^{-1}. \qquad (11.7.2)$$

That is, the matrix of sources S given the matrix of Regression coefficients B, the source covariance matrix R, the mixing coefficients Λ, the error covariance matrix Σ the matrix observable sources U, and the matrix of observed data X is Matrix Normally distributed.

The conditional posterior distribution of the source covariance matrix R is found by considering only the terms in the joint posterior distribution which involve R and is given by

$$p(R|C,S,\Sigma,U,X) \propto p(R)p(S|R)$$
$$\propto |R|^{-\frac{\nu}{2}}e^{-\frac{1}{2}trR^{-1}V}|R|^{-\frac{n}{2}}e^{-\frac{1}{2}trR^{-1}(S-S_0)'(S-S_0)}$$
$$\propto |R|^{-\frac{(n+\nu)}{2}}e^{-\frac{1}{2}trR^{-1}[(S-S_0)'(S-S_0)+V]}. \qquad (11.7.3)$$

That is, the posterior conditional distribution of the source covariance matrix R given the matrix of Regression/mixing coefficients C, the matrix of sources S, the error covariance matrix Σ, the matrix of observable sources U, and the matrix of data X has an Inverted Wishart distribution.

The conditional posterior distribution of the vector c containing the Regression coefficients B and the matrix of mixing coefficients Λ is found by considering only the terms in the joint posterior distribution which involve c or C and is given by

$$p(c|S,R,\Sigma,U,X) \propto p(c)p(X|S,C,\Sigma,U)$$
$$\propto |\Delta|^{-\frac{1}{2}}e^{-\frac{1}{2}(c-c_0)'\Delta^{-1}(c-c_0)}$$
$$\times |\Sigma|^{-\frac{n}{2}}e^{-\frac{1}{2}tr\Sigma^{-1}(X-ZC')'(X-ZC')}, \qquad (11.7.4)$$

which after performing some algebra in the exponent becomes

$$p(c|S,R,\Sigma,U,X) \propto e^{-\frac{1}{2}(c-\tilde{c})'(\Delta^{-1}+Z'Z\otimes\Sigma^{-1})(c-\tilde{c})}, \qquad (11.7.5)$$

where the vector $\tilde{c}$ has been defined to be

$$\tilde{c} = (\Delta^{-1}+Z'Z\otimes\Sigma^{-1})^{-1}[\Delta^{-1}c_0+(Z'Z\otimes\Sigma^{-1})\hat{c}] \qquad (11.7.6)$$

and the vector $\hat{c}$ has been defined to be

$$\hat{c} = vec[X'Z(Z'Z)^{-1}]. \qquad (11.7.7)$$

Note that the vector $\tilde{c}$ can be written as a weighted combination of the prior mean c_0 from the prior distribution and the data mean $\hat{c}$ from the likelihood.

The conditional posterior distribution of the vector c containing the matrix of Regression coefficients B and the matrix of mixing coefficients Λ given the matrix of sources S, the source covariance matrix R, the error covariance matrix Σ, the matrix of observable sources U, and the matrix of observed data X is Multivariate Normally distributed.

The conditional posterior distribution of the error covariance matrix Σ is found by considering only the terms in the joint posterior distribution which involve Σ and is given by

$$p(\Sigma|S,R,C,U,X) \propto p(\Sigma)p(X|S,C,\Sigma,U)$$
$$\propto |\Sigma|^{-\frac{(n+\nu)}{2}} e^{-\frac{1}{2}tr\Sigma^{-1}[(X-ZC')'(X-ZC')+Q]}. \qquad (11.7.8)$$

That is, the conditional posterior distribution of the error covariance matrix Σ given the matrix of sources S, the source covariance matrix R, the matrix of coefficients C, the matrix of observable sources U, and the matrix of observable data X has an Inverted Wishart distribution.

The modes of these posterior conditional distributions are as described in Chapter 2 and given by $\tilde{S}$, $\tilde{c}$, (both as defined above)

$$\tilde{R} = \frac{(S-S_0)'(S-S_0)+V}{n+\eta}, \qquad (11.7.9)$$

and

$$\tilde{\Sigma} = \frac{(X-ZC')'(X-ZC')+Q}{n+\nu}, \qquad (11.7.10)$$

respectively.

11.7.2   Gibbs Sampling

To find marginal mean estimates of the parameters from the joint posterior distribution using the Gibbs sampling algorithm, start with initial values for the matrix of sources S and the error covariance matrix Σ, say $\bar{S}_{(0)}$ and $\bar{\Sigma}_{(0)}$, and then cycle through

$$\bar{c}_{(l+1)} = \text{a random variate from } p(c|\bar{S}_{(l)},\bar{\Sigma}_{(l)},\bar{R}_{(l+1)},U,X)$$
$$= A_c Y_c + M_c, \tag{11.7.11}$$
$$\bar{\Sigma}_{(l+1)} = \text{a random variate from } p(\Sigma|\bar{S}_{(l)},\bar{R}_{(l+1)},\bar{c}_{(l+1)},U,X)$$
$$= A_\Sigma (Y'_\Sigma Y_\Sigma)^{-1} A'_\Sigma, \tag{11.7.12}$$
$$\bar{R}_{(l+1)} = \text{a random variate from } p(R|\bar{S}_{(l)},\bar{c}_{(l+1)},\bar{\Sigma}_{(l+1)},U,X)$$
$$= A_R (Y'_R Y_R)^{-1} A'_R, \tag{11.7.13}$$
$$\bar{S}_{(l+1)} = \text{a random variate from } p(S|\bar{R}_{(l+1)},\bar{c}_{(l+1)},\bar{\Sigma}_{(l+1)},U,X)$$
$$= Y_S B'_S + M_S, \tag{11.7.14}$$

where

$$\hat{c}_{(l)} = vec[X'\bar{Z}_{(l)}(\bar{Z}'_{(l)}\bar{Z}_{(l)})^{-1}],$$
$$\bar{c}_{(l+1)} = [\Delta^{-1} + \bar{Z}'_{(l)}\bar{Z}_{(l)} \otimes \bar{\Sigma}^{-1}_{(l)}]^{-1}[\Delta^{-1}c_0 + (\bar{Z}'_{(l)}\bar{Z}_{(l)} \otimes \bar{\Sigma}^{-1}_{(l)})\hat{c}_{(l)}],$$
$$A_c A'_c = (\Delta^{-1} + \bar{Z}'_{(l)}\bar{Z}_{(l)} \otimes \bar{\Sigma}^{-1}_{(l)})^{-1},$$
$$M_c = [\Delta^{-1} + \bar{Z}'_{(l)}\bar{Z}_{(l)} \otimes \bar{\Sigma}^{-1}_{(l)}]^{-1}[\Delta^{-1}c_0 + (\bar{Z}'_{(l)}\bar{Z}_{(l)} \otimes \bar{\Sigma}^{-1}_{(l)})\hat{c}],$$
$$A_\Sigma A'_\Sigma = (X - \bar{Z}_{(l)}\bar{C}'_{(l+1)})'(X - \bar{Z}_{(l)}\bar{C}'_{(l+1)}) + Q,$$
$$A_R A'_R = (\bar{S}_{(l)} - S_0)'(\bar{S}_{(l)} - S_0) + V,$$
$$B_S B'_S = (\bar{R}^{-1}_{(l+1)} + \bar{\Lambda}'_{(l+1)}\bar{\Sigma}^{-1}_{(l+1)}\bar{\Lambda}_{(l+1)})^{-1},$$
$$M_S = [S_0 \bar{R}^{-1}_{(l+1)} + (X - U\bar{B}'_{(l+1)})\bar{\Sigma}^{-1}_{(l+1)}\bar{\Lambda}_{(l+1)}]$$
$$\times (\bar{R}^{-1}_{(l+1)} + \bar{\Lambda}'_{(l+1)}\bar{\Sigma}^{-1}_{(l+1)}\bar{\Lambda}_{(l+1)})^{-1}$$

while Y_c, Y_Σ, Y_R, and Y_S are $p(m+1) \times 1$, $(n+\nu+p+1) \times p$, $(n+\eta+m+1) \times m$, and $n \times m$ dimensional matrices whose respective elements are random variates from the standard Scalar Normal distribution. The formulas for the generation of random variates from the conditional posterior distributions is easily found from the methods in Chapter 6.

The first random variates called the "burn in" are discarded and after doing so, compute from the next L variates means of each of the parameters

$$\bar{S} = \frac{1}{L}\sum_{l=1}^{L}\bar{S}_{(l)} \qquad \bar{R} = \frac{1}{L}\sum_{l=1}^{L}\bar{R}_{(l)} \qquad \bar{c} = \frac{1}{L}\sum_{l=1}^{L}\bar{c}_{(l)} \qquad \bar{\Sigma} = \frac{1}{L}\sum_{l=1}^{L}\bar{\Sigma}_{(l)}$$

which are the exact sampling-based marginal posterior mean estimates of the parameters. Exact sampling-based estimates of other quantities can also be found. Similar to Regression, Factor Analysis, and Source Separation, there is interest in the estimate of the marginal posterior variance of the matrix containing the Regression and mixing coefficients

$$\overline{var}(c|\bar{c}, X, U) = \frac{1}{L} \sum_{l=1}^{L} \bar{c}_{(l)} \bar{c}'_{(l)} - \bar{c}\bar{c}'$$

$$= \bar{\Delta},$$

where $c = vec(C)$.

The covariance matrices of the other parameters follow similarly. With a specification of Normality for the marginal posterior distribution of the vector containing the Regression and mixing coefficients, their distribution is

$$p(c|\bar{c}, X, U) \propto |\bar{\Delta}|^{-\frac{1}{2}} e^{-\frac{1}{2}(c-\bar{c})' \bar{\Delta}^{-1}(c-\bar{c})}, \tag{11.7.15}$$

where $\bar{c}$ and $\bar{\Delta}$ are as previously defined.

To evaluate statistical significance with the Gibbs sampling approach, use the marginal distribution of the vector c containing the Regression and mixing coefficients given above. General simultaneous hypotheses can be evaluated regarding the entire coefficient vector of Regression and mixing coefficients, a subset of it, or the coefficients for a particular independent variable or source by computing marginal distributions. It can be shown that the marginal distribution of the kth column of the matrix containing the Regression and mixing coefficients C, C_k is Multivariate Normal

$$p(C_k|\bar{C}_k, X, U) \propto |\bar{\Delta}_k|^{-\frac{1}{2}} e^{-\frac{1}{2}(C_k-\bar{C}_k)' \bar{\Delta}_k^{-1}(C_k-\bar{C}_k)}, \tag{11.7.16}$$

where $\bar{\Delta}_k$ is the covariance matrix of C_k found by taking the kth $p \times p$ submatrix along the diagonal of $\bar{\Delta}$.

Significance can be evaluated for a subset of means or coefficients of the kth column of C by determining the marginal distribution of the subset within C_k which is also Multivariate Normal. With the subset being a singleton set, significance can be evaluated for a particular mean or coefficient with the marginal distribution of the scalar coefficient which is

$$p(C_{kj}|\bar{C}_{kj}, X, U) \propto (\bar{\Delta}_{kj})^{-\frac{1}{2}} e^{-\frac{(C_{kj}-\bar{C}_{kj})^2}{2\bar{\Delta}_{kj}}}, \tag{11.7.17}$$

where $\bar{\Delta}_{kj}$ is the jth diagonal element of $\bar{\Delta}_k$. Note that $\bar{C}_{kj} = \bar{c}_{jk}$ and that

$$z = \frac{(C_{kj} - \bar{C}_{kj})}{\sqrt{\bar{\Delta}_{kj}}} \tag{11.7.18}$$

follows a Normal distribution with a mean of zero and variance of one.

11.7.3   Maximum a Posteriori

The joint posterior distribution can also be maximized with respect to the vector of coefficients c, the matrix of sources S, the source covariance matrix R, and the error covariance matrix Σ using the ICM algorithm. To jointly maximize the joint posterior distribution using the ICM algorithm, start with initial values for the matrix of sources $\tilde{S}$ and the error covariance matrix Σ, say $\tilde{S}_{(0)}$ and $\tilde{\Sigma}_{(0)}$, and then cycle through

$$\hat{c}_{(l)} = vec[X'\tilde{Z}_{(l)}(\tilde{Z}'_{(l)}\tilde{Z}_{(l)})^{-1}],$$

$$\tilde{c}_{(l+1)} = \overset{\text{Arg Max}}{c} \; p(c|\tilde{S}_{(l)},\tilde{R}_{(l)},\tilde{\Sigma}_{(l)},X,U)$$
$$= (\Delta^{-1}+\tilde{Z}'_{(l)}\tilde{Z}_{(l)}\otimes\tilde{\Sigma}^{-1}_{(l)})^{-1}[\Delta^{-1}c_0 + (\tilde{Z}'_{(l)}\tilde{Z}_{(l)}\otimes\tilde{\Sigma}^{-1}_{(l)})\hat{c}_{(l)}],$$

$$\tilde{\Sigma}_{(l+1)} = \overset{\text{Arg Max}}{\Sigma} \; p(\Sigma|\tilde{C}_{(l+1)},\tilde{R}_{(l)},\tilde{S}_{(l)},X,U)$$
$$= \frac{(X-\tilde{Z}_{(l)}\tilde{C}'_{(l+1)})'(X-\tilde{Z}_{(l)}\tilde{C}'_{(l+1)})+Q}{n+\nu},$$

$$\tilde{R}_{(l+1)} = \overset{\text{Arg Max}}{R} \; p(R|\tilde{S}_{(l)},\tilde{C}_{(l+1)},\tilde{\Sigma}_{(l+1)},X,U)$$
$$= \frac{(\tilde{S}_{(l)}-S_0)'(\tilde{S}_{(l)}-S_0)+V}{n+\eta},$$

$$\tilde{S}_{(l+1)} = \overset{\text{Arg Max}}{S} \; p(S|\tilde{C}_{(l+1)},\tilde{R}_{(l+1)},\tilde{\Sigma}_{(l+1)},X,U)$$
$$= [S_0\tilde{R}^{-1}_{(l+1)} + (X-U\tilde{B}'_{(l+1)})\tilde{\Sigma}^{-1}_{(l+1)}\tilde{\Lambda}_{(l+1)}]$$
$$\times(\tilde{R}^{-1}_{(l+1)} + \tilde{\Lambda}'_{(l+1)}\tilde{\Sigma}^{-1}_{(l+1)}\tilde{\Lambda}_{(l+1)})^{-1},$$

where the matrix $\tilde{Z}_{(l)} = (U,\tilde{S}_{(l)})$ has been defined until convergence is reached with the joint modal (maximum a posteriori) estimates for the unknown model parameters $(\tilde{S},\tilde{R},\tilde{c},\tilde{\Sigma})$. Conditional maximum a posteriori variance estimates can also be found. The conditional modal variance of the matrix containing the Regression and mixing coefficients is

$$var(c|\tilde{c},\tilde{S},\tilde{R},\tilde{\Sigma},X,U) = [\Delta^{-1}+\tilde{Z}'\tilde{Z}\otimes\tilde{\Sigma}]^{-1}$$
$$= \tilde{\Delta},$$

where $c = vec(C)$, while $\tilde{S}$, $\tilde{R}$, and $\tilde{\Sigma}$ are the converged value from the ICM algorithm.

Conditional modal intervals may be computed by using the conditional distribution for a particular parameter given the modal values of the others. The posterior conditional distribution of the matrix containing the Regression and mixing coefficients C given the modal values of the other parameters and the data is

$$p(c|\tilde{c},\tilde{S},\tilde{\Sigma},X,U) \propto |\tilde{\Delta}|^{-\frac{1}{2}} e^{-\frac{1}{2}(c-\tilde{c})'\tilde{\Delta}^{-1}(c-\tilde{c})}. \qquad (11.7.19)$$

To evaluate statistical significance with the ICM approach, use the marginal conditional distribution of the matrix containing the Regression and mixing coefficients given above. General simultaneous hypotheses can be evaluated regarding the entire vector, a subset of it, or the coefficients of a particular independent variable or source by computing marginal distributions. It can be shown that the marginal conditional distribution of the kth column C_k of the matrix C containing the Regression and mixing coefficients is Multivariate Normal

$$p(C_k|\tilde{C}_k,\tilde{\Sigma},X,U) \propto |\tilde{\Delta}_k|^{-\frac{1}{2}} e^{-\frac{1}{2}(C_k-\tilde{C}_k)'\tilde{\Delta}_k^{-1}(C_k-\tilde{C}_k)}, \qquad (11.7.20)$$

where $\tilde{\Delta}_k$ is the covariance matrix of C_k found by taking the kth $p \times p$ submatrix along the diagonal of $\tilde{\Delta}$.

Significance can be evaluated for a subset of Regression or mixing coefficients of the kth column of C by determining the marginal distribution of the subset within C_k which is also Multivariate Normal. With the subset being a singleton set, significance can be evaluated for a particular coefficient with the marginal distribution of the scalar coefficient which is

$$p(C_{kj}|\tilde{C}_{kj},\tilde{S},\tilde{\Sigma}_{jj},X) \propto (\tilde{\Delta}_{kj})^{-\frac{1}{2}} e^{-\frac{(C_{kj}-\tilde{C}_{kj})^2}{2\tilde{\Delta}_{kj}}}, \qquad (11.7.21)$$

where $\tilde{\Delta}_{kj}$ is the jth diagonal element of $\tilde{\Delta}_k$. Note that $\tilde{C}_{kj} = \tilde{c}_{jk}$ and that

$$z = \frac{(C_{kj}-\tilde{C}_{kj})}{\sqrt{\tilde{\Delta}_{kj}}}$$

follows a Normal distribution with a mean of zero and variance of one.

11.8 Interpretation

Although the main focus after having performed a Bayesian Source Separation is the separated sources, there are others. One focus as in Bayesian Regression is on the estimate of the Regression coefficient matrix B which defines a "fitted" line. Coefficients are evaluated to determine whether they are statistically "large" meaning that the associated independent variable contributes to the dependent variable or statistically "small" meaning that the associated independent variable does not contribute to the dependent variable.

The coefficient matrix also has the interpretation that if all of the independent variables were held fixed except for one u_{ij} which if increased to u_{ij}^*, the dependent variable x_{ij} increases to an amount x_{ij}^* given by

$$x_{ij}^* = \beta_{i0} + \cdots + \beta_{ij} u_{ij}^* + \cdots + \beta_{iq} u_{iq}. \qquad (11.8.1)$$

Another focus after performing a Bayesian Source Separation is in the estimated mixing coefficients. The mixing coefficients are the amplitudes which determine the relative contribution of the sources. A particular mixing coefficient which is relatively "small" indicates that the corresponding source does not significantly contribute to the associated observed mixed signal. If a particular mixing coefficient is relatively "large," this indicates that the corresponding source does significantly contribute to the associated observed mixed signal.

11.9   Discussion

Returning to the cocktail party problem, the matrix of Regression coefficients B where $B = (\mu, B_\star)$ contains the matrix of mixing coefficients $B_\star$ for the observed conversation (sources) U, and the population mean μ which is a vector of the overall background mean level at each microphone.

After having estimated the model parameters, the estimates of the sources as well as the mixing matrix are now available. The estimated matrix of sources corresponds to the unobservable signals or conversations emitted from the mouths of the speakers at the cocktail party. Row i of the estimated source matrix is the estimate of the unobserved source vector at time i and column j of the estimated source matrix is the estimate of the unobserved conversation of speaker j at the party for all n time increments.

Exercises

1. Specify that B and Λ are independent with the prior distribution for Regression coefficients B being the vague prior

$$p(B) \propto (\text{a constant}),$$

the distribution for the mixing matrix being

$$p(\Lambda|\Sigma) \propto |A|^{-\frac{p}{2}}|\Sigma|^{-\frac{m}{2}}e^{-\frac{1}{2}tr\Sigma^{-1}(\Lambda-\Lambda_0)A^{-1}(\Lambda-\Lambda_0)'},$$

and the others as in Equations 11.4.2-11.4.4.

Combine these prior distributions with the likelihood in Equation 11.3.2 to obtain a posterior distribution. Derive Gibbs sampling and ICM algorithms for marginal posterior mean and joint maximum a posteriori parameter estimates [52, 59].

2. Specify that B and Λ are independent with the prior distribution for the vector of Regression coefficients β to be the Conjugate prior

$$p(B|\Sigma) \propto |\Sigma|^{-\frac{1}{2}}e^{-\frac{1}{2}tr\Sigma^{-1}(B-B_0)H^{-1}(B-B_0)'},$$

the distribution for the vector of mixing coefficients λ to be

$$p(\Lambda|\Sigma) \propto |A|^{-\frac{p}{2}}|\Sigma|^{-\frac{m}{2}}e^{-\frac{1}{2}tr\Sigma^{-1}(\Lambda-\Lambda_0)A^{-1}(\Lambda-\Lambda_0)'},$$

and the others to be as in Equations 11.4.2–11.4.4.

Combine these prior distributions with the likelihood in Equation 11.3.2 to obtain a posterior distribution. Derive Gibbs sampling and ICM algorithms for marginal mean and joint maximum a posteriori parameter estimates [57].

3. Specify that B and Λ are independent with the prior distribution for the overall mean μ being the generalized Conjugate prior

$$p(\beta) \propto |\Gamma|^{-\frac{1}{2}}e^{-\frac{1}{2}(\beta-\beta_0)'\Gamma^{-1}(\beta-\beta_0)},$$

the distribution for the mixing matrix being

$$p(\lambda) \propto |\Delta|^{-\frac{1}{2}}e^{-\frac{1}{2}(\lambda-\lambda_0)'\Delta^{-1}(\lambda-\lambda_0)},$$

and the others as in Equations 11.4.2–11.4.4.

Combine these prior distributions with the likelihood in Equation 11.3.2 to obtain a posterior distribution. Derive Gibbs sampling and ICM algorithms for marginal mean and joint maximum a posteriori parameter estimates [57].

4. Show that by (a) setting the number of observable sources q to be equal to zero, the resulting model is the Source Separation model of Chapter 10 and (b) by setting the number of unobservable sources m to be equal to zero, the resulting model is the Regression model of Chapter 8.

12

FMRI Case Study

12.1 Introduction

Functional magnetic resonance imaging (FMRI) is a designed experiment
[15] which often consists of a patient being given a sequence of stimuli AB
or ABC. Imaging takes place while the patient is responding either passively
or actively to these stimuli. A model is used which views the observed time
courses as being made up of a linear (polynomial) trend, responses (possibly
zero valued) due to the presentation of the stimuli, and other cognitive ac-
tivites that are typically termed random and grouped into the error term. The
association between the observed time course in each voxel and the sequence
of stimuli is determined. Different levels of activation (association) for the
stimuli are colored accordingly. This chapter focuses on block designs but is
readily adapted to event-related designs.

In computing the activation level in a given voxel, a standard method [68]
is to assume known reference functions (independent variable) corresponding
to the responses of the different stimuli, often square waves based on the experimental sequence (but sometimes sine waves and functions which mimic
the "hemodynamic" response), and then to perform a multiple Regression of
the observed time courses on them, a linear trend, and any other independent
variables. In the multiple Regression, t or F Statistics are computed for the
coefficient associated with the reference function and voxels colored accordingly. But the most important question is: How do we choose the reference
functions? What if they change (possibly nonlinearly) over the course of the
experiment? What if we're interested in observing an "ah ha" moment? An a
priori fixed reference functions are not capable of showing an "ah ha" moment.

The choice of the reference functions in computing the activation of FMRI
has been somewhat arbitrary and subjective. This chapter uses a coherent
Bayesian statistical approach to determine the underlying responses (or reference functions) to the presentation of the stimuli and determine statistically
significant activation. In this approach, all the voxels contribute to "telling
us" the underlying responses due to the experimental stimuli. The model
is presented and applied to a simulated FMRI data in which available prior
information is quantified and dependent contributions (components) to the
observed hemodynamic response are determined (separated).

The utility of the Bayesian Source Separation model for FMRI can be motivated by returning to the classic "cocktail party" problem [27, 28, 52]. At the cocktail party, there are microphones scattered about that record partygoers or speakers at a given number of time increments. The observed conversations consist of mixtures of true unobservable conversations. The objective is to separate these observed signal vectors into true unobservable source signal vectors. As previously mentioned, there may be instances where some sources may be specified to be observable while others are not. This situation in which some sources are observable while others are unobservable is exactly the problem we are addressing in FMRI.

The Bayesian Source Separation model decomposes the observed time course in a voxel into a linear (or polynomial) trend and a linear combination of unobserved component sequences. The linear (or polynomial) trend corresponds to the observable sources and the (unobservable) sources that make up the observed time course corresponds to the unobservable speakers. If the sources were assumed to be known and no priors specified for the remaining parameters, then the Bayesian approach reduces to the standard model and activations determined accordingly. In practice we do not know the true underlying hemodynamic time response (source reference) functions due to the presentation of the experimental stimuli.

The Bayesian Source Separation model assesses a prior distribution for the response functions as well as for the other parameters, and combines them with the data to form a joint posterior distribution. From the posterior distribution, values for the source response functions as well as for the other parameters are computed and statistically significant activation determined. The Bayesian Source Separation model allows the source reference functions to be correlated and can incorporate incidental cognitive processes (blood flow) such as that due to cardiac activity and respiration. Modeling them instead of grouping them into the error term could prove useful.

12.2   Model

In describing the model, sometimes it will be parameterized in terms of rows while other times in terms of columns. Considering the observed time course in voxel j at time t, the model is

$$x_{tj} = \beta_{j0} + \beta_{j1}u_{t1} + \cdots + \beta_{jm}u_{tq} + \lambda_{j1}s_{t1} + \cdots + \lambda_{jm}s_{tm} + \epsilon_{tj}, \quad (12.2.1)$$

in which the observed signal in voxel j at time t, x_{tj} is made up of an overall mean β_{j0} (the intercept); a linear combination of q observed source reference functions $u_{t1} + \cdots + u_{tq}$ (the time trend and other conative processes) which characterize a change in the observed time course over time and includes other observable source reference functions; in addition to a linear combination of the m unobserved source reference functions $s_{t1}, \ldots, s_{tm}$ which characterize the contributions due to the presentation of the stimuli that make up the observed time course; and random error ϵ_{tj}. Coefficients of the observed source reference functions are called Regression coefficients and those of the unobserved source reference functions called mixing coefficients. This model can be written in terms of vectors as

$$x_{tj} = \beta'_j u_t + \lambda'_j s_t + \epsilon_{tj}, \qquad (12.2.2)$$

where a linear trend is specified to be observable so that $u_t = (1, t)'$, $\beta_j = (\beta_{j0}, \beta_{j1})'$, $\lambda_j = (\lambda_{j1}, \ldots, \lambda_{jm})'$, and $s_t = (s_{t1}, \ldots, s_{tm})'$. If any or all of the sources were assumed to be observable, they could be grouped into the u's and their coefficients computed.

Each voxel has its own slope and intercept in addition to a set of mixing coefficients that do not change over time. In contrast, the unobserved underlying source reference functions are the same for all voxels (with possibly zero-valued coefficients) at a given time but do change over time.

Now, considering p voxels at time t, the model can be written as

$$x_t = B u_t + \Lambda s_t + \epsilon_t, \qquad (12.2.3)$$

where x_t is a $p \times 1$ vector of observed values at time t, $B = (\beta_1, \ldots, \beta_p)' = (B_0, \ldots, B_q)$ is the $p \times (q+1)$ matrix of Regression coefficients (slopes and intercepts), and $\Lambda = (\lambda_1, \ldots, \lambda_p)' = (\Lambda_1, \ldots, \Lambda_m)$ is the $p \times m$ dimensional matrix of mixing coefficients.

Alternatively, considering a given voxel j, at all n time points, the model can be written as

$$X_j = U \beta_j + S \lambda_j + E_j, \qquad (12.2.4)$$

where X_j is an $n \times 1$ vector of observed values for voxel j, $U = (e_n, c_n) = (u_1, \ldots, u_n)' = (e_n, U_1)$, e_n is a $n \times 1$ vector of ones, $c_n = (1, \ldots, n)'$, $S = (s_1, \ldots, s_n)' = (S_1, \ldots, S_m)$, E_j is an $n \times 1$ vector of errors, while β_j and λ_j are as previously defined.

The model which considers all of the voxels at all time points can be written in terms of matrices as

$$X = U B' + S \Lambda' + E, \qquad (12.2.5)$$

where $X = (x_1, \ldots, x_n)' = (X_1, \ldots, X_p)$, $E = (\epsilon_1, \ldots, \epsilon_n)' = (E_1, \ldots, E_p)$ while B, Λ, U, and S are as before.

Motivated by the central limit theorem, the errors of observation at each time increment are taken to be Multivariate Normally distributed, as $(\epsilon_t | \Sigma) \sim N(0, \Sigma)$; thus the observations are also Normally distributed as

$$p(X|U,B,S,\Lambda,\Sigma) \propto |\Sigma|^{-\frac{n}{2}} e^{-\frac{1}{2}tr\Sigma^{-1}(X-UB'-S\Lambda')'(X-UB'-S\Lambda')}, \qquad (12.2.6)$$

where Σ is the error covariance matrix and the remaining variables are as previously defined.

By letting $Z = (U,S) = (z_1,\ldots,z_n)' = (Z_0,\ldots,Z_{m+q})$ and $C = (B,\Lambda) = (c_1,\ldots,c_n)' = (C_0,\ldots,C_{m+q})$, the model and likelihood become

$$X = ZC' + E \qquad (12.2.7)$$

and

$$p(X|U,B,S,\Lambda,\Sigma) \propto |\Sigma|^{-\frac{n}{2}} e^{-\frac{1}{2}tr\Sigma^{-1}(X-ZC')'(X-ZC')}. \qquad (12.2.8)$$

12.3   Priors and Posterior

The method of subjectively assigning source reference functions and performing a multiple Regression is equivalent to assigning degenerate prior distributions for them and vague priors for the remaining parameters. That is equivalent to assuming that the probability distribution for the source reference functions is equal to unity at these assigned values and zero otherwise.

Instead of subjectively choosing the source reference functions as being fixed, prior information as to their values in the form of prior distributions are assessed as are priors for any other observable contributing source reference functions to the observed signal. These prior distributions are combined with the data and the source reference functions are determined statistically using information contributed from every voxel. In addition, prior distributions are assessed for the covariance matrix for the (dependent) source reference functions, the Regression coefficients (slopes and intercepts), the mixing coefficients, and the covariance matrix for the observation error. The prior distribution for the source reference functions, the covariance matrix for source reference functions, the Regression coefficients, the mixing coefficients, and the error covariance matrix are taken to be Normal, Inverted Wishart, Normal, Normal, and Inverted Wishart distributed respectively.

The prior distributions for the parameters are the Normal and Inverted Wishart distributions

$$p(S|R) \propto |R|^{-\frac{n}{2}} e^{-\frac{1}{2}trR^{-1}(S-S_0)'(S-S_0)}, \qquad (12.3.1)$$

$$p(R) \propto |R|^{-\frac{\eta}{2}} e^{-\frac{1}{2}trR^{-1}V}, \qquad (12.3.2)$$

$$p(C|\Sigma) \propto |D|^{-\frac{p}{2}} |\Sigma|^{-\frac{(m+q+1)}{2}} e^{-\frac{1}{2}tr\Sigma^{-1}(C-C_0)D^{-1}(C-C_0)'}, \qquad (12.3.3)$$

$$p(\Sigma) \propto |\Sigma|^{-\frac{\nu}{2}} e^{-\frac{1}{2} tr \Sigma^{-1} Q}, \tag{12.3.4}$$

where the prior mean for the source component reference functions is $S_0 = (s_{01}, \ldots, s_{0n})' = (S_{01}, \ldots, S_{0m})$ and R is the covariance matrix of the source reference functions. The hyperparameters S_0, V, η, M, $C_0 = (B_0, \Lambda_0)$, Q, and ν which uniquely define the remaining prior distributions are to be assessed (see Appendix B).

Note that both Σ and R are full covariance matrices allowing the observed mixed signals (the voxels) and also the unobserved source reference functions to be correlated. The Regression and mixing coefficients are also allowed to be correlated by specifying a joint distribution for them and not constraining them to be independent.

Upon using Bayes' rule the posterior distribution for the unknown parameters is written as being proportional to the product of the aforementioned priors and likelihood

$$p(B, S, R, \Lambda, \Sigma | U, X) \propto |\Sigma|^{-\frac{(n+\nu+m+q+1)}{2}} e^{-\frac{1}{2} tr \Sigma^{-1} G}$$
$$\times |R|^{-\frac{(n+\eta)}{2}} e^{-\frac{1}{2} tr R^{-1} [(S-S_0)'(S-S_0)+V]}, \tag{12.3.5}$$

where

$$G = (X - ZC')'(X - ZC') + (C - C_0)D^{-1}(C - C_0)' + Q. \tag{12.3.6}$$

This posterior distribution must now be evaluated in order to obtain parameter estimates of the source reference functions, the covariance matrix for the source reference functions, the matrix of mixing coefficients, the matrix of Regression coefficients, and the error covariance matrix. In addition, statistically significant activation is determined from the posterior distribution.

12.4 Estimation and Inference

As stated in the estimation section of the Unobservable and Observable source Separation model, the above posterior distribution cannot be integrated or differentiated analytically to obtain marginal distributions for marginal estimates or maxima for maximum a posteriori estimates. Marginal and maximum a posteriori estimates can be obtained via Gibbs sampling and iterated conditional modes (ICM) algorithms [13, 14, 36, 40, 52, 57]. These algorithms use the posterior conditional distributions and either generate random variates or cycle through their modes.

The posterior conditional distributions are as in Chapter 11 with Conjugate prior distributions.

From Chapter 11, the estimates

$$\bar{S} = \frac{1}{L}\sum_{l=1}^{L}\bar{S}_{(l)} \qquad \bar{R} = \frac{1}{L}\sum_{l=1}^{L}\bar{R}_{(l)} \qquad \bar{C} = \frac{1}{L}\sum_{l=1}^{L}\bar{C}_{(l)} \qquad \bar{\Sigma} = \frac{1}{L}\sum_{l=1}^{L}\bar{\Sigma}_{(l)}$$

are sampling-based marginal posterior mean estimates of the parameters which converge almost surely to their population values.

Interval estimates can also be obtained by computing marginal covariance matrices from the sample variates. The marginal covariance for the matrix of Regression/mixing coefficients is

$$\overline{var}(c|X,U) = \frac{1}{L}\sum_{l=1}^{L}\bar{c}_{(l)}\bar{c}'_{(l)} - \bar{c}\bar{c}'$$
$$= \bar{\Delta},$$

where $c = vec(C)$, $\bar{c} = vec(\bar{C})$, and $\bar{c}_{(l)} = vec(\bar{C}_{(l)})$.

The covariance matrices of the other parameters follow similarly. With a specification of Normality for the marginal posterior distribution of the mixing coefficients, their distribution is

$$p(c|X,U) \propto |\bar{\Delta}|^{-\frac{1}{2}}e^{-\frac{1}{2}(c-\bar{c})\bar{\Delta}^{-1}(c-\bar{c})'}, \qquad (12.4.1)$$

where $\bar{c}$ and $\bar{\Delta}$ are as previously defined.

The marginal posterior distribution of the mixing coefficients is

$$p(\lambda|X,U) \propto |\bar{\Upsilon}|^{-\frac{1}{2}}e^{-\frac{1}{2}(\lambda-\bar{\lambda})\bar{\Upsilon}^{-1}(\lambda-\bar{\lambda})'}, \qquad (12.4.2)$$

where $\bar{\lambda} = vec(\bar{\Lambda})$ and $\bar{\Upsilon}$ is the lower right $pm \times pm$ (covariance) submatrix in $\bar{\Delta}$.

From Chapter 11, the ICM estimates found by cycling through the modes

$$\tilde{C} = [X'\tilde{Z} + C_0 D^{-1}](D^{-1} + \tilde{Z}'\tilde{Z})^{-1},$$
$$\tilde{\Sigma} = [(X - \tilde{Z}\tilde{C}')'(X - \tilde{Z}\tilde{C}') + (\tilde{C} - C_0)D^{-1}(\tilde{C} - C_0)'$$
$$+ Q]/(n + \nu + m + q + 1),$$
$$\tilde{R} = \frac{(\tilde{S} - S_0)'(\tilde{S} - S_0) + V}{n + \eta},$$
$$\tilde{S} = (X - U\tilde{B}')\tilde{\Sigma}^{-1}\tilde{\Lambda}(\tilde{R}^{-1} + \tilde{\Lambda}'\tilde{\Sigma}^{-1}\tilde{\Lambda})^{-1}$$

are maximum a posteriori estimates via the ICM estimation procedure [36, 40].

Conditional modal intervals may be computed by using the conditional distribution for a particular parameter given the modal values of the others. The posterior conditional distribution of the matrix of Regression/mixing coefficients C given the modal values of the other parameters and the data is

$$p(C|\tilde{C},\tilde{S},\tilde{R},\tilde{\Sigma},U,X) \propto |(D^{-1}+\tilde{Z}'\tilde{Z})|^{\frac{p}{2}}|\tilde{\Sigma}|^{-\frac{m}{2}}$$
$$\times e^{-\frac{1}{2}tr\tilde{\Sigma}^{-1}(C-\tilde{C})(D^{-1}+\tilde{Z}'\tilde{Z})(C-\tilde{C})'}, \quad (12.4.3)$$

which may be also written in terms of vectors as

$$p(c|\tilde{c},\tilde{S},\tilde{R},\tilde{\Sigma},U,X) \propto |(D^{-1}+\tilde{Z}'\tilde{Z})^{-1}\otimes\tilde{\Sigma}|^{-\frac{1}{2}}$$
$$\times e^{-\frac{1}{2}(c-\tilde{c})'[(D^{-1}+\tilde{Z}'\tilde{Z})^{-1}\otimes\tilde{\Sigma}]^{-1}(c-\tilde{c})}. \quad (12.4.4)$$

The marginal posterior conditional distribution of the mixing coefficients $\lambda = vec(\Lambda)$ which is the last mp rows of c written in terms of vectors as

$$p(\lambda|\tilde{\lambda},\tilde{S},\tilde{R},\tilde{\Sigma},U,X) \propto |\tilde{\Gamma}\otimes\tilde{\Sigma}|^{-\frac{1}{2}}e^{-\frac{1}{2}(\lambda-\tilde{\lambda})'(\tilde{\Gamma}\otimes\tilde{\Sigma})^{-1}(\lambda-\tilde{\lambda})} \quad (12.4.5)$$

and in terms of the matrix Λ is

$$p(\Lambda|\tilde{\Lambda},\tilde{S},\tilde{R},\tilde{\Sigma},U,X) \propto |\tilde{\Gamma}|^{-\frac{p}{2}}|\tilde{\Sigma}|^{-\frac{m}{2}}e^{-\frac{1}{2}tr\tilde{\Sigma}^{-1}(\Lambda-\tilde{\Lambda})\tilde{\Gamma}^{-1}(\Lambda-\tilde{\Lambda})'}, \quad (12.4.6)$$

where $\tilde{\Gamma}$ is the lower right $m \times m$ portion of $(D^{-1}+\tilde{Z}'\tilde{Z})^{-1}$.

After determining the test statistics, a threshold or significance level is set and a one to one color mapping is performed. The image of the colored voxels is superimposed onto an anatomical image.

12.5   Simulated FMRI Experiment

For an example, data were generated to mimic a scaled down version of a real FMRI experiment. The simulated experiment was designed to have three stimuli, A, B, and C, but this method is readily adapted to any number of stimuli and to event related designs. Stimuli A, B, and C were 22, 10, and 32 seconds in length respectively, and eight trials were performed for a total of 512 seconds as illustrated in Figure 12.1.

A single trial is focused upon in Figure 12.2 which shows stimulus A lasting 22 seconds, stimulus B lasting 10 seconds, and stimulus C lasting 32 seconds. As in the real FMRI experiment in the next section, observations are

FIGURE 12.1

Experimental design: white task A, gray task B, and black task C in seconds.

taken every 4 seconds so that there are $n = 128$ in each voxel. The simulated functional data are created with true source reference functions. These true reference functions, one trial of which is given in Figure 12.3, both start at -1 when the respective stimulus is first presented, increasing to $+1$ just as the first quarter of a sinusoid with a period of 16 seconds and amplitude of 2. The reference functions are at $+1$ when each of the respective stimuli are removed and decrease to -1 just as the third quarter of a sinusoid with a period of 16 seconds and amplitude of 2. This sinusoidal increase and decrease take 4 seconds and is assumed to simulate hemodynamic responses to the presentation and removal of each of the stimuli.

FIGURE 12.2

One trial: $+1$ when presented, -1 when not presented.

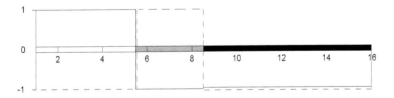

A simulated anatomical 4×4 image is determined as in Figure 12.4. Statistically significant activation is to be superimposed onto the anatomical image.

The voxels are numbered from 1 to 16 starting from the top left, proceeding across and then down. The functional data for these 16 voxels was created according to the Source Separation model

$$
\begin{array}{ccccccc}
x_{tj} & = & \beta'_{Tj} & u_{Tt} & + & \lambda'_{Tj} & s_{Tt} & + & \epsilon_{tj}, \\
(1 \times 1) & & (1 \times 2) & (2 \times 1) & + & (1 \times 3) & (3 \times 1) & & (1 \times 1)
\end{array}
\tag{12.5.1}
$$

where j denotes the voxel, t denotes the time increment, ϵ_{tj} denotes the ran-

FIGURE 12.3

True source reference functions.

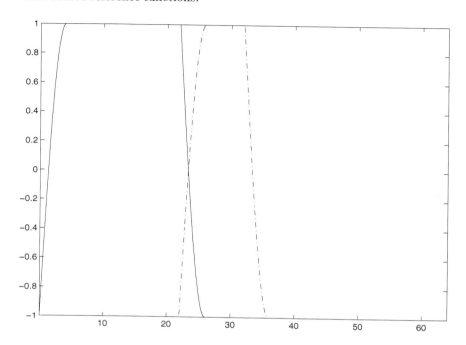

dom error term, and the T subscript denotes that these are the true values. In each voxel, the simulated observed data at each time increment was generated according to the above model with random error added to each of the parameters as follows. The random errors were generated according to $\epsilon_{tj} \sim N(0, 10)$. Noise was added to the sources reference functions and the mixing coefficients according to $s_t \sim N(s_{Tt}, 0.2I_2)$, $\lambda_j \sim N(\lambda_{Tj}, 0.25I_2)$, and $\beta_j \sim N(\beta_{Tj}, 0.1I_2)$ where I_2 is the 2×2 identity matrix. The true sources were sampled every 4 seconds from those in Figure 12.3.

TABLE 12.1

True Regression and mixing coefficients.

B_T	1	2	3	4	Λ_T	1	2	3	4
1	.2, .5	.7, .1	.4, .9	.3, .2	1	15, 5	2, 1	1, 15	2, 15
2	.9, .6	.4, .8	.5, .3	.2, .7	2	1, 2	15, 1	-2, 1	2, 15
3	.9, .1	.1, .3	.5, .5	.1, .6	3	1, 15	-1, 2	15, 1	2, 2
4	.6, .4	.4, .2	.4, .5	.8, .9	4	1, 15	2, 15	1, 2	15, 1

The true slopes and intercepts for the voxels along with the true source amplitudes (mixing coefficients) are displayed in their voxels location as in

FIGURE 12.4
Simulated anatomical image.

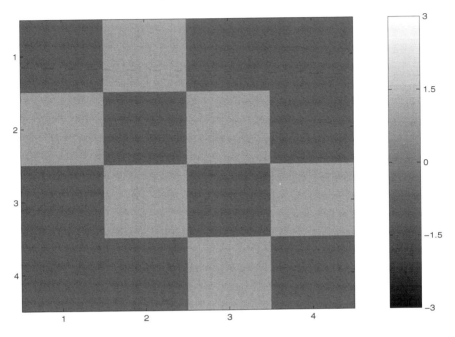

Table 12.1.

All hyperparameters were assessed according to the methods in Appendix B and an empirical Bayes, approach was taken that uses the current data as the prior data. For presentation purposes, all values have been rounded to either one or two digits.

For the prior means of the source reference functions, square functions are assessed with unit amplitude. Note that observations are taken at 20 and 24 seconds while the point where stimulus A ends and stimulus B begins is at 22 seconds. The prior means for the source reference function associated with stimulus A is at $+1$ until 20 seconds (observation 5), 0 at 24 seconds (observation 6) and then at -1 thereafter; and the one associated with stimulus B is at -1 until 20 seconds (observation 5), 0 at 24 seconds (observation 6), and then $+1$ thereafter. Both are at -1 between 32 seconds and 64 seconds. The prior, Gibbs, and ICM sources are displayed in Figure 12.5.

The assessed prior values for D were as in Table 12.2 while Q was as in Table 12.6 and $\nu = n$. The assessed values for η and v_0 were $\eta = 6$ and $v_0 = 0.13$.

The Regression and mixing coefficients are as in Tables 12.3 and 12.4. The top set of values are the prior means, the middle set are the Gibbs sampling estimates, and the bottom set are the ICM estimates.

FIGURE 12.5

Prior —, Gibbs estimated −−, and ICM estimated ·· reference functions for

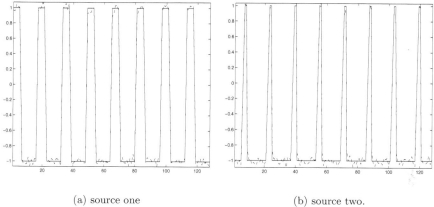

(a) source one   (b) source two.

TABLE 12.2

Covariance hyperparameter for coefficients.

D	1	2	3	4
1	0.0405	-0.0004	0.0036	0.0124
2		0.0000	0.0000	0.0000
3			0.0100	0.0033
4				0.0179

TABLE 12.3

Prior, Gibbs, and ICM Regression coefficients.

B_0	1	2	3	4
1	-3.95, 0.50	0.95, 0.04	1.81, 0.83	-2.20, 0.23
2	3.56, 0.58	-4.09, 0.83	3.99, 0.26	1.90, 0.68
3	2.50, 0.08	2.00, 0.26	-6.38, 0.55	1.81, 0.58
4	2.31, 0.39	3.59, 0.17	-3.68, 0.57	-3.46, 0.89

$\bar{B}$	1	2	3	4
1	-3.93, 0.50	0.95, 0.04	1.81, 0.83	-2.19, 0.22
2	3.58, 0.57	-4.08, 0.83	3.99, 0.26	1.90, 0.68
3	2.52, 0.08	2.00, 0.26	-6.40, 0.55	1.82, 0.58
4	2.31, 0.39	3.60, 0.17	-3.68, 0.57	-3.45, 0.89

$\tilde{B}$	1	2	3	4
1	-3.95, 0.50	0.95, 0.04	1.82, 0.83	-2.19, 0.22
2	3.57, 0.57	-4.08, 0.83	3.98, 0.26	1.91, 0.68
3	2.51, 0.08	2.00, 0.26	-6.38, 0.55	1.81, 0.58
4	2.32, 0.38	3.60, 0.17	-3.67, 0.57	-3.46, 0.89

TABLE 12.4
Prior, Gibbs, and ICM mixing coefficients.

Λ_0	1	2	3	4
1	10.32, -1.63	0.48, 1.28	-2.28, 12.28	0.40, 12.61
2	2.65, 3.35	12.69, -0.07	-1.10, 2.09	0.52, 12.75
3	-1.45, 13.50	-1.47, 1.24	11.53, -1.84	1.82, 2.69
4	-0.83, 14.27	1.58, 15.18	1.39, 0.71	12.13, 0.37

$\bar{\Lambda}$	1	2	3	4
1	10.35, -1.62	0.48, 1.29	-2.28, 12.30	0.39, 12.60
2	2.66, 3.36	12.71, -0.07	-1.10, 2.10	0.54, 12.78
3	-1.45, 13.53	-1.47, 1.24	11.55, -1.85	1.82, 2.70
4	-0.83, 14.29	1.59, 15.21	1.39, 0.70	12.15, 0.397

$\tilde{\Lambda}$	1	2	3	4
1	10.32, -1.63	0.48, 1.29	-2.28, 12.29	0.40, 12.62
2	2.65, 3.35	12.70, -0.07	-1.10, 2.09	0.53, 12.76
3	-1.45, 13.51	-1.47, 1.24	11.54, -1.84	1.82, 2.69
4	-0.83, 14.28	1.58, 15.52	1.39, 0.71	12.14, 0.38

The (left) prior mode along with the Bayesian (center) Gibbs sampling marginal and Bayesian (right) ICM maximum a posteriori source covariance matrices for the source reference functions are displayed in Table 12.5.

TABLE 12.5
Prior, Gibbs, and ICM source covariances.

V/η	1	2	$\bar{R}$	1	2	$\tilde{R}$	1	2
1	0.02	0	1	0.04	0.00	1	0.35	0.05
2		0.02	2		0.03	2		0.30

The prior mode along with the Bayesian Gibbs sampling marginal and Bayesian ICM maximum a posteriori error covariance matrices are displayed in Tables 12.6-12.8.

TABLE 12.6
Error covariance prior mode.

Q/ν	1	2	3	4	5	6	7	8	9	10	11	12	13	14	15	16
1	159.3	25.2	12.7	29.0	3.1	35.8	-11.2	19.0	15.6	10.2	45.7	66.1	13.8	6.1	-0.7	26.6
2		148.8	-.1	13.7	15.8	5.3	-7.3	8.9	3.8	-9.7	10.4	3.4	-10.9	7.8	-9.3	25.9
3			193.4	44.3	-18.3	29.1	12.2	23.9	18.9	7.4	29.8	32.9	21.2	18.9	-30.6	-9.8
4				174.7	-.7	3.5	-6.4	31.6	30.2	15.5	35.3	19.8	23.6	9.7	8.4	2.4
5					135.7	-23.8	1.9	8.3	23.2	-3.8	-2.2	-3.0	2.5	9.7	29.2	16.0
6						186.3	19.7	12.9	5.7	1.2	60.9	17.4	-30.0	.0	-22.6	19.4
7							105.7	-1.9	5.4	15.6	5.5	-11.3	.5	2.7	-9.9	-13.4
8								126.9	14.8	7.8	11.4	37.3	32.9	4.6	17.7	20.6
9									192.4	17.7	0.5	26.8	45.9	37.7	-15.0	-3.0
10										145.4	20.7	6.5	3.4	23.2	32.7	-2.1
11											188.3	16.8	1.7	14.0	11.9	7.3
12												170.8	14.3	-10.6	-3.1	-8.6
13													193.5	45.4	37.2	-4.7
14														208.2	7.2	14.1
15															145.9	5.1
16																191.2

TABLE 12.7
Gibbs estimates of covariances.

$\bar{\Psi}$	1	2	3	4	5	6	7	8	9	10	11	12	13	14	15	16
1	158.6	25.3	13.3	29.4	2.8	34.4	-11.1	19.1	16.2	10.4	44.3	66.1	14.1	6.12	-0.9	25.2
2		149.3	-0.3	13.6	15.9	5.2	-7.3	8.72	3.7	-9.7	10.4	3.4	-11.1	7.6	-9.3	26.0
3			192.5	42.7	-18.8	29.6	12.1	22.3	17.1	7.23	30.4	32.7	19.3	16.9	-30.9	-9.6
4				173.5	-1.1	3.5	-6.7	29.9	28.5	15.4	35.6	19.6	21.6	7.6	8.34	2.3
5					136.0	-24.3	1.9	7.8	22.8	-3.8	-2.5	-3.2	2.0	9.1	29.3	15.78
6						185.0	20.0	12.8	6.0	1.3	59.3	17.2	-30.0	-0.3	-23.0	17.6
7							106.0	-2.1	5.0	15.6	5.7	-11.3	0.3	2.4	-9.9	-13.2
8								125.5	13.1	7.6	11.4	37.0	31.0	2.4	17.6	20.4
9									191.2	17.5	1.0	26.6	44.0	35.5	-15.2	-2.8
10										145.8	20.9	6.5	3.34	23.1	32.9	-1.9
11											187.3	16.6	2.0	14.2	11.7	5.7
12												171.3	14.0	-11.2	-3.1	-9.0
13													191.9	43.1	37.24	-4.7
14														206.4	7.0	13.7
15															146.5	4.8
16																190.2

TABLE 12.8
ICM estimates of covariances.

$\tilde{\Psi}$	1	2	3	4	5	6	7	8	9	10	11	12	13	14	15	16
1	157.4	24.9	12.6	28.7	3.0	35.3	-11.1	18.8	15.5	10.1	45.0	65.3	13.6	6.1	-0.7	26.1
2		147.0	-0.2	13.6	15.6	5.2	-7.2	8.8	3.7	-9.6	10.3	3.4	-10.8	7.7	-9.2	25.6
3			191.0	43.6	-18.1	28.8	12.1	23.5	18.5	7.3	29.5	32.5	20.8	18.5	-30.3	-9.7
4				172.5	-0.7	3.5	-6.4	31.1	29.7	15.3	34.9	19.5	23.1	9.4	8.3	2.3
5					134.2	-23.5	1.9	8.1	22.9	-3.8	-2.2	-3.0	2.4	9.6	28.9	15.8
6						184.0	19.5	12.8	5.6	1.2	60.1	17.2	-29.6	-0.0	-22.4	19.1
7							104.5	-1.9	5.3	15.4	5.4	-11.1	0.5	2.6	-9.8	-13.3
8								125.3	14.4	7.7	11.3	36.8	32.3	4.3	17.5	20.3
9									190.0	17.5	0.6	26.4	45.2	37.1	-14.9	-3.0
10										143.7	20.4	6.4	3.4	22.9	32.3	-2.0
11											186.0	16.6	1.7	13.9	11.7	7.1
12												168.8	14.1	-10.5	-3.0	-8.6
13													191.0	44.7	36.8	-4.6
14														205.6	7.1	13.9
15															144.2	5.0
16																188.8

Activation as shown in Table 12.9 was determined using (top) Regression with the prior source reference functions in addition to using (middle) Gibbs sampling and (bottom) ICM estimation methods.

TABLE 12.9
Prior, Gibbs, and ICM coefficient statistics.

t_{Reg}	1	2	3	4
1	8.04, -0.95	0.38, 0.77	-1.61, 6.49	0.30, 7.02
2	2.24, 2.11	9.14, -0.04	-1.06, 1.50	0.46, 8.33
3	-1.03, 7.16	-1.20, 0.76	8.26, -0.99	1.37, 1.51
4	-0.59, 7.54	1.08, 7.74	1.13, 0.43	8.63, 0.20
z_{Gibbs}	1	2	3	4
1	11.46, -1.36	0.55, 1.12	-2.34, 9.26	0.43, 10.08
2	3.23, 3.04	13.11, -0.05	-1.52, 2.16	0.67, 11.94
3	-1.47, 10.35	-1.73, 1.09	11.89, -1.42	1.97, 2.20
4	-0.84, 10.78	1.56, 11.15	1.62, 0.61	12.34, 0.29
z_{ICM}	1	2	3	4
1	11.64, -1.37	0.55, 1.12	-2.33, 9.40	0.43, 10.16
2	3.23, 3.06	13.22, -0.05	-1.53, 2.16	0.67, 12.06
3	-1.48, 10.36	-1.74, 1.10	11.95, -1.42	1.98, 2.190
4	-0.85, 10.92	1.56, 11.20	1.63, 0.62	12.48, 0.29

The Regression activations follow Scalar Student t-distributions with $n - m - q - 1 = 124$ degrees of freedom which is negligibly different than the corresponding Normal distributions. As by design, positive activations for source reference function 1 are along the diagonal from upper left to lower right and those for source reference function 2 are on the upper right and lower left. A threshold was set at 5 and illustrated in Figure 12.6.

The same threshold is used for the Gibbs sampling and ICM activations. For Gibbs sampling, all diagonal activations in Figure 12.7 are present and more pronounced than those by the standard Regression method. The activation along the diagonal of the image increased by an average of 3.62 from the standard Regression method while those for the corners increased by an average of 3.13. For the ICM activations in Figure 12.8, all activations are present and more pronounced than those by the standard Regression method. The activation along the diagonal of the image increased by an average of 5.08 from the standard Regression method while those for the corners increased by an average of 4.03.

These functional activations are to be superimposed onto the previously shown anatomical image. In this example, the Bayesian statistical Source Separation model outperformed the common method of multiple Regression for both estimation methods and ICM was the best.

FIGURE 12.6
Activations thresholded at 5 for prior reference functions.

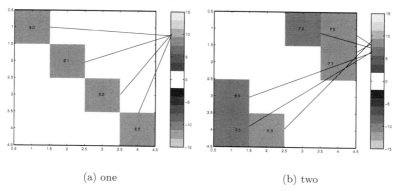

(a) one   (b) two

FIGURE 12.7
Activations thresholded at 5 for Bayesian Gibbs reference functions.

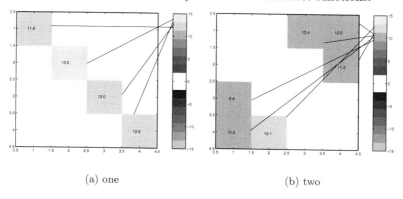

(a) one   (b) two

FIGURE 12.8
Activations thresholded at 5 Bayesian ICM reference functions.

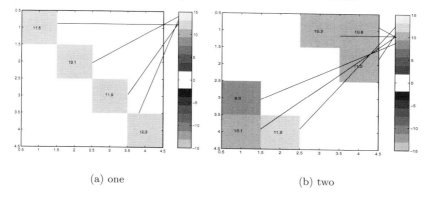

(a) one   (b) two

It can be seen that the Bayesian method of determining the reference function for computation of voxel activation performed well especially with only sixteen voxels.

12.6   Real FMRI Experiment

Due to the fact that the number of voxels is large and that the ICM and Gibbs sampling procedures requires the inversion of the voxels large covariance matrix and additionally the Gibbs sampling procedure requires a matrix factorization, spatial independence is assumed for computational simplicity. The Gibbs sampling procedure is also very computationally intensive and since the ICM procedure produced nearly identical simulation results, only the ICM procedure is implemented. The software package AFNI [7] is used to display the results.

The current FMRI data [59] provides the motivation for using Bayesian Source Separation to determine the true underling unobserved source reference functions. These reference functions are the underlying responses due to the presentation of the experimental stimuli. The data were collected from an experiment in which a subject was given eight trials of stimuli A, B, and C. The timing and trials were exactly the same as in the simulated example. Experimental task A was an implementation of a common experimental economic method for determining participants' valuation of objects in the setting of an auction [4]. The participant was given an item and told that the item can be kept or sold. If the item is kept, the participant retains ownership at the end of the experiment and is paid the items stated value.

Task A consisted of the participant reading text from a screen, determining a number, and entering the number using button response unit all in 22 seconds. Task B consisted of the subject receiving feedback displayed on a screen for 10 seconds. Task C was a control stimulus which consisted of a blank screen for 32 seconds.

For the functional data, 24 axial slices of size 64×64 were taken. Each voxel has dimensions of $3 \times 3 \times 5$ mm. Scanning was performed using a 1.5 Tesla Siemens Magneton with $TE = 40$ ms. Observations were taken every 4 seconds so that there are 128 in each voxel. All hyperparameters were assessed according to the Regression technique in Appendix A with an empirical Bayes' approach. For the prior mean, a square function was assessed with unit amplitude as discussed in the simulation example which mimics the experiment. Due to space limitations, the prior and posterior parameter values for the 98,304 voxel's have been omitted.

In Figure 12.9 are the prior square and ICM Bayesian source reference functions corresponding to the (a) first and (b) second source reference functions.

FIGURE 12.9

Prior $-\cdot$ and Bayesian ICM — reference functions.

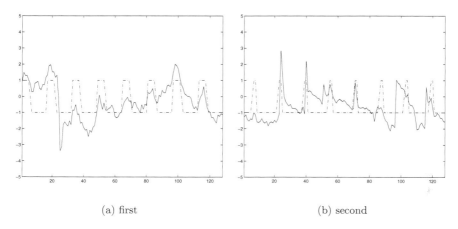

(a) first   (b) second

The activations corresponding to the first prior square reference function was computed and displayed in Figure 12.10 (a). It is evident that the activation in Figure 12.10 (a) is buried in the noise. The threshold is set at 1.885 and if raised, the activation begins to disappear while noise remains.

The activations corresponding to the first Bayesian ICM reference function was computed and displayed in Figure 12.10 (b). It is evident that the activation in Figure 12.10 (b) is larger and is no longer buried in the noise. The activations stand out above the noise. The threshold is set at 12.63 and if raised, the activation begins to disappear.

The activations corresponding to the second prior square reference function was computed and displayed in Figure 12.11(a). It is evident that the activation in Figure 12.11 (a) is buried in the noise. The threshold is set at 3.205 and if raised, the activation begins to disappear while noise remains.

The activations corresponding to the second ICM Bayesian square reference function was computed and displayed in Figure 12.11 (b). It is evident that the activation in Figure 12.11 (b) is much larger and is no longer buried in the noise. Further, the activation is more localized. The activations stand out above the noise. The threshold is set at 20.58 and if raised, the activation begins to disappear.

The activations that were computed using the underlying reference functions from Bayesian Source Separation were much larger and more distinct than those using the prior square reference functions.

FIGURE 12.10
Activations for first reference functions.

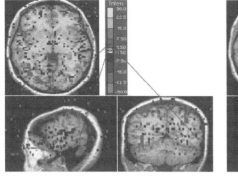

(a) prior thresholded at 1.885   (b) Bayesian thresholded at 12.63

FIGURE 12.11
Activations for second reference functions.

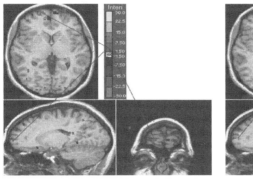

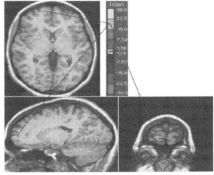

(a) prior thresholded at 3.205   (b) Bayesian thresholded at 20.58

12.7   FMRI Conclusion

In computing the activations in FMRI, the choice of the source reference function is subjective. It has been shown that the reference function need not be assigned but may be determined statistically using Bayesian methods. A dynamic (nonstatic or fixed) source reference function can be determined for each FMRI participant. It was further found in the simulation example, that when computing activations, the iterated conditional modes and Gibbs sampling algorithms performed similarly.

Part III

Generalizations

13

Delayed Sources and Dynamic Coefficients

13.1   Introduction

In Part II, the mixing of the sources was specified to be instantaneous and the Regression/mixing coefficients were specified to be constant over time. In this Chapter, the sources are allowed to be delayed through the Regression/mixing coefficients and the Regression/mixing coefficients are allowed to change over time. Delayed sources and nonconstant or dynamic mixing coefficients take into account the facts that speakers at a cocktail party are a physical distance from the microphones, thus the sound from them taking time to travel to the various microphones at different distances, and that the speakers at the party may be moving around.

13.2   Model

The observation and source vectors x_i and s_i are stacked into single vectors x and s which are $np \times 1$ and $nm \times 1$ respectively. The Source Separation model is written

$$(x|\mathcal{B}, u, \mathcal{L}, s) = \underset{(np \times 1)}{\mathcal{B}} \underset{[np \times n(q+1)]}{} \underset{[n(q+1) \times 1]}{u} + \underset{(np \times nm)}{\mathcal{L}} \underset{(nm \times 1)}{s} + \underset{(np \times 1)}{\epsilon},$$
$$(13.2.1)$$

where the observation vector x, the observed source vector u, the unobserved source vector s, and the error vector ϵ given by

$$x = \begin{pmatrix} x_1 \\ \vdots \\ x_n \end{pmatrix}, \quad u = \begin{pmatrix} u_1 \\ \vdots \\ u_n \end{pmatrix}, \quad s = \begin{pmatrix} s_1 \\ \vdots \\ s_n \end{pmatrix}, \quad \epsilon = \begin{pmatrix} \epsilon_1 \\ \vdots \\ \epsilon_n \end{pmatrix}, \qquad (13.2.2)$$

have been defined. The matrices $\mathcal{B}$ and $\mathcal{L}$ are generalized Regression and mixing coefficients. The vectors x, u, s, and ϵ contain each of the observation, observed source, unobserved source, and error vectors stacked in time order into single vectors.

Taking a closer look at the general mixing matrix $\mathcal{L}$ (similarly for $\mathcal{B}$), it is evident that it can be partitioned into $p \times m$ blocks and has the lower triangular form

$$\mathcal{L} = \begin{pmatrix} \mathcal{L}_{11} & 0 & \cdots & 0 \\ \mathcal{L}_{21} & \mathcal{L}_{22} & & \\ & & \ddots & \vdots \\ \vdots & & & 0 \\ \mathcal{L}_{n1} & \cdots & & \mathcal{L}_{nn} \end{pmatrix}, \tag{13.2.3}$$

where the blocks above the diagonal are $p \times m$ zero matrices and each $\mathcal{L}_{ii'}$ are $p \times m$ mixing matrices. Only blocks below the diagonal are nonzero because only current and past sources that are delayed can enter into the mixing and not future sources. Upon multiplying the observed source vector u by the generalized Regression coefficient matrix $\mathcal{B}$ and the observed source vector s by the generalized mixing coefficient matrix $\mathcal{L}$ or upon mixing the observed and unobserved sources, the observed mixed signal vectors are

$$x_i = \sum_{i'=1}^{i} (\mathcal{B}_{ii'} u_{i'} + \mathcal{L}_{ii'} s_{i'}) + \epsilon_i. \tag{13.2.4}$$

The matrix $\mathcal{L}_{ii}$ is the instantaneous mixing matrix at time i and the matrix $\mathcal{L}_{i,(i-d)}$ is the mixing matrix at time i for signals delayed d time units. The same is true for the generalized Regression coefficient matrix $\mathcal{B}$ for the observed sources.

For example, assume that there are $m = 2$ unobserved speakers' and $p = 1$ microphones. Let D_s be a matrix of time delays for the unobservable sources where d_{jk} is element (j, k) indicating the delay of unobserved source k to microphone j. The general mixing matrix $\mathcal{L}$ for the case where the delays are known to be described by $D_s = (0, 2)$ and the mixing process for the unobserved sources which is allowed to vary over time is

$$\mathcal{L} = \begin{pmatrix} \mathcal{L}_{11} & 0 & 0 & \cdots & & 0 \\ 0 & \mathcal{L}_{22} & 0 & & & \\ \mathcal{L}_{31} & 0 & \mathcal{L}_{33} & 0 & & \vdots \\ 0 & \mathcal{L}_{42} & 0 & \mathcal{L}_{44} & 0 & \vdots \\ 0 & 0 & \mathcal{L}_{53} & 0 & \mathcal{L}_{55} & 0 \\ \vdots & & & & & \ddots \end{pmatrix}, \tag{13.2.5}$$

where at time increment i, $\mathcal{L}_{ii} = (l_{ii}, 0)$ indicating that the signal from speaker 1 is instantaneously mixed and the signal from speaker 2 has not yet reached the microphone (thus not mixed), $\mathcal{L}_{i,(i-1)} = (0, 0)$ indicating that neither of the speakers signals are delayed 1 time increment, and $\mathcal{L}_{i,(i-2)} = (0, l_{i,(i-2)})$

indicating that the signal from speaker 2 takes 2 time units to enter the mixing process, $\mathcal{L}_{ii'} = 0$ for $(i - i') > 2$. The same procedure for delayed mixing of observed sources is true for the generalized Regression coefficient matrix $\mathcal{B}$.

The above very general delayed nonconstant mixing model can be divided into three general cases. The first is a delayed constant mixing process, the second is a delayed nonconstant mixing process, and the third is an instantaneous nonconstant mixing process.

13.3 Delayed Constant Mixing

It is certainly reasonable that in many instances a delayed nonconstant mixing process can be well approximated by a delayed constant one over "short" periods of time. The mixing matrix $\mathcal{L}$ (similarly for $\mathcal{B}$) for the delayed constant mixing process is given by the matrix with n row and column blocks

$$\mathcal{L} = \begin{pmatrix} \mathcal{L}_0 & 0 & 0 & \cdots & & 0 \\ \mathcal{L}_1 & \mathcal{L}_0 & 0 & & & \\ \mathcal{L}_2 & \mathcal{L}_1 & \mathcal{L}_0 & 0 & & \vdots \\ \mathcal{L}_3 & \mathcal{L}_2 & \mathcal{L}_1 & \mathcal{L}_0 & 0 & \vdots \\ \mathcal{L}_4 & \mathcal{L}_3 & \mathcal{L}_2 & \mathcal{L}_1 & \mathcal{L}_0 & 0 \\ \vdots & & & & & \ddots \end{pmatrix}, \tag{13.3.1}$$

where $\mathcal{L}_d$, $d = 0, 1, 2, ..., n-1$ describes the delayed mixing process at d time increments. The observed mixed vectors x_i at time i are represented as linear combinations of the observed and unobserved source vectors $u_{i'}$ and $s_{i'}$ multiplied by the appropriate Regression and mixing coefficient matrices $\mathcal{B}_{i'-1}$ and $\mathcal{L}_{i'-1}$ (where i' ranges from 1 to i) plus a random error vector ϵ_i which is given by

$$x_i = \sum_{i'=1}^{i} (\mathcal{B}_{i'-1}\, u_{i'} + \mathcal{L}_{i'-1}\, s_{i'}) + \epsilon_i. \tag{13.3.2}$$

In the above example with $(m = 2)$ two unobserved speakers and $(p = 1)$ one microphone, the generalized mixing matrix $\mathcal{L}$ for the unobservabe sources with a delayed constant mixing process is given by

$$
\mathcal{L} = \begin{pmatrix}
\mathcal{L}_0 & 0 & 0 & \cdots & & 0 \\
\mathcal{L}_1 & \mathcal{L}_0 & 0 & & & \\
\mathcal{L}_2 & \mathcal{L}_1 & \mathcal{L}_0 & 0 & & \vdots \\
0 & \mathcal{L}_2 & \mathcal{L}_1 & \mathcal{L}_0 & 0 & \vdots \\
0 & 0 & \mathcal{L}_2 & \mathcal{L}_1 & \mathcal{L}_0 & 0 \\
\vdots & & & & & \ddots
\end{pmatrix}.
\tag{13.3.3}
$$

Note that at any given time increment i, a given source s_{ik} enters into the mixing through only one mixing coefficient l_k in $\mathcal{L}_k$, $k = 1, \ldots, m$. For the above example with $p = 1$ microphone and $m = 2$ speakers, the general mixing matrix $\mathcal{L}$ has the form

$$
\mathcal{L} = \begin{pmatrix}
(l_1,0) & 0 & 0 & \cdots & & 0 \\
(0,0) & (l_1,0) & 0 & & & \\
(0,l_2) & (0,0) & (l_1,0) & 0 & & \vdots \\
0 & (0,l_2) & (0,0) & (l_1,0) & 0 & \vdots \\
0 & 0 & (0,l_2) & (0,0) & (l_1,0) & 0 \\
\vdots & & & & & \ddots
\end{pmatrix},
\tag{13.3.4}
$$

(similarly for $\mathcal{B}$) and the contribution of the unobserved source signals to the observed mixed signals is

$$
\mathcal{L}s = \begin{pmatrix}
(l_1,l_2)(s_{1,1}, s_{1-2,2})' \\
(l_1,l_2)(s_{2,1}, s_{2-2,2})' \\
(l_1,l_2)(s_{3,1}, s_{3-2,2})' \\
\vdots \\
(l_1,l_2)(s_{i,1}, s_{i-2,2})' \\
\vdots \\
(l_1,l_2)(s_{n,1}, s_{n-2,2})'
\end{pmatrix},
\tag{13.3.5}
$$

where $s_{1-2,2}$ and $s_{2-2,2}$ are zero. With general delays, the elements are

$$
(l_1,l_2)'(s_{i-d_{11},1}, s_{i-d_{12},2})',
\tag{13.3.6}
$$

where d_{jk} is the delay of source k to microphone j.

13.4 Delayed Nonconstant Mixing

The mixing matrix $\mathcal{L}$ for the delayed nonconstant mixing process can be divided a part that contains the mixing coefficients $I_n \otimes \Lambda$ and a part that contains the delays P_s as $\mathcal{L} = (I_n \otimes \Lambda)P_s$. The part that contains the delays P_s is a permutation-like matrix. For the example, the permutation-like matrix P_s

$$P_s = \begin{pmatrix} \begin{cases} 1\,0\,0\,0\,0\,0\,0\,0\,0\,0 \cdot 0 \\ 0\,0\,0\,0\,0\,0\,0\,0\,0\,0 \cdot 0 \\ 0\,0\,1\,0\,0\,0\,0\,0\,0\,0 \cdot 0 \\ 0\,0\,0\,0\,0\,0\,0\,0\,0\,0 \cdot 0 \\ 0\,0\,0\,0\,1\,0\,0\,0\,0\,0 \cdot 0 \\ 0\,1\,0\,0\,0\,0\,0\,0\,0\,0 \cdot 0 \\ 0\,0\,0\,0\,0\,0\,1\,0\,0\,0 \cdot 0 \\ 0\,0\,0\,1\,0\,0\,0\,0\,0\,0 \cdot 0 \\ \cdots\cdots\cdots\cdots \end{cases} \end{pmatrix}, \tag{13.4.1}$$

where for the ith set of m rows, the rows are shifted (from the identity matrix) kd_k columns to the left, where k is the row number, m is the number of sources, and d_k is the delay for the kth source, so that

$$P_s s = \begin{pmatrix} s_{11} \\ 0 \\ s_{21} \\ 0 \\ s_{31} \\ s_{32} \\ s_{41} \\ s_{42} \\ \vdots \end{pmatrix}. \tag{13.4.2}$$

This can easily be generalized to p microphones and m speakers.

With the generalized mixing coefficient matrix $\mathcal{L}$ written as $(I_n \otimes \Lambda)P_s$ and the generalized Regression coefficient matrix $\mathcal{B}$ written as $(I_n \otimes B)P_u$, the linear synthesis model becomes

$$\begin{aligned} (x|B,u,\Lambda,s) = & \quad (I_n \otimes B) & P_u u & \quad + \;(I_n \otimes \Lambda) & P_s s & \quad + & \epsilon, \\ (np \times 1) & \quad [np \times n(q+1)]\;[n(q+1) \times 1] & (np \times nm)\;(nm \times 1) & (np \times 1) \end{aligned}$$
$$\tag{13.4.3}$$

where P_u and P_s are $[n(q+1) \times n(q+1)]$ and $(nm \times nm)$ matrices respectively.

If the delays were known or specified to be well explained by a given constant delayed process, then the delayed constant mixing process is the instantaneous constant one where $P_s s$ is generically replaced by s, $P_u u$ by u, and written as

$$
\begin{array}{cccccccc}
(x|B,u,\Lambda,s) = & (I_n \otimes B) & u & + & (I_n \otimes \Lambda) & s & + & \epsilon. \\
(np \times 1) & [np \times n(q+1)] & [n(q+1) \times 1] & & (np \times nm) & (nm \times 1) & & (np \times 1)
\end{array}
\tag{13.4.4}
$$

Then the model can be written in the matrix formulation

$$
\begin{array}{ccccccc}
(X|U,B,\Lambda,S) = & U & B' & + & S & \Lambda' & + & E, \\
(n \times p) & [n \times (q+1)] & [(q+1) \times p] & & (n \times m) & (m \times p) & & (n \times p)
\end{array}
\tag{13.4.5}
$$

and also the combined formulation

$$
\begin{array}{cccccc}
(X|C,Z) = & Z & C' & + & E, \\
(n \times p) & n \times (m+q+1) & (m+q+1) \times p & & (n \times p)
\end{array}
\tag{13.4.6}
$$

which is the previously model described from Part II.

The matrix of sources for the previous example with two sources and one microphone is

$$
S = \begin{pmatrix}
s_{11} & 0 \\
s_{21} & 0 \\
s_{31} & s_{32} \\
s_{41} & s_{42} \\
\vdots & \vdots
\end{pmatrix}.
\tag{13.4.7}
$$

For this model, the parameters are estimated as before; then the columns of U and S are shifted down by the appropriate amounts. If the delays were unknown, an instantaneous constant mixing process could be assumed. Then the estimates of the sources s would actually be $P_s s$ and similarly for the mixing process for $P_u u$, observed sources u. The mixing process is estimated correctly except for shifts in the columns of U and S.

A model could constrain the appropriate number of rows in a column to be zero and shift the columns, but this is not necessary with the current robust model. Shifts are not necessary because we incorporate our prior beliefs regarding the values of the sources and the shifts by assessing a prior mean for the sources S_0 with the appropriate number of zeros in each column representing our prior belief regarding the sources delays.

It has been shown that the delayed constant mixing process can be transformed into the instantaneous constant one. The likelihood, prior distributions, joint posterior distribution, and estimation methods are as in Chapter 11.

13.5 Instantaneous Nonconstant Mixing

A delayed nonconstant mixing process can sometimes be approximated by an instantaneous constant one due to the length of time it takes for a given source component (signal from a speaker) to be observed (recorded by the microphone). Let r be the distance from a source (speaker) to an observing device (microphone), v the speed that the source signal travels in the medium (air for instance), and t be the number of time increments per second (sampling rate). Then, the distance a signal can be from the recorder before there is a delay of 1 time increment is v/t. For example, if the microphone sampled at 100 times per second with the speed of sound $v = 343$ m/s, then the speaker can be 3.43 m or 11.25 ft from the source before there is a delay of 1 time increment.

The generalized mixing matrix $\mathcal{L}$ (similarly for the generalized Regression coefficient matrix $\mathcal{B}$) for the instantaneous nonconstant mixing process is

$$
\mathcal{L} =
\begin{pmatrix}
\mathcal{L}_1 & 0 & 0 & \cdots & & 0 \\
0 & \mathcal{L}_2 & 0 & & & \\
0 & 0 & \mathcal{L}_3 & 0 & & \vdots \\
0 & 0 & 0 & \mathcal{L}_4 & 0 & \vdots \\
0 & 0 & 0 & 0 & \mathcal{L}_5 & 0 \\
\vdots & & & & & \ddots
\end{pmatrix},
\tag{13.5.1}
$$

which yields observed mixed signals x_i at time increment i of the form

$$
x_i = \mathcal{B}_i u_i + \mathcal{L}_i s_i + \epsilon_i
\tag{13.5.2}
$$
$$
x_i = \mathcal{C}_i z_i + \epsilon_i,
\tag{13.5.3}
$$

where $\mathcal{C}_i = (\mathcal{B}_i, \mathcal{L}_i)$, $z_i = (u_i', s_i')'$. For the usual instantaneous constant mixing process of the unobserved and observed Source Separation model, described in Part II, $B = \mathcal{B}_i$ and $\Lambda = \mathcal{L}_i$ for all i.

The general Source Separation model describing the instantaneous nonconstant mixing process for all observations at all time increments is

$$
\begin{array}{ccccc}
(x|c,z) & = & \mathcal{C} & z & + & \epsilon, \\
(np \times 1) & & [np \times n(m+q+1)] & [n(m+q+1) \times 1] & & (np \times 1)
\end{array}
\tag{13.5.4}
$$

where the general Regression/mixing coefficient matrix $\mathcal{C}$ is given by

$$
\mathcal{C} =
\begin{pmatrix}
\mathcal{C}_1 & 0 & 0 & \cdots & & 0 \\
0 & \mathcal{C}_2 & 0 & & & \\
0 & 0 & \mathcal{C}_3 & 0 & & \vdots \\
0 & 0 & 0 & \mathcal{C}_4 & 0 & \vdots \\
0 & 0 & 0 & 0 & \mathcal{C}_5 & 0 \\
\vdots & & & & & \ddots
\end{pmatrix}, \tag{13.5.5}
$$

$\mathcal{C}_i = (\mathcal{B}_i, \mathcal{L}_i)$, $z = (z_1, \ldots, z_n)$, and the remaining variables are as previously defined in this Chapter.

13.6 Likelihood

As in Part II, the observation vectors errors ϵ_i are specified to be independent and Multivariate Normally distributed with mean zero and covariance matrix Σ. The Multivariate Normal distributional likelihood for the instantaneous nonconstant model is given by

$$
p(x|\mathcal{C}, z, \Sigma) \propto |I_n \otimes \Sigma|^{-\frac{1}{2}} e^{-\frac{1}{2}(x - \mathcal{C}z)'(I_n \otimes \Sigma)^{-1}(x - \mathcal{C}z)}, \tag{13.6.1}
$$

which can be written by performing some algebra in the exponent and using a determinant property of Kroneker products as

$$
p(x|\mathcal{C}, z, \Sigma) \propto |\Sigma|^{-\frac{n}{2}} e^{-\frac{1}{2} tr \Sigma^{-1} \sum_{i=1}^{n} (x_i - \mathcal{C}_i z_i)(x_i - \mathcal{C}_i z_i)'}, \tag{13.6.2}
$$

where the variables are as previously defined.

To quantify available prior knowledge regarding our prior beliefs about the model parameter values, both Conjugate and generalized Conjugate prior distributions are utilized.

13.7 Conjugate Priors and Posterior

When quantifying available prior information regarding the parameters of interest, Conjugate prior distributions are specified as described in Chapter 4. The joint prior distribution for the model parameters which are the matrix of Regression/mixing coefficients C, the matrix of sources S, the source covariance matrix R, and the error covariance matrix Σ is given by

$$p(S,R,C,\Sigma) = p(S|R)p(R)p(C|\Sigma)p(\Sigma), \qquad (13.7.1)$$

where the prior distribution for the model parameters from the Conjugate procedure outlined in Chapter 4 are given by

$$p(S|R) \propto |R|^{-\frac{n}{2}} e^{-\frac{1}{2}tr(S-S_0)R^{-1}(S-S_0)'}, \qquad (13.7.2)$$

$$p(\Sigma) \propto |\Sigma|^{-\frac{\nu}{2}} e^{-\frac{1}{2}tr\Sigma^{-1}Q}, \qquad (13.7.3)$$

$$p(R) \propto |R|^{-\frac{\eta}{2}} e^{-\frac{1}{2}trR^{-1}V}, \qquad (13.7.4)$$

$$p(C|\Sigma) \propto |\Sigma|^{-\frac{n(m+q+1)}{2}} e^{-\frac{1}{2}tr\Sigma^{-1}(C-C_0)(I_n \otimes D^{-1})(C-C_0)'}, \qquad (13.7.5)$$

where the $p \times n(m+q+1)$ matrix of coefficients is $C = (\mathcal{C}_1,\dots,\mathcal{C}_n)$. The matrices Σ, R, Q, V, and D are positive definite. The hyperparameters S_0, ν, Q, η, V, D, and C_0 are to be assessed and having done so completely determine the joint prior distribution.

The prior distributions for the parameters are Matrix Normal for the matrix of Regression/mixing coefficients C, Matrix Normal for the matrix of sources S, Inverted Wishart distributed for the source covariance matrix R, and Inverted Wishart distributed for the error covariance matrix.

Upon using Bayes' rule, the likelihood of the observed mixed signals and the joint prior distribution for the unknown model parameters are combined and their product is proportional to the joint posterior distribution

$$p(S,R,C,\Sigma|u,x) \propto |\Sigma|^{-\frac{\nu+n(m+q+2)}{2}} e^{-\frac{1}{2}tr\Sigma^{-1}G}$$
$$\times |R|^{-\frac{(n+\eta)}{2}} e^{-\frac{1}{2}trR^{-1}[V+(S-S_0)'(S-S_0)]}, \qquad (13.7.6)$$

where the $p \times p$ matrix G has been defined to be

$$G = \sum_{i=1}^{n}(x_i - \mathcal{C}_i z_i)(x_i - \mathcal{C}_i z_i)' + (C - C_0)(I_n \otimes D^{-1})(C - C_0)' + Q. \qquad (13.7.7)$$

Note that the second term in G can be rewritten as

$$(C - C_0)(I_n \otimes D^{-1})(C - C_0)' = \sum_{i=1}^{n}(\mathcal{C}_i - \mathcal{C}_{0i})D^{-1}(\mathcal{C}_i - \mathcal{C}_{0i})'. \qquad (13.7.8)$$

This joint posterior distribution must now be evaluated in order to obtain our parameter estimates of the matrix of sources S, the matrix of instantaneous nonconstant Regression/mixing coefficients C, the source covariance matrix R, and the error covariance matrix Σ. Marginal posterior mean and joint maximum a posteriori estimates of the parameters S, R, C, and Σ are found by the Gibbs sampling and ICM algorithms.

13.8 Conjugate Estimation and Inference

With the above posterior distribution, it is not possible to obtain marginal distributions and thus marginal estimates for any of the parameters in an analytic closed form or explicit maximum a posteriori estimates from differentiation. It is possible to use both Gibbs sampling to obtain marginal posterior parameter estimates and the ICM algorithm for joint modal or maximum a posteriori estimates. For both estimation procedures, the posterior conditional distributions are required.

13.8.1 Posterior Conditionals

From the joint posterior distribution we can obtain the posterior conditional distribution for each of the model parameters.

The conditional posterior distributions for the matrix of instantaneous nonconstant Regression/mixing coefficients C is found by considering only the terms in the joint posterior distribution which involve C and is given by

$$p(C|S,R,\Sigma,U,X) \propto p(C|\Sigma)p(X|S,C,\Sigma,U)$$

$$\propto |\Sigma|^{-\frac{n(m+q+1)}{2}} e^{-\frac{1}{2}tr\Sigma^{-1}\sum_{i=1}^{n}(\mathcal{C}_i-\mathcal{C}_{0i})D^{-1}(\mathcal{C}_i-\mathcal{C}_{0i})'}$$

$$\times |\Sigma|^{-\frac{n}{2}} e^{-\frac{1}{2}tr\Sigma^{-1}\sum_{i=1}^{n}(x_i-\mathcal{C}_i z_i)(x_i-\mathcal{C}_i z_i)'}, \qquad (13.8.1)$$

which after performing some algebra in the exponent can be written as

$$p(C|S,R,\Sigma,U,X) \propto e^{-\frac{1}{2}\sum_{i=1}^{n} tr\Sigma^{-1}(\mathcal{C}_i-\tilde{\mathcal{C}}_i)(D^{-1}+z_i z_i')(\mathcal{C}_i-\tilde{\mathcal{C}}_i)'}, \quad (13.8.2)$$

where the variable $\tilde{C}$, the posterior conditional mean and mode, has been defined and is given by

$$\tilde{\mathcal{C}}_i = [\mathcal{C}_{0i}D^{-1} + x_i z_i'](D^{-1} + z_i z_i')^{-1} \qquad (13.8.3)$$

for all i, $i = 1,\ldots,n$.

The posterior conditional distribution for the matrix of combined instantaneous nonconstant Regression/mixing coefficients $\mathcal{C}_i$ can be written as

$$p(\mathcal{C}_i|S_i,R,\Sigma,u_i,x_i) \propto |\Sigma|^{-\frac{m+q+1}{2}} e^{-\frac{1}{2}tr\Sigma^{-1}(\mathcal{C}_i-\tilde{\mathcal{C}}_i)(D^{-1}+z_i z_i')(\mathcal{C}_i-\tilde{\mathcal{C}}_i)'}. \quad (13.8.4)$$

for all i, $i = 1,\ldots,n$.

That is, the matrix of instantaneous nonconstant Regression/mixing $\mathcal{C}_i$ coefficients given the source vector s_i, the source covariance matrix R, the error covariance matrix Σ, the observed source vector u_i, and the observed data vector x_i is Multivariate Normally distributed.

The conditional posterior distribution of the observation error covariance matrix Σ is found by considering only the terms in the joint posterior distribution which involve Σ and is given by

$$
\begin{aligned}
p(\Sigma|S,R,C,U,X) &\propto p(\Sigma)p(C|\Sigma)p(X|Z,C,\Sigma) \\
&\propto |\Sigma|^{-\frac{\nu}{2}}e^{-\frac{1}{2}tr\Sigma^{-1}Q} \\
&\times |\Sigma|^{-\frac{n(m+q+1)}{2}}e^{-\frac{1}{2}tr\sum_{i=1}^{n}\Sigma^{-1}(C_i-C_{0i})D^{-1}(C_i-C_{0i})'} \\
&\times |\Sigma|^{-\frac{n}{2}}e^{-\frac{1}{2}\sum_{i=1}^{n}(x_i-C_iz_i)'\Sigma^{-1}(x_i-C_iz_i)} \\
&\propto |\Sigma|^{-\frac{\nu+n(m+q+2)}{2}}e^{-\frac{1}{2}tr\Sigma^{-1}G},
\end{aligned}
\tag{13.8.5}
$$

where the $p \times p$ matrix G has been defined to be

$$
G = \sum_{i=1}^{n}[(x_i-C_iz_i)(x_i-C_iz_i)' + (C_i-C_{0i})D^{-1}(C_i-C_{0i})'] + Q \tag{13.8.6}
$$

with a mode as described in Chapter 2 given by

$$
\tilde{\Sigma} = \frac{G}{\nu+n(m+q+2)} \ . \tag{13.8.7}
$$

The conditional posterior distribution of the observation error covariance matrix Σ given the matrix of sources S, the source covariance matrix R, the matrix of instantaneous nonconstant Regression/mixing coefficients C, the matrix of observed sources U, and the matrix of data X is an Inverted Wishart distribution.

The conditional posterior distribution for the matrix of sources S is found by considering only the terms in the joint posterior distribution which involve S and is given by

$$
\begin{aligned}
p(S|\mathcal{B},\mathcal{L},R,\Sigma,U,X) &\propto p(S|R)p(X|\mathcal{B},\mathcal{L},S,\Sigma,U) \\
&\propto |R|^{-\frac{n}{2}}e^{-\frac{1}{2}tr(S-S_0)R^{-1}(S-S_0)'} \\
&\times |\Sigma|^{-\frac{n}{2}}e^{-\frac{1}{2}\sum_{i=1}^{n}(x_i-\mathcal{B}_iu_i-\mathcal{L}_is_i)'\Sigma^{-1}(x_i-\mathcal{B}_iu_i-\mathcal{L}_is_i)},
\end{aligned}
\tag{13.8.8}
$$

which after performing some algebra in the exponent can be written as

$$
p(S|\mathcal{B},\mathcal{L},R,\Sigma,U,X) \propto e^{-\frac{1}{2}\sum_{i=1}^{n}(s_i-\tilde{s}_i)'(R^{-1}+\mathcal{L}_i'\Sigma^{-1}\mathcal{L}_i)(s_i-\tilde{s}_i)}, \tag{13.8.9}
$$

where the vectors s_i have been defined which are the posterior conditional mean and mode as described in Chapter 2 and given by

$$\tilde{s}_i = (R^{-1} + \mathcal{L}_i'\Sigma^{-1}\mathcal{L}_i)^{-1}[\mathcal{L}_i'\Sigma^{-1}(x_i - \mathcal{B}_i u_i) + R^{-1}s_{0i}]. \tag{13.8.10}$$

The posterior conditional distribution for the vectors of sources s_i can also be written as

$$p(s_i|\mathcal{B}_i,\mathcal{L}_i,R,\Sigma,u_i,x_i) \propto |R^{-1} + \mathcal{L}_i'\Sigma^{-1}\mathcal{L}_i|^{-\frac{1}{2}} e^{-\frac{1}{2}(s_i-\tilde{s}_i)'(R^{-1}+\mathcal{L}_i'\Sigma^{-1}\mathcal{L}_i)(s_i-\tilde{s}_i)} \tag{13.8.11}$$

for all i, $i = 1,\ldots,n$.

The conditional posterior distribution for the sources s_i given the instantaneous nonconstant Regression coefficients $\mathcal{B}_i$, the instantaneous nonconstant mixing coefficients $\mathcal{L}_i$, the source covariance matrix R, the error covariance matrix Σ, the observed source vector u_i, and the vector of data x_i is Multivariate Normally distributed.

The conditional posterior distribution for the source covariance matrix R is found by considering only the terms in the joint posterior distribution which involve R and is given by

$$p(R|C,S,\Sigma,U,X) \propto p(R)p(S|R)p(X|\mathcal{B},\mathcal{L},S,\Sigma,U)$$
$$\propto |R|^{-\frac{\eta}{2}} e^{-\frac{1}{2}trR^{-1}V}|R|^{-\frac{n}{2}} e^{-\frac{1}{2}tr(S-S_0)R^{-1}(S-S_0)'}$$
$$\propto |R|^{-\frac{(n+\eta)}{2}} e^{-\frac{1}{2}trR^{-1}[(S-S_0)'(S-S_0)+V]}, \tag{13.8.12}$$

with the posterior conditional mode as described in Chapter 2 given by

$$\tilde{R} = \frac{(S-S_0)'(S-S_0)+V}{n+\eta}. \tag{13.8.13}$$

The conditional posterior distribution for the source covariance matrix R given the instantaneous nonconstant matrix of Regression/mixing coefficients C, the matrix of sources S, the error covariance matrix Σ, the matrix of observed sources U, and the matrix of data X is Inverted Wishart distributed.

13.8.2 Gibbs Sampling

To find the marginal posterior mean estimates of the parameters from the joint posterior distribution using the Gibbs sampling algorithm, start with initial values for the matrix of sources S and error covariance matrix Σ, say $\bar{S}_{(0)}$ and $\bar{\Sigma}_{(0)}$, and then cycle through

$$\bar{C} = \text{a random variate from } p(C|\bar{S}_{(l)},\bar{R}_{(l)},\bar{\Sigma}_{(l)},U,X)$$

$$\bar{\mathcal{C}}_{i,(l+1)} = A_{\mathcal{C}_i} Y_{\mathcal{C}_i} B'_{\mathcal{C}_i} + M_{\mathcal{C}_i}, \quad i = 1,\dots,n, \tag{13.8.14}$$

$$\bar{\Sigma}_{(l+1)} = \text{a random variate from } p(\Sigma|\bar{S}_{(l)},\bar{R}_{(l)},\bar{C}_{(l+1)},U,X)$$

$$= A_\Sigma (Y'_\Sigma Y_\Sigma)^{-1} A'_\Sigma, \tag{13.8.15}$$

$$\bar{R}_{(l+1)} = \text{a random variate from } p(R|\bar{S}_{(l)},\bar{C}_{(l+1)},\bar{\Sigma}_{(l+1)},U,X)$$

$$= A_R (Y'_R Y_R)^{-1} A'_R, \tag{13.8.16}$$

$$\bar{S}_{(l+1)} = \text{a random variate from } p(S|\bar{C}_{(l+1)},\bar{\Sigma}_{(l+1)},\bar{R}_{(l+1)},U,X),$$

$$s_{i,(l+1)} = A_{s_i} Y_{s_i} + M_{s_i}, \quad i = 1,\dots,n, \tag{13.8.17}$$

where

$$A_{\mathcal{C}_i} A'_{\mathcal{C}_i} = \bar{\Sigma}_{(l)},$$

$$B_{\mathcal{C}_i} B'_{\mathcal{C}_i} = (D^{-1} + \bar{z}_{i,(l)} \bar{z}'_{i,(l)})^{-1},$$

$$\bar{z}_{i,(l)} = (u'_{i,(l)},\bar{s}'_{i,(l)})',$$

$$M_{\mathcal{C}_i} = [\mathcal{C}_{0i} D^{-1} + x_i \bar{z}'_{i,(l)}](D^{-1} + \bar{z}_{i,(l)} \bar{z}'_{i,(l)})^{-1},$$

$$A_\Sigma A'_\Sigma = \sum_{i=1}^{n} [(x_i - \bar{\mathcal{C}}_{i,(l+1)} \bar{z}_{i,(l)})(x_i - \bar{\mathcal{C}}_{i,(l+1)} \bar{z}_{i,(l)})'$$

$$+ (\bar{\mathcal{C}}_{i,(l+1)} - \mathcal{C}_{0i}) D^{-1} (\bar{\mathcal{C}}_{i,(l+1)} - \mathcal{C}_{0i})'] + Q,$$

$$A_R A'_R = (\bar{S}_{(l)} - S_0)'(\bar{S}_{(l)} - S_0) + V,$$

$$A_{s_i} A'_{s_i} = (\bar{R}_{(l+1)}^{-1} + \bar{\mathcal{L}}'_{i,(l+1)} \bar{\Sigma}_{(l+1)}^{-1} \bar{\mathcal{L}}_{i,(l+1)})^{-1},$$

$$M_{s_i} = (\bar{R}_{(l+1)}^{-1} + \bar{\mathcal{L}}'_{i,(l+1)} \bar{\Sigma}_{(l+1)}^{-1} \bar{\mathcal{L}}_{i,(l+1)})^{-1}$$

$$\times [\bar{\mathcal{L}}'_{i,(l+1)} \bar{\Sigma}_{(l+1)}^{-1} (x_i - \bar{\mathcal{B}}_{i,(l+1)} u_i) + \bar{R}_{(l+1)}^{-1} s_{0i}],$$

while $Y_{\mathcal{C}_i}$, Y_Σ, Y_R, and Y_{s_i} are $p \times (m+q+1)$, $(\nu+n(m+q+2)+p+1) \times p$, $(n+\eta+m+1) \times m$, and $(m \times 1)$-dimensional matrices respectively, whose elements are random variates from the standard Scalar Normal distribution. The formulas for the generation of random variates from the conditional posterior distributions are easily found from the methods in Chapter 6.

The first random variates called the "burn in" are discarded and after doing so, compute from the next L variates means of the parameters

$$\bar{S} = \frac{1}{L} \sum_{l=1}^{L} \bar{S}_{(l)} \qquad \bar{R} = \frac{1}{L} \sum_{l=1}^{L} \bar{R}_{(l)} \qquad \bar{C} = \frac{1}{L} \sum_{l=1}^{L} \bar{C}_{(l)} \qquad \bar{\Sigma} = \frac{1}{L} \sum_{l=1}^{L} \bar{\Sigma}_{(l)}$$

which are the exact sampling-based marginal posterior mean estimates of the parameters. In the above, $C = (\mathcal{C}_1,\dots,\mathcal{C}_n)$. Exact sampling-based estimates of other quantities can also be found. Similar to Bayesian Regression and Bayesian Factor Analysis there is interest in the estimate of the marginal posterior variance of the matrix containing the Regression and mixing coefficients

$$\overline{var}(c|X,U) = \frac{1}{L}\sum_{l=1}^{L}\bar{c}_{(l)}\bar{c}'_{(l)} - \bar{c}\bar{c}',$$

$$= \bar{\Delta},$$

where $c = vec(C)$ and $\bar{c} = vec(\bar{C})$.

The covariance matrices of the other parameters follow similarly.

13.8.3   Maximum a Posteriori

The joint posterior distribution can also be jointly maximized with respect to the matrix of instantaneous nonconstant coefficients C, the error covariance matrix Σ, the matrix of sources S, and the source covariance matrix R by using the ICM algorithm. To maximize the joint posterior distribution using the ICM algorithm, start with an initial value for the matrix of sources S, say $\tilde{S}_{(0)}$, and then cycle through

$$\tilde{C}_{(l+1)} = \underset{C}{\text{Arg Max}}\ p(C|\tilde{S}_{(l)},\tilde{R}_{(l)},\tilde{\Sigma}_{(l)},X,U),$$

$$\tilde{C}_{i,(l+1)} = (D^{-1}+\tilde{z}_{i,(l)}\tilde{z}'_{i,(l)})^{-1}[D^{-1}\mathcal{C}_{0i}+\tilde{z}_{i,(l)}x'_i],\quad i=1,\dots,n,$$

$$\tilde{\Sigma}_{(l+1)} = \underset{\Sigma}{\text{Arg Max}}\ p(\Sigma|\bar{C}_{(l+1)},R_{(l)},\tilde{S}_{(l)},X,U)$$

$$= \sum_{i=1}^{n}[(x_i-\tilde{\mathcal{C}}_{i,(l+1)}\tilde{z}_{i,(l)})(x_i-\tilde{\mathcal{C}}_{i,(l+1)}\tilde{z}_{i,(l)})'$$

$$+ (\tilde{\mathcal{C}}_{i,(l+1)}-\mathcal{C}_{0i})D^{-1}(\tilde{\mathcal{C}}_{i,(l+1)}-\mathcal{C}_{0i})']+Q]/[\nu+n(m+q+2)],$$

$$\tilde{R}_{(l+1)} = \underset{R}{\text{Arg Max}}\ p(R|\tilde{S}_{(l)},\bar{C}_{(l+1)},\tilde{\Sigma}_{(l+1)},X,U)$$

$$= \frac{(\tilde{S}_{(l)}-S_0)'(\tilde{S}_{(l)}-S_0)+V}{n+\eta},$$

$$\tilde{S}_{(l+1)} = \underset{S}{\text{Arg Max}}\ p(S|\tilde{C}_{(l+1)},\tilde{R}_{(l+1)},\tilde{\Sigma}_{(l+1)},X,U),$$

$$\tilde{s}_{i,(l+1)} = (\tilde{R}_{(l+1)}^{-1}+\tilde{\mathcal{L}}'_{i,(l+1)}\tilde{\Sigma}_{(l+1)}^{-1}\tilde{\mathcal{L}}_{i,(l+1)})^{-1}$$

$$\times[\tilde{\mathcal{L}}'_{i,(l+1)}\tilde{\Sigma}_{(l+1)}^{-1}(x_i-\tilde{\mathcal{B}}_{i,(l+1)}u_i)+\tilde{R}_{(l+1)}^{-1}s_{0i}],\quad i=1,\dots,n,$$

where the vector $\tilde{z}_i = (u'_i,\tilde{s}'_i)'$ has been defined, until convergence is reached. The converged values $(\tilde{S},\tilde{R},\tilde{C},\tilde{\Sigma})$ are joint posterior modal (maximum a posteriori) estimators of the parameters. Conditional maximum a posteriori variance estimates can also be found. The conditional modal variance of the matrix containing the Regression and mixing coefficients is

$$var(\mathcal{C}_i|\tilde{\mathcal{C}}_i,\tilde{S},\tilde{R},\tilde{\Sigma},X,U) = \tilde{\Sigma}\otimes(D^{-1}+\tilde{z}_i\tilde{z}'_i)^{-1},\quad i=1,\dots,n$$

or equivalently

$$var(c_i|\tilde{c}_i, \tilde{S}, \tilde{R}, \tilde{\Sigma}, X, U) = (D^{-1} + \tilde{z}_i \tilde{z}_i')^{-1} \otimes \tilde{\Sigma}, \quad i = 1, \ldots, n$$
$$= \tilde{\Delta}_i,$$

where $c_i = vec(\mathcal{C}_i)$, $\tilde{c}$, $\tilde{S}$, $\tilde{R}$, and $\tilde{\Sigma}$ are the converged value from the ICM algorithm.

To determine statistical significance with the ICM approach, use the conditional distribution of the matrix containing the Regression and mixing coefficients which is

$$p(\mathcal{C}_i|\tilde{\mathcal{C}}_i, \tilde{S}, \tilde{R}, \tilde{\Sigma}, X, U) \propto |D^{-1} + \tilde{z}_i \tilde{z}_i'|^{\frac{1}{2}} |\tilde{\Sigma}|^{-\frac{1}{2}} e^{-\frac{1}{2} tr \tilde{\Sigma}^{-1} (\mathcal{C}_i - \tilde{\mathcal{C}}_i)(D^{-1} + \tilde{z}_i \tilde{z}_i')(\mathcal{C}_i - \tilde{\mathcal{C}}_i)'}.$$
$$(13.8.18)$$

That is,

$$\mathcal{C}_i|\tilde{\mathcal{C}}_i, \tilde{S}, \tilde{R}, \tilde{\Sigma}, X, U \sim N\left(\tilde{\mathcal{C}}_i, \tilde{\Sigma} \otimes (D^{-1} + \tilde{z}_i \tilde{z}_i')^{-1}\right). \qquad (13.8.19)$$

General simultaneous hypotheses can be evaluated regarding the entire matrix containing the Regression/mixing coefficients, a submatrix, or the coefficients of a particular independent variable or source, or an element by computing marginal conditional distributions.

It can be shown [17, 41] that the marginal conditional distribution of any column of the matrix containing the Regression and mixing coefficients $\mathcal{C}_i$, $\mathcal{C}_{i,k}$ is Multivariate Normal

$$p(\mathcal{C}_{i,k}|\tilde{\mathcal{C}}_{i,k}, \tilde{S}, \tilde{\Sigma}, U, X) \propto |W_{i,kk}\tilde{\Sigma}|^{-\frac{1}{2}} e^{-\frac{1}{2}(\mathcal{C}_{i,k} - \tilde{\mathcal{C}}_{i,k})'(W_{i,kk}\tilde{\Sigma})^{-1}(\mathcal{C}_{i,k} - \tilde{\mathcal{C}}_{i,k})},$$
$$(13.8.20)$$

where $W_i = (D^{-1} + \tilde{z}_i \tilde{z}_i')^{-1}$ and $W_{i,kk}$ is its kth diagonal element.

With the marginal distribution of a column of $\mathcal{C}_i$, significance can be determined for a particular independent variable or source. Significance can be determined for a subset of coefficients by determining the marginal distribution of the subset within $\mathcal{C}_{i,k}$ which is also Multivariate Normal. With the subset being a singleton set, significance can be determined for a particular coefficient with the marginal distribution of the scalar coefficient which is

$$p(\mathcal{C}_{i,kj}|\tilde{\mathcal{C}}_{i,kj}, \tilde{S}, \tilde{\Sigma}_{jj}, U, X) \propto (W_{i,kk}\tilde{\Sigma}_{jj})^{-\frac{1}{2}} e^{-\frac{(\mathcal{C}_{i,kj} - \tilde{\mathcal{C}}_{i,kj})^2}{2W_{i,kk}\tilde{\Sigma}_{jj}}}, \qquad (13.8.21)$$

where $\tilde{\Sigma}_{jj}$ is the jth diagonal element of $\tilde{\Sigma}$. Note that $\tilde{\mathcal{C}}_{i,kj} = \tilde{c}_{i,jk}$ and that

$$z = \frac{(\mathcal{C}_{i,kj} - \tilde{\mathcal{C}}_{i,kj})}{\sqrt{W_{i,kk}\tilde{\Sigma}_{jj}}}$$

$$(13.8.22)$$

follows a Normal distribution with a mean of zero and variance of one.

13.9   Generalized Priors and Posterior

Generalized Conjugate prior distributions are assessed in order to quantify available prior information regarding values of the model parameters. The joint prior distribution for the sources S, the source covariance matrix R, the vector of instantaneous nonconstant Regression/mixing coefficients $c = vec(C)$, and the error covariance matrix Σ is given by

$$p(S,R,\Sigma,c) = p(S|R)p(R)p(\Sigma)p(c), \qquad (13.9.1)$$

where the prior distribution for the parameters from the generalized Conjugate procedure outlined in Chapter 4 are as follows

$$p(S|R) \propto |R|^{-\frac{n}{2}} e^{-\frac{1}{2}tr(S-S_0)R^{-1}(S-S_0)'}, \qquad (13.9.2)$$

$$p(\Sigma) \propto |\Sigma|^{-\frac{\nu}{2}} e^{-\frac{1}{2}tr\Sigma^{-1}Q}, \qquad (13.9.3)$$

$$p(R) \propto |R|^{-\frac{\eta}{2}} e^{-\frac{1}{2}trR^{-1}V}, \qquad (13.9.4)$$

$$p(c) \propto |\Delta|^{-\frac{1}{2}} e^{-\frac{1}{2}(c-c_0)'\Delta^{-1}(c-c_0)}, \qquad (13.9.5)$$

where $c = vec(C)$, R, Σ, Q, V, and Δ are positive definite matrices. The hyperparameters S_0, ν, Q, η, V, Δ, and c_0 are to be assessed. Upon assessing the hyperparameters, the joint prior distribution is completely determined.

The prior distribution for the matrix of sources S is Matrix Normally distributed, the prior distribution for the source vector covariance matrix R is Inverted Wishart distributed, the vector of combined Regression/mixing coefficients $c = vec(C)$, $C = (B,\Lambda)$ is Multivariate Normally distributed, and the prior distribution for the error covariance matrix Σ is Inverted Wishart distributed.

Using Bayes' rule the joint posterior distribution for the unknown parameters with generalized Conjugate prior distributions for the model parameters is given by

$$p(c,S,R,\Sigma|U,X) = p(S|R)p(\Sigma)p(R)p(c)p(X|Z,C,\Sigma), \qquad (13.9.6)$$

which is

$$p(c,S,R,\Sigma|u,x) \propto |\Sigma|^{-\frac{(n+\nu)}{2}} |R|^{-\frac{(n+\eta)}{2}} e^{-\frac{1}{2}tr\Sigma^{-1}G} e^{-\frac{1}{2}(c-c_0)'\Delta^{-1}(c-c_0)}$$
$$\times e^{-\frac{1}{2}trR^{-1}[V+(S-S_0)'(S-S_0)]}, \qquad (13.9.7)$$

where the $p \times p$ matrix G has been defined to be

$$G = \sum_{i=1}^{n}(x_i - C_i z_i)(x_i - C_i z_i)' + Q \tag{13.9.8}$$

after inserting the prior distributions and the likelihood.

This joint posterior distribution must now be evaluated in order to obtain estimates of the sources S, the vector of instantaneous nonconstant Regression/mixing coefficients c, the source covariance matrix R, and the error covariance matrix Σ.

13.10 Generalized Estimation and Inference

With the above joint posterior distribution that uses generalized Conjugate prior distributions, it is not possible to obtain marginal distributions and thus marginal estimates for all or any of the parameters in an analytic closed form or explicit maximum a posteriori estimates from differentiation. It is possible to use both Gibbs sampling, to obtain marginal mean parameter estimates and the ICM algorithm for maximum a posteriori estimates. For both estimation procedures, the posterior conditional distributions are required.

13.10.1 Posterior Conditionals

Both the Gibbs sampling and ICM algorithms require the posterior conditional distributions. Gibbs sampling requires them for the generation of random variates while the ICM algorithm requires them for maximization by cycling through their modes or maxima.

The conditional posterior distribution of the observation error covariance matrix Σ is found by considering only those terms in the joint posterior distribution which involve Σ and is given by

$$\begin{aligned}
p(\Sigma|S,R,C,U,X) &\propto p(\Sigma)p(X|S,C,\Sigma,U) \\
&\propto |\Sigma|^{-\frac{\nu}{2}} e^{-\frac{1}{2}tr\Sigma^{-1}Q} \\
&\quad \times |\Sigma|^{-\frac{n}{2}} e^{-\frac{1}{2}\sum_{i=1}^{n}(x_i - C_i z_i)'\Sigma^{-1}(x_i - C_i z_i)} \\
&\propto |\Sigma|^{-\frac{(n+\nu)}{2}} e^{-\frac{1}{2}tr\Sigma^{-1}G},
\end{aligned} \tag{13.10.1}$$

where the $p \times p$ matrix G has been defined to be

$$G = \sum_{i=1}^{n} (x_i - \mathcal{C}_i z_i)(x_i - \mathcal{C}_i z_i)' + Q \tag{13.10.2}$$

and the mode of this posterior conditional distribution is as described in Chapter 2 and is given by

$$\tilde{\Sigma} = \frac{G}{n+\nu}. \tag{13.10.3}$$

The posterior conditional distribution of the observation error covariance matrix Σ given the matrix of sources S, the source covariance matrix R, the matrix of instantaneous nonconstant Regression/mixing coefficients $\mathcal{C}$, the matrix of observable sources U, and the matrix of data X is an Inverted Wishart distribution.

The conditional posterior distribution for the matrix of sources S is found by considering only those terms in the joint posterior distribution which involve S and is given by

$$p(S|\mathcal{B},R,\mathcal{L},\Sigma,U,X) \propto p(S|R)p(X|\mathcal{B},\mathcal{L},S,\Sigma,U)$$
$$\propto |R|^{-\frac{n}{2}} e^{-\frac{1}{2}tr(S-S_0)R^{-1}(S-S_0)'}$$
$$\times |\Sigma|^{-\frac{n}{2}} e^{-\frac{1}{2}\sum_{i=1}^{n}(x_i-\mathcal{B}_i u_i - \mathcal{L}_i s_i)'\Sigma^{-1}(x_i-\mathcal{B}_i u_i-\mathcal{L}_i s_i)}$$
$$\propto e^{-\frac{1}{2}\sum_{i=1}^{n}(s_i-\tilde{s}_i)'(R^{-1}+\mathcal{L}_i'\Sigma^{-1}\mathcal{L}_i)(s_i-\tilde{s}_i)}, \tag{13.10.4}$$

where the vector $\tilde{s}_i$ has been defined to be

$$\tilde{s}_i = (R^{-1}+\mathcal{L}_i'\Sigma^{-1}\mathcal{L}_i)^{-1}[\mathcal{L}_i'\Sigma^{-1}(x_i-\mathcal{B}_i u_i)+R^{-1}s_{0i}] \tag{13.10.5}$$

which is the posterior conditional mean and mode.

The posterior conditional distribution for the vectors of sources s_i can also be written as

$$p(s_i|\mathcal{B}_i,\mathcal{L}_i,R,\Sigma,u_i,x_i) \propto |R^{-1}+\mathcal{L}_i'\Sigma^{-1}\mathcal{L}_i|^{-\frac{1}{2}}$$
$$\times e^{-\frac{1}{2}(s_i-\tilde{s}_i)'(R^{-1}+\mathcal{L}_i'\Sigma^{-1}\mathcal{L}_i)(s_i-\tilde{s}_i)} \tag{13.10.6}$$

for all i, $i = 1,\ldots,n$.

The conditional posterior distribution for the sources s_i given the instantaneous nonconstant Regression coefficients $\mathcal{B}_i$, the instantaneous nonconstant mixing coefficients $\mathcal{L}_i$, the source covariance matrix R, the error covariance matrix Σ, the observed source vector u_i, and the vector of data x_i is Multivariate Normally distributed.

The conditional posterior distribution for the source covariance matrix R is found by considering only those terms in the joint posterior distribution which involve R and is given by

$$p(R|\mathcal{B},\mathcal{L},S,\Sigma,U,X) \propto p(R)p(S|R)p(X|\mathcal{B},\mathcal{L},S,\Sigma,U)$$
$$\propto |R|^{-\frac{\eta}{2}}e^{-\frac{1}{2}trR^{-1}V}|R|^{-\frac{n}{2}}e^{-\frac{1}{2}tr(S-S_0)R^{-1}(S-S_0)'}$$
$$\propto |R|^{-\frac{(n+\eta)}{2}}e^{-\frac{1}{2}trR^{-1}[(S-S_0)'(S-S_0)+V]}, \qquad (13.10.7)$$

with the posterior conditional mode as described in Chapter 2 given by

$$\tilde{R} = \frac{(S-S_0)'(S-S_0)+V}{n+\eta}. \qquad (13.10.8)$$

The conditional posterior distribution for the source covariance matrix R given the matrix of sources S, the instantaneous nonconstant Regression coefficients $\mathcal{B}$, the matrix of instantaneous nonconstant mixing coefficients $\mathcal{L}$, the error covariance matrix Σ, the matrix of observable sources U, and the matrix of data X is Inverted Wishart distributed.

The conditional posterior distribution of the vector of instantaneous nonconstant Regression/mixing coefficients is found by considering only those terms in the joint posterior distribution which involve c, C, or $\mathcal{C}$ and is given by

$$p(c|S,R,\Sigma,U,X) \propto p(c)p(X|C,S,\Sigma,U)$$
$$\propto |\Delta|^{-\frac{1}{2}}e^{-\frac{1}{2}(c-c_0)'\Delta^{-1}(c-c_0)}$$
$$\times|\Sigma|^{-\frac{n}{2}}e^{-\frac{1}{2}\sum_{i=1}^{n}(x_i-\mathcal{C}_iz_i)'\Sigma^{-1}(x_i-\mathcal{C}_iz_i)}, \quad (13.10.9)$$

which after performing some algebra in the exponent can be written as

$$p(c|S,R,\Sigma,U,X) \propto e^{-\frac{1}{2}(c-\tilde{c})'(\Delta^{-1}+I\otimes\Sigma^{-1})(c-\tilde{c})}, \qquad (13.10.10)$$

where the vector $\tilde{c}$ has been defined to be

$$\tilde{c} = (\Delta^{-1}+I\otimes\Sigma^{-1})^{-1}(\Delta^{-1}c_0+I\otimes\Sigma^{-1}\hat{c}), \qquad (13.10.11)$$

and the vector $\hat{c}$ has been defined by first defining $\hat{C}$ to be

$$\hat{C} = (\hat{c}_1',\ldots,\hat{c}_n')', \qquad (13.10.12)$$

where

$$\hat{c}_i = x_iz_i'(z_iz_i')^{-1} \qquad (13.10.13)$$

and then $\hat{c} = vec(\hat{C})$. That is, the vector of instantaneous nonconstant Regression/mixing coefficients c given the matrix of sources S, the source covariance matrix R, the error covariance matrix Σ, the matrix of observed sources U, and the matrix of data X is Multivariate Normally distributed.

13.10.2   Gibbs Sampling

To find marginal mean estimates of the parameters from the joint posterior distribution using the Gibbs sampling algorithm, start with initial values for the matrix of sources S and the error covariance matrix Σ, say $\bar{S}_{(0)}$ and $\bar{\Sigma}_{(0)}$, and then cycle through

$$\bar{c}_{(l+1)} = \text{a random variate from } p(c|\bar{S}_{(l)}, \bar{R}_{(l)}, \bar{\Sigma}_{(l)}, U, X)$$
$$= A_c Y_c + M_c, \tag{13.10.14}$$
$$\bar{\Sigma}_{(l+1)} = \text{a random variate from } p(\Sigma|\bar{S}_{(l)}, \bar{R}_{(l)}, \bar{C}_{(l+1)}, U, X)$$
$$= A_\Sigma (Y'_\Sigma Y_\Sigma)^{-1} A'_\Sigma, \tag{13.10.15}$$
$$\bar{R}_{(l+1)} = \text{a random variate from } p(R|\bar{S}_{(l)}, \bar{C}_{(l+1)}, \bar{\Sigma}_{(l+1)}, U, X)$$
$$= A_R (Y'_R Y_R)^{-1} A'_R, \tag{13.10.16}$$
$$\bar{S}_{(l+1)} = \text{a random variate from } p(S|\bar{R}_{(l+1)}, \bar{C}_{(l+1)}, \bar{\Sigma}_{(l+1)}, U, X)$$
$$s_{i,(l+1)} = A_{s_i} Y_{s_i} + M_{s_i}, \tag{13.10.17}$$

where

$$A_c A'_c = (\Delta^{-1} + I \otimes \bar{\Sigma}_{(l)}^{-1})^{-1},$$
$$M_c = (\Delta^{-1} + I \otimes \bar{\Sigma}_{(l)}^{-1})^{-1}(\Delta^{-1} c_0 + I \otimes \bar{\Sigma}_{(l)}^{-1} \hat{c}_{(l)}),$$
$$\hat{c}_{i,(l)} = x_i \bar{z}'_{i,(l)} (\bar{z}_{i,(l)} \bar{z}_{i,(l)})^{-1}, \ i = 1, \ldots, n,$$
$$\hat{c}_{(l)} = (\hat{c}_{1,(l)}, \ldots, \hat{c}_{i,(l)})',$$
$$A_\Sigma A'_\Sigma = \sum_{i=1}^{n} [(x_i - \bar{C}_{i,(l+1)} \bar{z}_{i,(l)})(x_i - \bar{C}_{i,(l+1)} \bar{z}_{i,(l)})' + Q,$$
$$A_R A'_R = (\bar{S}_{(l)} - S_0)'(\bar{S}_{(l)} - S_0) + V,$$
$$A_{s_i} A'_{s_i} = (\bar{R}_{(l+1)}^{-1} + \bar{\mathcal{L}}'_{i,(l+1)} \bar{\Sigma}_{(l+1)}^{-1} \bar{\mathcal{L}}_{i,(l+1)})^{-1},$$
$$M_{s_i} = (\bar{R}_{(l+1)}^{-1} + \bar{\mathcal{L}}'_{i,(l+1)} \bar{\Sigma}_{(l+1)}^{-1} \bar{\mathcal{L}}_{i,(l+1)})^{-1}$$
$$\times [\bar{\mathcal{L}}'_{i,(l+1)} \bar{\Sigma}_{(l+1)}^{-1} (x_i - \bar{\mathcal{B}}_{i,(l+1)} u_i) + \bar{R}_{(l+1)}^{-1} s_{0i}],$$

while Y_c, Y_Σ, Y_R, and Y_{s_i} are $np(m+q+1) \times 1$, $(\nu + n(m+q+2) + p+1) \times p$, $(n + \eta + m + 1) \times m$, and $(m \times 1)$-dimensional matrices respectively, whose elements are random variates from the standard Scalar Normal distribution. The formulas for the generation of random variates from the conditional posterior distributions is easily found from the methods in Chapter 6. The formulas for the generation of random variates from the conditional posterior distributions are easily found from the methods in Chapter 6.

The first random variates called the "burn in" are discarded and after doing so, compute from the next L variates means of each of the parameters

$$\bar{S} = \frac{1}{L}\sum_{l=1}^{L}\bar{S}_{(l)} \qquad \bar{R} = \frac{1}{L}\sum_{l=1}^{L}\bar{R}_{(l)} \qquad \bar{c} = \frac{1}{L}\sum_{l=1}^{L}\bar{c}_{(l)} \qquad \bar{\Sigma} = \frac{1}{L}\sum_{l=1}^{L}\bar{\Sigma}_{(l)}$$

which are the exact sampling-based marginal posterior mean estimates of the parameters. Exact sampling-based estimates of other quantities can also be found. Similar to Bayesian Regression and Bayesian Factor Analysis, there is interest in the estimate of the marginal posterior variance of the matrix containing the Regression and mixing coefficients

$$\overline{var}(c|X,U) = \frac{1}{L}\sum_{l=1}^{L}\bar{c}_{(l)}\bar{c}'_{(l)} - \bar{c}\bar{c}'$$
$$= \bar{\Delta},$$

where $c = vec(C)$ and $\bar{c} = vec(\bar{C})$.

The covariance matrices of the other parameters follow similarly.

13.10.3 Maximum a Posteriori

The distribution can be maximized with respect to vector of instantaneous nonconstant Regression/mixing coefficients c, the matrix of sources S, the source covariance matrix R, and the error covariance Σ using the ICM algorithm. To jointly maximize the posterior distribution using the ICM algorithm, start with initial values for the matrix of sources $\tilde{S}$, and the error covariance matrix Σ, say $\tilde{S}_{(0)}$, and $\tilde{\Sigma}_{(0)}$, and then cycle through

$$\tilde{c}_{(l+1)} = \overset{\text{Arg Max}}{c}\, p(c|\tilde{S}_{(l)},\tilde{R}_{(l)},\tilde{\Sigma}_{(l)},X,U),$$
$$\hat{c}_{i,(l)} = x_i\tilde{z}'_{i,(l)}(\tilde{z}_{i,(l)}\tilde{z}_{i,(l)})^{-1},\quad i=1,\ldots,n,$$
$$\hat{c}_{(l)} = (\hat{c}_{1,(l)},\ldots,\hat{c}_{i,(l)})',$$
$$\tilde{c}_{(l+1)} = \left(\Delta^{-1} + I\otimes\tilde{\Sigma}_{(l)}^{-1}\right)^{-1}\left[\Delta^{-1}c_0 + \left(I\otimes\tilde{\Sigma}_{(l)}^{-1}\right)\hat{c}_{(l)}\right],$$
$$\tilde{\Sigma}_{(l+1)} = \overset{\text{Arg Max}}{\Sigma}\, p(\Sigma|\tilde{C}_{(l+1)},\tilde{R}_{(l)},\tilde{S}_{(l)},X,U)$$
$$= \frac{\sum_{i=1}^{n}(x_i - \tilde{C}_{i,(l+1)}\tilde{z}_{i,(l)})(x_i - \tilde{C}_{i,(l+1)}\tilde{z}_{i,(l)})' + Q}{n+\nu},$$
$$\tilde{R}_{(l+1)} = \overset{\text{Arg Max}}{R}\, p(R|\tilde{S}_{(l)},\tilde{C}_{(l+1)},\tilde{\Sigma}_{(l+1)},X,U)$$

$$= \frac{(\tilde{S}_{(l)} - S_0)'(\tilde{S}_{(l)} - S_0) + V}{n + \eta},$$

$$\tilde{S}_{(l+1)} = \overset{\text{Arg Max}}{S} p(S|\tilde{C}_{(l+1)}, \tilde{R}_{(l+1)}, \tilde{\Sigma}_{(l+1)}, X, U),$$

$$\tilde{s}_{i,(l+1)} = (\tilde{R}_{(l+1)}^{-1} + \tilde{\mathcal{L}}_{i,(l+1)}' \tilde{\Sigma}_{(l+1)}^{-1} \tilde{\mathcal{L}}_{(l+1)})^{-1}$$

$$\times [\tilde{\mathcal{L}}_{i,(l+1)}' \tilde{\Sigma}_{(l+1)}^{-1}(x_i - \tilde{\mathcal{B}}_{i,(l+1)} u_i) + \tilde{R}_{(l+1)}^{-1} s_{0i}], \quad i = 1, \dots, n$$

until convergence is reached. The converged values $(\tilde{S}, \tilde{R}, \tilde{c}, \tilde{\Sigma})$ are joint posterior modal (maximum a posteriori) estimates of the unknown model parameters. Conditional maximum a posteriori variance estimates can also be found. The conditional modal variance of the matrix containing the Regression and mixing coefficients is

$$var(c|\tilde{c}, \tilde{S}, \tilde{R}, \tilde{\Sigma}, X, U) = [\Delta^{-1} + \tilde{Z}'\tilde{Z} \otimes \tilde{\Sigma}]^{-1}$$
$$= \tilde{\Delta},$$

where $c = vec(C)$, while $\tilde{S}$, $\tilde{R}$, and $\tilde{\Sigma}$ are the converged value from the ICM algorithm.

Conditional modal intervals may be computed by using the conditional distribution for a particular parameter given the modal values of the others.

13.11   Interpretation

Although the main focus after having performed a Bayesian Source Separation is the separated sources, there are others. One focus as in Bayesian Regression is on the estimate of the Regression coefficient matrix B which defines a "fitted" line. Coefficients are evaluated to determine whether they are statistically "large" meaning that the associated independent variable contributes to the dependent variable or statistically "small" meaning that the associated independent variable does not contribute to the dependent variable. The coefficient matrix also has the interpretation that if all of the independent variables were held fixed except for one u_{ij} which if increased to u_{ij}^*, the dependent variable x_{ij} increases to an amount x_{ij}^* given by

$$x_{ij}^* = \beta_{i0} + \cdots + \beta_{ij} u_{ij}^* + \cdots + \beta_{iq} u_{iq}. \tag{13.11.1}$$

Another focus after performing a Bayesian Source Separation is in the estimated mixing coefficients. The mixing coefficients are the amplitudes which determine the relative contribution of the sources. A particular mixing coefficient which is relatively "small" indicates that the corresponding source does not significantly contribute to the associated observed mixed signal. If a particular mixing coefficient is relatively "large," this indicates that the corresponding source does significantly contribute to the associated observed mixed signal.

A useful way to visualize the changing mixing coefficients is to plot their value over time for each of the sources.

13.12 Discussion

Returning to the cocktail party problem, the matrix of Regression coefficients B where $B = (\mu, B_\star)$ contains the matrix of mixing coefficients $B_\star$ for the observed conversation (sources) U, and the population mean μ which is a vector of the overall background mean level at each microphone.

After having estimated the model parameters, the estimates of the sources as well as the mixing matrix are now available. The estimated matrix of sources corresponds to the unobservable signals or conversations emitted from the mouths of the speakers at the cocktail party. Row i of the estimated source matrix is the estimate of the unobserved source vector at time i and column j of the estimated source matrix is the estimate of the unobserved conversation of speaker j at the party for all n time increments.

Exercises

1. Show that in the instantaneous nonconstant mixing process model for both Conjugate and generalized Conjugate prior distributions that by setting $\mathcal{B}_i = B$ and $\mathcal{C}_i = \Lambda$, the resulting model is the unobservable/observable Source Separation model of Chapter 11.

2. Derive a model in which up to time i_0 the Regression and mixing coefficients are constant and after time i_0, they are also constant but different. Assume an instantaneous constant mixing process.

14

Correlated Observation and Source Vectors

14.1 Introduction

In Part II, the observed mixed source as well as the observed and unobserved source vectors were specified to allow correlation within the vectors at a given time, but to be uncorrelated over time. In the previous Chapter, the sources were allowed to be delayed; in addition, the Regression and mixing coefficients were allowed to change over time. In this Chapter, the observed mixed vectors as well as the unobserved source vectors are allowed to be correlated over time; in addition, the mixing is assumed to be instantaneous and constant over time.

14.2 Model

Just as in the previous Chapter, the observed mixed vectors, the observed source vectors, and unobserved source vectors are stacked into singleton vectors and the correlated vector Source Separation model is written as

$$
\begin{array}{cccccc}
(x|B,u,\Lambda,s) = & (I_n \otimes B) & u & + & (I_n \otimes \Lambda) & s & + & \epsilon, \\
(np \times 1) & [np \times n(q+1)] & [n(q+1) \times 1] & & (np \times nm) & (nm \times 1) & & (np \times 1)
\end{array}
$$

$$(14.2.1)$$

where the vector of observations x, the vector of observed sources u, the vector of unobserved sources s, and the vector of errors ϵ are

$$
x = \begin{pmatrix} x_1 \\ \vdots \\ x_n \end{pmatrix}, \quad u = \begin{pmatrix} u_1 \\ \vdots \\ u_n \end{pmatrix}, \quad s = \begin{pmatrix} s_1 \\ \vdots \\ s_n \end{pmatrix}, \quad \epsilon = \begin{pmatrix} \epsilon_1 \\ \vdots \\ \epsilon_n \end{pmatrix}, \quad (14.2.2)
$$

B and Λ are Regression and mixing coefficients respectively, and an instantaneous constant mixing process has been assumed.

The errors of observation ϵ are specified to have distribution $p(\epsilon|\Omega)$ with mean zero and covariance matrix Ω which induces a distribution on the observations x, namely, $p(x|B,u,\Lambda,s)$ with mean $(I_n \otimes B)u + (I_n \otimes \Lambda)s$ and co-variance matrix Ω.

The above is a very general model. The covariance matrix Ω defines a model in which all observations (x_{ij} and $x_{i'j'}$) are distinctly correlated. With this model in its full covariance generality, there are an enormous number of distinct covariance parameters. To be exact, there are

$$\frac{np(np+1)}{2} \tag{14.2.3}$$

distinct error covariance parameters [50]. The full error covariance matrix for the observation vector can be represented by the block partitioned matrix

$$\Omega = \begin{pmatrix} \Omega_{11} & \Omega_{12} & \cdots & \Omega_{1n} \\ & \Omega_{22} & & \\ & & \ddots & \vdots \\ & & & \Omega_{n-1,n} \\ & & & \Omega_{nn} \end{pmatrix}, \tag{14.2.4}$$

where only the diagonal and superdiagonal submatrices Ω_{ii} and $\Omega_{ii'}$ are displayed. The subdiagonal submatrices are found by reflection. The variance of observation vector i is given by the $p \times p$ submatrix

$$var(x_i|B,u_i,s_i,\Lambda,\Omega_{ii}) = \Omega_{ii} \tag{14.2.5}$$

and the covariance between observation vectors i and i' is given by the $p \times p$ submatrix

$$cov(x_i,x_{i'}|B,u_i,u_{i'},s_i,s_{i'},\Lambda,\Omega_{ii'}) = \Omega_{ii'}. \tag{14.2.6}$$

This generality is rarely needed. A simplified error covariance matrix is usually rich enough to capture the covariance structure. The error covariance matrix can be simplified by specifying a particular structure, thereby reducing the number of distinct parameters and the required computation. In the context of Bayesian Factor Analysis, separable and matrix intraclass covariance matrices have been considered [50].

A separable covariance matrix for the errors is

$$\Omega = \begin{pmatrix} \phi_{11}\Sigma & \phi_{12}\Sigma & \cdots & \phi_{1n}\Sigma \\ & \phi_{22}\Sigma & & \\ & & \ddots & \vdots \\ & & & \\ & & & \phi_{nn}\Sigma \end{pmatrix} = \Phi \otimes \Sigma \tag{14.2.7}$$

which is exactly the structure of a Matrix Normal distribution.

The covariance matrix Φ will be referred to as the between vector covariance matrix and Σ the within vector covariance matrix. Separable covariance matrices have a wide variety of applications such as in time series. In applications such as the previous FMRI example, spatial vector valued observations x_i are observed over time. In such applications, Φ could be referred to as the "temporal" covariance matrix and Σ as the "spatial" covariance. Previous work has also considered matrix intraclass covariance structures, but points out that the matrix intraclass covariance can be transformed to an independent covariance model by an orthogonal transformation [50]. The covariance matrix Φ has

$$\frac{n(n+1)}{2} \tag{14.2.8}$$

distinct covariance parameters. The number of distinct covariance parameters has been reduced to

$$\frac{n(n+1)}{2} + \frac{p(p+1)}{2}. \tag{14.2.9}$$

Covariance structures such as intraclass and first order Markov (also known as an AR(1)) which depend on a single parameter have been considered for Φ to further reduce the number of parameters and computation [50].

The motivation to model covariation among the observations and also the source vectors is that they are often taken in a time or spatial order, thus possibly not independent. This covariance structure allows for both spatial and temporal correlation.

The correlated vector Bayesian Source Separation model with the separable matrix specifications can be written in the matrix form

$$
\begin{array}{cccccccc}
(X|U,B,\Lambda,S) = & U & B' & + & S & \Lambda' & + & E, \\
(n \times p) & [n \times (q+1)] & [(q+1) \times p] & & (n \times m) & (m \times p) & & (n \times p)
\end{array}
\tag{14.2.10}
$$

which is the previously model described from Part II where $X' = (x_1,\ldots,x_n)$, $U' = (u_1,\ldots,u_n)$, $S' = (s_1,\ldots,s_n)$, and $E' = (\epsilon_1,\ldots,\epsilon_n)$.

14.3   Likelihood

The variability among the observations is specified to have arisen from a separable Multivariate Normal distribution. A Multivariate Normal distribution with a separable covariance structure $\Omega = \Phi \otimes \Sigma$ can be simplified to the Matrix Normal distribution form (as in Chapter 2)

$$p(X|U,B,\Lambda,S,\Phi,\Sigma) \propto |\Phi|^{-\frac{p}{2}}|\Sigma|^{-\frac{n}{2}}e^{-\frac{1}{2}tr\Phi^{-1}(X-UB-S\Lambda')\Sigma^{-1}(X-UB'-S\Lambda')'}, \tag{14.3.1}$$

where $X' = (x_1, \ldots, x_n)$, $S' = (s_1, \ldots, s_n)$, and $E' = (\epsilon_1, \ldots, \epsilon_n)$. If the between vector covariance matrix Φ were specified to be the identity matrix I_n, then this likelihood corresponds to the uncorrelated observation vector one as discussed in Chapter 11.

With the separable covariance matrix structure which is a Matrix Normal distribution, the marginal likelihood of any row of the matrix of data X, say x_i', is Multivariate Normally distributed with mean

$$E(x_i|B, u_i, \phi_{ii}, \Sigma, s_i, \Lambda) = Bu_i + \Lambda s_i, \qquad (14.3.2)$$

and the variance is given by

$$var(x_i|B, u_i, \phi_{ii}, \Sigma, s_i, \Lambda) = \phi_{ii}\Sigma, \qquad (14.3.3)$$

while the covariance between any two rows (observation vectors) x_i and $x_{i'}$ is given by

$$cov(x_i, x_{i'}|B, u_i, u_{i'}, s_i, s_{i'}, \Lambda, \phi_{ii'}, \Sigma) = \phi_{ii'}\Sigma, \qquad (14.3.4)$$

where $\phi_{ii'}$ is the ii'^{th} element of Φ.

The model may also be written in terms of columns as in Chapter 8 by parameterizing the data matrix X, the observable source matrix U, the matrix of Regression coefficients B, and the matrix of errors E in terms of columns as

$$X = (X_1, \ldots X_p), \qquad U = (e_n, U_1, \ldots U_q),$$
$$B = (B_0, B_1, \ldots B_q), \qquad E = (E_1, \ldots E_p). \qquad (14.3.5)$$

This leads to the Bayesian Source Separation model being also written as

$$(X_j|U, B_j, S, \Lambda_j) = \underset{(n \times 1)}{U} \underset{[n \times (q+1)]}{B_j} + \underset{[(q+1) \times 1]}{S} \underset{(n \times m)}{\Lambda_j} + \underset{(m \times 1)}{E_j}, \qquad (14.3.6)$$

which describes all the observations for a single microphone j at all n time points.

With the column representation of the Source Separation model, the marginal likelihood of any column of the matrix of data X, say X_j, is also Multivariate Normally distributed with mean

$$E(X_j|U, B_j, \Phi, \sigma_{jj}, S, \Lambda_j) = UB_j + S\Lambda_j, \qquad (14.3.7)$$

and variance given by

$$var(X_j|U, B_j, \Phi, \sigma_{jj}, S, \Lambda_j) = \sigma_{jj}\Phi, \qquad (14.3.8)$$

while the covariance between any two columns (observation vectors) X_j and $X_{j'}$ is given by

$$cov(x_j, x_{j'}|B, u_j, u_{j'}, s_j, s_{j'}, \Lambda, \Phi, \sigma_{jj'}) = \sigma_{jj'}\Phi, \qquad (14.3.9)$$

where $\sigma_{jj'}$ is the jj'^{th} element of Σ.

The Regression and the mixing coefficient matrices are joined into a single matrix as $C = (B, \Lambda)$. The observable and unobservable source matrices are also joined as $Z = (U, S)$.

Having joined these matrices, the correlated vector Source Separation model is now

$$\begin{array}{cccccc}
(X|C,Z) = & Z & C' & + & E, \\
n \times p & n \times (m+q+1) & (m+q+1) \times p & & (n \times p)
\end{array} \qquad (14.3.10)$$

and the corresponding likelihood is

$$p(X|C,Z,\Sigma) \propto |\Sigma|^{-\frac{n}{2}}|\Phi|^{-\frac{p}{2}}e^{-\frac{1}{2}tr\Phi^{-1}(X-ZC')\Sigma^{-1}(X-ZC')'}, \qquad (14.3.11)$$

where all variables are as previously defined.

Available prior knowledge regarding the parameter values are quantified through both Conjugate and generalized Conjugate prior distributions. For the Conjugate prior distribution model, different structures will be specified for Φ along with prior distributions.

14.4   Conjugate Priors and Posterior

When quantifying available prior information regarding the parameters of interest, Conjugate prior distributions can be specified. For the between vector covariance matrix Φ, structures will be considered along with the corresponding prior distribution. The joint posterior distribution for the model parameters which are the matrix of Regression/mixing coefficients C, the matrix of sources S, the source covariance matrix R, the within observation vector covariance matrix Σ, and the between vector covariance matrix Φ is given by

$$p(S,R,C,\Sigma,\Phi) = p(S|R,\Phi)p(R)p(C|\Sigma)p(\Sigma)p(\Phi), \qquad (14.4.1)$$

where the prior distributions for the model parameters are from the Conjugate procedure outlined in Chapter 4 and are given by

$$p(S|R,\Phi) \propto |R|^{-\frac{n}{2}} |\Phi|^{-\frac{m}{2}} e^{-\frac{1}{2}tr\Phi^{-1}(S-S_0)R^{-1}(S-S_0)'}, \qquad (14.4.2)$$

$$p(\Sigma) \propto |\Sigma|^{-\frac{\nu}{2}} e^{-\frac{1}{2}tr\Sigma^{-1}Q}, \qquad (14.4.3)$$

$$p(R) \propto |R|^{-\frac{\eta}{2}} e^{-\frac{1}{2}trR^{-1}V}, \qquad (14.4.4)$$

$$p(C|\Sigma) \propto |D|^{-\frac{p}{2}} |\Sigma|^{-\frac{m+q+1}{2}} e^{-\frac{1}{2}tr\Sigma^{-1}(C-C_0)D^{-1}(C-C_0)'}, \qquad (14.4.5)$$

$$p(\Phi) \quad \text{as below}, \qquad (14.4.6)$$

where Σ, Φ, R, Q, V, D, and Ψ, are positive definite matrices. The hyperparameters S_0, ν, Q, η, V, D, C_0, and any for $p(\Phi)$ are to be assessed. Upon assessing the hyperparameters, the joint prior distribution is completely determined.

The Conjugate prior distributions are Matrix Normal for the combined matrix of Regression/mixing coefficients C, Matrix Normal for the matrix of sources S, where the source vectors and components are free to be correlated, while the observation within Σ and between Φ, as well as source covariance matrices R are taken to be Inverted Wishart distributed.

Upon using Bayes' rule, the joint posterior distribution for the unknown model parameters is proportional to the product of the joint prior distribution and the likelihood and given by

$$p(S,C,\Sigma,\Phi|X) \propto |\Sigma|^{-\frac{(n+\nu+m+q+1)}{2}} e^{-\frac{1}{2}tr\Sigma^{-1}G} |R|^{-\frac{(n+\eta)}{2}} e^{-\frac{1}{2}trR^{-1}V}$$
$$\times |\Phi|^{-\frac{(p+m)}{2}} e^{-\frac{1}{2}tr\Phi^{-1}(S-S_0)R^{-1}(S-S_0)'} p(\Phi), \qquad (14.4.7)$$

where the $p \times p$ matrix G has been defined to be

$$G = (X - ZC')'\Phi^{-1}(X - ZC') + (C - C_0)D^{-1}(C - C_0)' + Q. \qquad (14.4.8)$$

Again, the objective is to unmix the unobservable sources by estimating S and to obtain knowledge about the mixing process by estimating B, Λ, Σ, and Φ.

This joint posterior distribution must now be evaluated in order to obtain our parameter estimates of the matrix of sources S, the matrix of Regression/mixing coefficients C, the within vector source covariance matrix R, the within observation vector covariance matrix Σ, and the between vector covariance matrix Φ. Marginal posterior mean and joint maximum a posteriori estimated of the parameters S, R, Φ, C, and Σ are found by the Gibbs sampling and ICM algorithms.

There are four cases which will be considered for the between vector covariance matrix Φ: (1) Φ, a known general covariance matrix with no unknown parameters, (2) Φ, a general unknown covariance matrix with $n(n+1)/2$ unknown parameters (3) Φ, an intraclass structured covariance matrix with one unknown parameter, and (4) Φ, a first order Markov structured covariance matrix with one unknown parameter.

Φ Known

In some instances, we know Φ, are able to assess Φ, or can estimate Φ using previous data, so that

$$p(\Phi) = \begin{cases} 1, \text{ if } \Phi = \Phi_0 \\ 0, \text{ if } \Phi \neq \Phi_0, \end{cases} \tag{14.4.9}$$

a degenerate distribution. If the observation vectors were independent, then $\Phi_0 = I_n$.

For Gibbs sampling, we will need the conditional posterior distributions. When the covariance matrix Φ is known to be Φ_0, then the only change in posterior conditional distributions for the parameters Σ, S, and C is that Φ will be replaced by Φ_0.

Φ General Covariance

If we determine that the observations are correlated according to a general covariance matrix, then we assess the Conjugate Inverted Wishart prior distribution

$$p(\Phi) \propto |\Phi|^{-\frac{\kappa}{2}} e^{-\frac{1}{2} tr \Phi^{-1} \Psi}, \tag{14.4.10}$$

where Ψ and κ are hyperparameters to be assessed which completely determine the prior.

Φ Unknown Structured

It is often the case that Φ is unknown but structured. When Φ is unknown, the conditionals for Σ, S, and C do not change from when Φ is known or unknown and general. Once the structure is determined, we need to assess the prior distributions for the unknown parameters in Φ and calculate the posterior conditional distribution for the unknown parameters in Φ. We will specify that the observations are homoscedastic and consider Φ to be a structured correlation matrix.

There are many possible structures that we are able to specify for Φ that apply to a wide variety of situations. Given that we have homoscedasticity of the observation vectors, then

$$\Omega = \Phi \otimes \Sigma = \begin{pmatrix} \Sigma & \phi_{12}\Sigma & \cdots & \phi_{1n}\Sigma \\ & \Sigma & & \\ & & \ddots & \vdots \\ & & & \Sigma \end{pmatrix},$$

where Φ is a correlation matrix.

One possibility is that there is a structure in the correlation matrix Φ so that its elements only depend on a single parameter ρ; then the covariance matrix becomes

$$
\Omega = \Phi \otimes \Sigma = \begin{pmatrix} \Sigma & \phi_{12}(\rho)\Sigma & \cdots & \phi_{1n}(\rho)\Sigma \\ & \Sigma & & \\ & & \ddots & \vdots \\ & & & \Sigma \end{pmatrix}.
$$

Two well-known examples of possible correlation structures for Φ are intraclass and first order Markov. We will state these correlation structures and derive the posterior conditionals for both of these correlations assuming a Generalized Scalar Beta prior distribution.

Φ Intraclass Correlation

It could be determined that the observations are correlated according to an intraclass correlation. An intraclass correlation is used when we have a set of variables and we believe that any two are related in the same way. Any two variables have the same correlation. Then the between observation correlation matrix Φ is

$$
\Phi = \begin{pmatrix} 1 & \rho & \rho & \cdots & \rho \\ & 1 & \rho & \cdots & \rho \\ & & \ddots & & \vdots \\ & & & & \rho \\ & & & & 1 \end{pmatrix} = (1-\rho)I_n + \rho e_n e_n', \qquad (14.4.11)
$$

where e_n is a column vector of ones and $-\frac{1}{n-1} < \rho < 1$.

If we determine that the observations are correlated according to an intraclass correlation matrix, then the Conjugate prior distribution

$$
p(\rho) \propto (1-\rho)^{-\frac{(n-1)(p+m)}{2}} [1+(n-1)\rho]^{-\frac{\beta}{2}} e^{-\frac{1}{2(1-\rho)}\left[k_1 - \frac{\rho k_2}{1+\alpha\rho}\right]} \qquad (14.4.12)
$$

is specified, where Ψ for

$$
k_1 = \sum_{i=1}^{n} \Psi_{ii} \quad \text{and} \quad k_2 = \sum_{i'=1}^{n}\sum_{i=1}^{n} \Psi_{ii'} \qquad (14.4.13)
$$

is a hyperparameter to be assessed which completely determines the prior.

Φ First Order Markov Correlation

It could be determined that the observations are correlated according to a first order Markov structure. In a first order Markov structure, the observations are related according to a vector auto regression with a time lag of one, $VAR(1)$. With this structure, the between observation correlation matrix Φ is

$$\Phi = \begin{pmatrix} 1 & \rho & \rho^2 & \cdots & \rho^{n-1} \\ \rho & 1 & \rho & \cdots & \rho^{n-2} \\ \vdots & \vdots & \vdots & & \vdots \\ \rho^{n-1} & \rho^{n-2} & & \cdots & 1 \end{pmatrix}, \tag{14.4.14}$$

where $0 < |\rho| < 1$.

If it is determined that the observations are correlated according to a first order Markov correlation matrix, then the Conjugate prior distribution

$$p(\rho) \propto (1-\rho^2)^{-\frac{(n-1)(p+m)}{2}} e^{-\frac{k_1 - \rho k_2 + \rho^2 k_3}{2(1-\rho^2)}} \tag{14.4.15}$$

is specified, where Ψ for

$$k_1 = \sum_{i=1}^{n} \Psi_{ii}, \quad k_2 = \sum_{i=1}^{n-1} (\Psi_{i,i+1} + \Psi_{i+1,i}), \quad \text{and} \quad k_3 = \sum_{i=2}^{n-1} \Psi_{ii} \tag{14.4.16}$$

is a hyperparameter to be assessed which completely determines the prior.

14.5   Conjugate Estimation and Inference

With the above posterior distribution, it is not possible to obtain marginal distributions and thus marginal estimates for all or any of the parameters in an analytic closed form or explicit maximum a posteriori estimates from differentiation. It is possible to use both Gibbs sampling to obtain marginal parameter estimates and the ICM algorithm to find maximum a posteriori estimates. For both estimation procedures which are described in Chapter 6, the posterior conditional distributions are required.

14.5.1   Posterior Conditionals

From the joint posterior distribution we can obtain the posterior conditional distribution for each of the model parameters.

The conditional posterior distributions for the Regression/mixing matrix C is found by considering only the terms in the joint posterior distribution which involve C and is given by

$$p(C|S,R,\Sigma,\Phi,U,X) \propto p(C|\Sigma)p(X|C,Z,\Sigma,\Phi)$$
$$\propto |\Sigma|^{-\frac{m+q+1}{2}} e^{-\frac{1}{2}tr\Sigma^{-1}(C-C_0)D^{-1}(C-C_0)'}$$
$$\times |\Sigma|^{-\frac{n}{2}} e^{-\frac{1}{2}tr\Sigma^{-1}(X-ZC')'\Phi^{-1}(X-ZC')}$$
$$\propto e^{-\frac{1}{2}tr\Sigma^{-1}[(C-C_0)D^{-1}(C-C_0)'+(X-ZC')'\Phi^{-1}(X-ZC')]}$$
$$\propto e^{-\frac{1}{2}tr\Sigma^{-1}(C-\tilde{C})(D^{-1}+Z'\Phi^{-1}Z)(C-\tilde{C})'}, \tag{14.5.1}$$

where the vector $\tilde{C}$, the posterior conditional mean and mode, has been defined and is given by

$$\tilde{C} = [C_0 D^{-1} + X'\Phi^{-1}Z](D^{-1}+Z'\Phi^{-1}Z)^{-1} \tag{14.5.2}$$
$$= [C_0 D^{-1} + \hat{C}(Z'\Phi^{-1}Z)](D^{-1}+Z'\Phi^{-1}Z)^{-1}. \tag{14.5.3}$$

Note that the matrix of coefficients C can be written as a weighted combination of the prior mean C_0 from the prior distribution and the data mean $\hat{C} = X'\Phi^{-1}Z(Z'\Phi^{-1}Z)^{-1}$ from the likelihood.

The conditional distribution for the combined Regression mixing matrix given the matrix of unobservable sources S, the within vector source covariance matrix R, the within observation vector covariance matrix Σ, the between vector covariance matrix Φ, the matrix of observed sources U, and the matrix of data X is Matrix Normally distributed.

The conditional posterior distribution of the within observation vector covariance matrix Σ is found by considering only the terms in the joint posterior distribution which involve Σ and is given by

$$p(\Sigma|B,S,R,\Lambda,\Phi,U,X) \propto p(\Sigma)p(\Lambda|\Sigma)p(X|Z,C,\Sigma,\Phi)$$
$$\propto |\Sigma|^{-\frac{m+q+1}{2}} e^{-\frac{1}{2}tr\Sigma^{-1}(C-C_0)D^{-1}(C-C_0)'}$$
$$\times |\Sigma|^{-\frac{\nu}{2}} e^{-\frac{1}{2}tr\Sigma^{-1}Q}$$
$$\times |\Sigma|^{-\frac{n}{2}} e^{-\frac{1}{2}tr\Sigma^{-1}(X-ZC')'\Phi^{-1}(X-ZC')}$$
$$\propto |\Sigma|^{-\frac{(n+\nu+m+q+1)}{2}} e^{-\frac{1}{2}tr\Sigma^{-1}G}, \tag{14.5.4}$$

where the $p \times p$ matrix G has been defined to be

$$G = (X-ZC')'\Phi^{-1}(X-ZC')+(C-C_0)D^{-1}(C-C_0)'+Q \tag{14.5.5}$$

with a mode as described in Chapter 2 given by

$$\tilde{\Sigma} = \frac{G}{n+\nu+m+q+1}. \tag{14.5.6}$$

The posterior conditional distribution of the within observation vector co-variance matrix Σ given the matrix of unobservable sources S, the within source vector covariance matrix R, the matrix of Regression/mixing coefficients C, the between vector covariance matrix Φ, the matrix of observable sources U, and the matrix of data X is an Inverted Wishart distribution.

The conditional posterior distribution for the matrix of sources S is found by considering only the terms in the joint posterior distribution which involve S and is given by

$$p(S|B,R,\Lambda,\Sigma,\Phi,U,X) \propto p(S|R,\Phi)p(X|B,\Lambda,S,\Sigma,\Phi,U)$$
$$\propto |\Phi|^{-\frac{m}{2}} |R|^{-\frac{n}{2}} e^{-\frac{1}{2}tr\Phi^{-1}(S-S_0)R^{-1}(S-S_0)'}$$
$$\times |\Sigma|^{-\frac{n}{2}} e^{-\frac{1}{2}tr\Sigma^{-1}(X-UB'-S\Lambda')'\Phi^{-1}(X-UB'-S\Lambda')}$$
$$\propto e^{-\frac{1}{2}tr\Phi^{-1}(S-\tilde{S})(R^{-1}+\Lambda'\Sigma^{-1}\Lambda)(S-\tilde{S})'}, \qquad (14.5.7)$$

where the matrix $\tilde{S}$ has been defined which is the posterior conditional mean and mode as described in Chapter 2 and is given by

$$\tilde{S} = [S_0 R^{-1} + (X - UB')\Sigma^{-1}\Lambda](R^{-1} + \Lambda'\Sigma^{-1}\Lambda)^{-1}. \qquad (14.5.8)$$

The conditional posterior distribution for the matrix of sources S given the matrix of Regression coefficients B, the within source vector covariance matrix R, the matrix of mixing coefficients Λ, the within observation vector covariance matrix Σ, the between vector covariance matrix Φ, the matrix of observable sources U, and the matrix of data X is Normally Matrix distributed.

The conditional posterior distribution for the within source vector covariance matrix R is found by considering only the terms in the joint posterior distribution which involve R and is given by

$$p(R|C,Z,\Sigma,\Phi,X) \propto p(R)p(S|R,\Phi)p(X|C,Z,\Sigma,\Phi)$$
$$\propto |R|^{-\frac{\eta}{2}} e^{-\frac{1}{2}trR^{-1}V} |\Phi|^{-\frac{m}{2}} |R|^{-\frac{n}{2}} e^{-\frac{1}{2}tr\Phi^{-1}(S-S_0)R^{-1}(S-S_0)'}$$
$$\propto |R|^{-\frac{(n+\eta)}{2}} e^{-\frac{1}{2}trR^{-1}[(S-S_0)'\Phi^{-1}(S-S_0)+V]}, \qquad (14.5.9)$$

with the posterior conditional mode as described in Chapter 2 given by

$$\tilde{R} = \frac{(S-S_0)'\Phi^{-1}(S-S_0)+V}{n+\eta}. \qquad (14.5.10)$$

The conditional posterior distribution for the within source vector covariance matrix R given the matrix of Regression/mixing coefficients C, the matrix of unobservable sources S, the within observation vector covariance matrix

Σ, the between vector covariance matrix Φ, the matrix of observable sources, and the matrix of X data is Inverted Wishart distributed.

Φ Known

The conditional posterior distribution for the between vector covariance matrix Φ is found by considering only the terms in the joint posterior distribution which involve Φ and is given by

$$p(\Phi|C, Z, R, \Sigma, X) \propto p(\Phi)p(S|R, \Phi)p(X|C, Z, \Sigma, \Phi)$$
$$\propto 1_{\Phi=\Phi_0}|\Phi|^{-\frac{m}{2}}e^{-\frac{1}{2}tr\Phi^{-1}(S-S_0)R^{-1}(S-S_0)'}$$
$$\times|\Phi|^{-\frac{p}{2}}e^{-\frac{1}{2}tr\Phi^{-1}(X-ZC')\Sigma^{-1}(X-ZC')'}$$
$$= 1_{\Phi=\Phi_0}, \qquad (14.5.11)$$

where $1_{\Phi=\Phi_0}$ is used to denote the prior distribution for Φ which is one at Φ_0 and zero otherwise.

Φ General Covariance

The conditional posterior distribution for the between vector covariance matrix Φ is found by considering only the terms in the joint posterior distribution which involve Φ and is given by

$$p(\Phi|C, Z, R, \Sigma, X) \propto p(\Phi)p(S|R, \Phi)p(X|C, Z, \Sigma, \Phi)$$
$$\propto |\Phi|^{-\frac{\kappa}{2}}e^{-\frac{1}{2}tr\Phi^{-1}\Psi}|\Phi|^{-\frac{m}{2}}e^{-\frac{1}{2}tr\Phi^{-1}(S-S_0)R^{-1}(S-S_0)'}$$
$$\times|\Phi|^{-\frac{p}{2}}e^{-\frac{1}{2}tr\Phi^{-1}(X-ZC')\Sigma^{-1}(X-ZC')'}, \qquad (14.5.12)$$

with the posterior conditional mode as described in Chapter 2 given by

$$\tilde{\Phi} = \frac{(X-ZC')\Sigma^{-1}(X-ZC')'+(S-S_0)R^{-1}(S-S_0)'+\Psi}{p+m+\kappa}. \qquad (14.5.13)$$

The conditional posterior distribution for the between vector covariance matrix Φ given the matrix of Regression/mixing coefficients C, the matrix of unobservable sources S, the matrix of observable sources U, the within source vector covariance matrix R, the within observation vector covariance matrix Σ, and the matrix of data X is Inverted Wishart distributed.

Φ Intraclass Correlation

As previously stated, the exact form of the conditional posterior distribution depends on which structure is determined for the correlation matrix Φ. If the intraclass structure that has the covariance between any two observations being the same is determined, then we can use the result that the determinant of Φ has the form

$$|\Phi| = (1-\rho)^{n-1}[1+\rho(n-1)] \tag{14.5.14}$$

and the result that the inverse of Φ has the form

$$\Phi^{-1} = \frac{I_n}{1-\rho} - \frac{\rho e_n e_n'}{(1-\rho)[1+(n-1)\rho]}, \tag{14.5.15}$$

which is again a matrix with intraclass correlation structure. Using the afore-mentioned likelihood, priors, and forms above the posterior conditional distribution is

$$p(\rho|S,C,\Psi,X,U) \propto p(\rho)p(S|\Phi)p(X|\Phi,S,C,\Sigma,U)$$

$$\propto (1-\rho)^{-\frac{(n-1)(p+m)}{2}}[1+(n-1)\rho]^{-\frac{(p+m)}{2}}$$

$$\propto e^{-\frac{1}{2(1-\rho)}\left[tr(\Psi)-\frac{\rho tr(e_n e_n'\Psi)}{1+(n-1)\rho}\right]}$$

$$\times |\Phi|^{-\frac{m}{2}}|R|^{-\frac{n}{2}}e^{-\frac{1}{2}tr\Phi^{-1}(S-S_0)R^{-1}(S-S_0)'}$$

$$\times |\Phi|^{-\frac{p}{2}}|\Sigma|^{-\frac{n}{2}}e^{-\frac{1}{2}tr\Phi^{-1}(X-ZC')\Sigma^{-1}(X-ZC')'}$$

$$\propto (1-\rho)^{-(n-1)(p+m)}[1+\rho(n-1)]^{-(p+m)}$$

$$\times e^{-\frac{1}{2}\left[\frac{k_1}{1-\rho}-\frac{k_2\rho}{(1-\rho)[1+(n-1)\rho]}\right]}, \tag{14.5.16}$$

where

$$\Xi = (X-ZC')\Sigma^{-1}(X-ZC')' + (S-S_0)R^{-1}(S-S_0)' + \Psi, \tag{14.5.17}$$

$$k_1 = tr(\Xi) = \sum_{i=1}^{n}\Xi_{ii}, \quad \text{and} \quad k_2 = tr(e_n e_n'\Xi) = \sum_{i'=1}^{n}\sum_{i=1}^{n}\Xi_{ii'}. \tag{14.5.18}$$

This is not recognizable as a common distribution.

Φ First Order Markov Correlation

If the first order Markov structure is determined, then the result that the determinant of a matrix with such structure has the form

$$|\Phi| = (1-\rho^2)^{n-1} \tag{14.5.19}$$

and the result that the inverse of such a patterned matrix has the form

$$\Phi^{-1} = \frac{1}{1-\rho^2}\begin{pmatrix} 1 & -\rho & & & 0 \\ -\rho & (1+\rho^2) & -\rho & & \\ & \ddots & \ddots & \ddots & \\ & & (1+\rho^2) & -\rho \\ 0 & & & -\rho & 1 \end{pmatrix}. \tag{14.5.20}$$

These results are used along with the aforementioned likelihood and prior distributions to obtain

$$p(\rho|S,C,\Sigma,X,U) = p(\rho)p(S|\Phi)p(X|\Phi,S,C,\Sigma,U)$$

$$\propto (1-\rho^2)^{-\frac{(n-1)(p+m)}{2}} e^{-\frac{k_1-\rho k_2+\rho^2 k_3}{2(1-\rho^2)}}$$

$$\times |\Phi|^{-\frac{m}{2}} |R|^{-\frac{n}{2}} e^{-\frac{1}{2}tr\Phi^{-1}(S-S_0)R^{-1}(S-S_0)'}$$

$$\times |\Phi|^{-\frac{p}{2}} |\Sigma|^{-\frac{n}{2}} e^{-\frac{1}{2}tr\Phi^{-1}(X-ZC')\Sigma^{-1}(X-ZC')'}$$

$$\propto (1-\rho^2)^{-(n-1)(p+m)} e^{-\frac{k_1-\rho k_2+\rho^2 k_3}{2(1-\rho^2)}} , \qquad (14.5.21)$$

where the matrices and constants used are Ξ as defined previously,

$$\Psi_1 = I_n, \quad \Psi_2 = \begin{pmatrix} 0 & 1 & & & 0 \\ 1 & 0 & 1 & & \\ & \ddots & \ddots & \ddots & \\ & & 0 & 1 \\ 0 & & 1 & 0 \end{pmatrix}, \quad \Psi_3 = \begin{pmatrix} 0 & & & 0 \\ & 1 & & \\ & & \ddots & \\ & & & 1 \\ 0 & & & 0 \end{pmatrix},$$

$$k_1 = tr(\Psi_1 \Xi) = \sum_{i=1}^{n} \Xi_{ii}, \qquad (14.5.22)$$

$$k_2 = tr(\Psi_2 \Xi) = \sum_{i=1}^{n-1} (\Xi_{i,i+1} + \Xi_{i+1,i}), \qquad (14.5.23)$$

and

$$k_3 = tr(\Psi_3 \Xi) = \sum_{i=2}^{n-1} \Xi_{ii}. \qquad (14.5.24)$$

Again, this is not recognizable as a common distribution.

These are two simple possible structures. There may be others that also depend on a single parameter or on several parameters. The rejection sampling technique is needed and is simple to carry out because one only needs to generate random variates from a univariate distribution.

14.5.2   Gibbs Sampling

To find marginal mean estimates of the model parameters from the joint posterior distribution using the Gibbs sampling algorithm, start with initial values for the matrix of sources S, the within observation vector covariance matrix Σ and the between vector covariance matrix Φ, say $\bar{S}_{(0)}$, $\bar{\Sigma}_{(0)}$, and $\bar{\Phi}_{(0)}$ or $\bar{\rho}_{(0)}$, then cycle through

$$
\begin{aligned}
\bar{C}_{(l+1)} &= \text{a random variate from } p(C|\bar{S}_{(l)}, \bar{R}_{(l)}, \bar{\Sigma}_{(l)}, \bar{\Phi}_{(l)}, U, X) \\
&= A_C Y_C B_C' + M_C, & (14.5.25) \\
\bar{\Sigma}_{(l+1)} &= \text{a random variate from } p(\Sigma|\bar{S}_{(l)}, \bar{R}_{(l)}, \bar{C}_{(l+1)}, \bar{\Phi}_{(l)}, U, X) \\
&= A_\Sigma (Y_\Sigma' Y_\Sigma)^{-1} A_\Sigma', & (14.5.26) \\
\bar{R}_{(l+1)} &= \text{a random variate from } p(R|\bar{S}_{(l)}, \bar{C}_{(l+1)}, \bar{\Sigma}_{(l+1)}, \bar{\Phi}_{(l)}, U, X) \\
&= A_R (Y_R' Y_R)^{-1} A_R', & (14.5.27) \\
\bar{S}_{(l+1)} &= \text{a random variate from } p(S|\bar{R}_{(l+1)}, \bar{C}_{(l+1)}, \bar{\Sigma}_{(l+1)}, \bar{\Phi}_{(l)}, U, X) \\
&= A_S Y_S B_S' + M_S, & (14.5.28) \\
\bar{\Phi}_{(l+1)} &= \text{a random variate from } p(\Phi|\bar{S}_{(l+1)}, \bar{R}_{(l+1)}, \bar{C}_{(l+1)}, \bar{\Sigma}_{(l+1)}, U, X)
\end{aligned}
$$

$$
\bar{\Phi}_{(l+1)} = \begin{cases} \Phi_0 & \text{if known} \\ A_\Phi (Y_\Phi' Y_\Phi)^{-1} A_\Phi' & \text{if general,} \end{cases} \qquad (14.5.29)
$$

or

$$
\bar{\rho}_{(i+1)} = \text{a random variate from } p(\rho|\bar{S}_{(l+1)}, \bar{R}_{(l+1)}, \bar{C}_{(l+1)}, \bar{\Sigma}_{(l+1)}, U, X),
$$

where

$$
\begin{aligned}
A_C A_C' &= \bar{\Sigma}_{(l)}, \\
B_C B_C' &= (D^{-1} + \bar{Z}_{(l)}' \bar{\Phi}_{(l)}^{-1} \bar{Z}_{(l)})^{-1}, \\
\bar{Z}_{(l)} &= (U, \bar{S}_{(l)}), \\
M_C &= (X' \bar{\Phi}_{(l)}^{-1} \bar{Z}_{(l)} + C_0 D^{-1})(D^{-1} + \bar{Z}_{(l)}' \bar{\Phi}_{(l)}^{-1} \bar{Z}_{(l)})^{-1}, \\
A_\Sigma A_\Sigma' &= (X - \bar{Z}_{(l)} \bar{C}_{(l+1)}')' \bar{\Phi}_{(l)}^{-1} (X - \bar{Z}_{(l)} \bar{C}_{(l+1)}') \\
&\quad + (\bar{C}_{(l+1)} - C_0) D^{-1} (\bar{C}_{(l+1)} - C_0)' + Q, \\
A_R A_R' &= (\bar{S}_{(l)} - S_0)' \bar{\Phi}_{(l)}^{-1} (\bar{S}_{(l)} - S_0) + V, \\
A_S A_S' &= \bar{\Phi}_{(l)}, \\
B_S B_S' &= (\bar{R}_{(l+1)}^{-1} + \bar{\Lambda}_{(l+1)}' \bar{\Sigma}_{(l+1)}^{-1} \bar{\Lambda}_{(l+1)})^{-1}, \\
M_S &= [S_0 \bar{R}_{(l+1)}^{-1} + (X - U \bar{B}_{(l+1)}') \bar{\Sigma}_{(l+1)}^{-1} \bar{\Lambda}_{(l+1)}] \\
&\quad \times (\bar{R}_{(l+1)}^{-1} + \bar{\Lambda}_{(l+1)}' \bar{\Sigma}_{(l+1)}^{-1} \bar{\Lambda}_{(l+1)})^{-1}, \\
A_\Phi A_\Phi' &= (X - \bar{Z}_{(l+1)} \bar{C}_{(l+1)}') \bar{\Sigma}_{(l+1)}^{-1} (X - \bar{Z}_{(l+1)} \bar{C}_{(l+1)}')' + \\
&\quad (\bar{S}_{(l+1)} - S_0) \bar{R}_{(l+1)}^{-1} (\bar{S}_{(l+1)} - S_0)' + \Psi,
\end{aligned}
$$

while Y_C, Y_Σ, Y_R, Y_S, and Y_Φ are $p \times (m+q+1)$, $(n+\nu+m+q+1+p+1) \times p$, $(n+\eta+m+1) \times m$, $n \times m$, and $(p+m+\kappa+n+1) \times n$ dimensional matrices respectively, whose elements are random variates from the standard Scalar Normal distribution. The formulas for the generation of random variates from the conditional posterior distributions are easily found from the methods in Chapter 6.

The first random variates called the "burn in" are discarded and after doing so, compute from the next L variates means of the parameters

$$\bar{S} = \frac{1}{L}\sum_{l=1}^{L}\bar{S}_{(l)} \qquad \bar{R} = \frac{1}{L}\sum_{l=1}^{L}\bar{R}_{(l)} \qquad \bar{C} = \frac{1}{L}\sum_{l=1}^{L}\bar{C}_{(l)}$$

$$\bar{\Sigma} = \frac{1}{L}\sum_{l=1}^{L}\bar{\Sigma}_{(l)} \qquad \bar{\Phi} = \frac{1}{L}\sum_{l=1}^{L}\bar{\Phi}_{(l)} \text{ or } \bar{\rho} = \frac{1}{L}\sum_{l=1}^{L}\bar{\rho}_{(l)}$$

which are the exact sampling-based marginal posterior mean estimates of the parameters. Exact sampling-based estimates of other quantities can also be found. Similar to Bayesian Regression, Bayesian Factor Analysis, and Bayesian Source Separation, there is interest in the estimate of the marginal posterior variance of the matrix containing the Regression and mixing coefficients

$$\overline{var}(c|X,U) = \frac{1}{L}\sum_{l=1}^{L}\bar{c}_{(l)}\bar{c}'_{(l)} - \bar{c}\bar{c}'$$
$$= \bar{\Delta},$$

where $c = vec(C)$ and $\bar{c} = vec(\bar{C})$. The covariance matrices of the other parameters follow similarly. With a specification of Normality for the marginal posterior distribution of the vector containing the Regression and mixing coefficients, their distribution is

$$p(c|X,U) \propto |\bar{\Delta}|^{-\frac{1}{2}} e^{-\frac{1}{2}(c-\bar{c})'\bar{\Delta}^{-1}(c-\bar{c})}, \tag{14.5.30}$$

where $\bar{c}$ and $\bar{\Delta}$ are as previously defined.

To determine statistical significance with the Gibbs sampling approach, use the marginal distribution of the matrix containing the Regression and mixing coefficients given above. General simultaneous hypotheses can be evaluated regarding the entire matrix containing the Regression and mixing coefficients, a submatrix, or a particular independent variable or source, or an element by computing marginal distributions. It can be shown that the marginal distribution of the kth column of the matrix containing the Regression and mixing coefficients C, C_k is Multivariate Normal

$$p(C_k|\bar{C}_k,X,U) \propto |\bar{\Delta}_k|^{-\frac{1}{2}} e^{-\frac{1}{2}(C_k-\bar{C}_k)'\bar{\Delta}_k^{-1}(C_k-\bar{C}_k)}, \tag{14.5.31}$$

where $\bar{\Delta}_k$ is the covariance matrix of C_k found by taking the kth $p \times p$ submatrix along the diagonal of $\bar{\Delta}$.

Significance can be determined for a subset of coefficients of the kth column of C by determining the marginal distribution of the subset within C_k which is also Multivariate Normal. With the subset being a singleton set, significance can be determined for a particular coefficient with the marginal distribution of the scalar coefficient which is

$$p(C_{kj}|\bar{C}_{kj}, X, U) \propto (\bar{\Delta}_{kj})^{-\frac{1}{2}} e^{-\frac{(C_{kj}-\bar{C}_{kj})^2}{2\bar{\Delta}_{kj}}}, \qquad (14.5.32)$$

where $\bar{\Delta}_{kj}$ is the jth diagonal element of $\bar{\Delta}_k$. Note that $\bar{C}_{kj} = \bar{c}_{jk}$ and that

$$z = \frac{(C_{kj} - \bar{C}_{kj})}{\sqrt{\bar{\Delta}_{kj}}} \qquad (14.5.33)$$

follows a Normal distribution with a mean of zero and variance of one.

14.5.3 Maximum a Posteriori

The joint posterior distribution can also be maximized with respect to the model parameters. To maximize the joint posterior distribution using the ICM algorithm, start with initial values for S, and Φ, say $\tilde{S}_{(0)}$ and $\tilde{\Phi}_{(0)}$, and then cycle through

$$\tilde{C}_{(l+1)} = \overset{\text{Arg Max}}{C} \; p(C|\tilde{S}_{(l)}, \tilde{R}_{(l)}, \tilde{\Sigma}_{(l)}, \tilde{\Phi}_{(l)}, U, X)$$
$$= [X'\tilde{\Phi}_{(l)}^{-1}\tilde{Z}_{(l)} + C_0 D^{-1}](D^{-1} + \tilde{Z}'_{(l)}\tilde{\Phi}_{(l)}^{-1}\tilde{Z}_{(l)})^{-1},$$

$$\tilde{\Sigma}_{(l+1)} = \overset{\text{Arg Max}}{\Sigma} \; p(\Sigma|\tilde{C}_{(l+1)}, \tilde{S}_{(l)}, \tilde{R}_{(l)}, \tilde{\Phi}_{(l)}, U, X)$$
$$= [(X - \tilde{Z}_{(l)}\tilde{C}'_{(l+1)})'\tilde{\Phi}_{(l)}^{-1}(X - \tilde{Z}_{(l)}\tilde{C}'_{(l+1)})$$
$$+ (\tilde{C}_{(l+1)} - C_0)D^{-1}(\tilde{C}_{(l+1)} - C_0)' + Q]/(n+\nu+m+q+1),$$

$$\tilde{R}_{(l+1)} = \overset{\text{Arg Max}}{R} \; p(R|\tilde{C}_{(l+1)}, \tilde{Z}_{(l)}, \tilde{\Sigma}_{(l+1)}, \tilde{\Phi}_{(l)}, X)$$
$$= \frac{(\tilde{S}_{(l)} - S_0)'\tilde{\Phi}_{(l)}^{-1}(\tilde{S}_{(l)} - S_0) + V}{n+\eta},$$

$$\tilde{S}_{(l+1)} = \overset{\text{Arg Max}}{S} \; p(S|\tilde{B}_{(l+1)}, \tilde{R}_{(l+1)}, \tilde{\Lambda}_{(l+1)}, \tilde{\Sigma}_{(l+1)}, \tilde{\Phi}_{(l)}, U, X),$$
$$= [S_0\tilde{R}_{(l+1)}^{-1} + (X - U\tilde{B}'_{(l+1)})\tilde{\Sigma}_{(l+1)}^{-1}\tilde{\Lambda}_{(l+1)}]$$
$$\times (\tilde{R}_{(l+1)}^{-1} + \tilde{\Lambda}'_{(l+1)}\tilde{\Sigma}_{(l+1)}^{-1}\tilde{\Lambda}_{(l+1)})^{-1},$$

$$\tilde{\Phi}_{(l+1)} = \overset{\text{Arg Max}}{\Phi} \; p(\Phi|\tilde{C}_{(l+1)}, \tilde{Z}_{(l+1)}, \tilde{R}_{(l+1)}, \tilde{\Sigma}_{(l+1)}, X),$$
$$\tilde{\Phi}_{(l+1)} = [(X - \tilde{Z}_{(l+1)}\tilde{C}'_{(l+1)})\tilde{\Sigma}_{(l+1)}^{-1}(X - \tilde{Z}_{(l+1)}\tilde{C}'_{(l+1)})'$$

$$+ (\tilde{S}_{(l+1)} - S_0)\tilde{R}_{(l+1)}^{-1}(\tilde{S}_{(l+1)} - S_0)' + \Psi]/(p+m+\kappa),$$

or

$$\tilde{\rho}_{(l+1)} = \overset{\text{Arg Max}}{\rho} \; p(\rho|\tilde{C}_{(l+1)}, \tilde{Z}_{(l+1)}, \tilde{R}_{(l+1)}, \tilde{\Sigma}_{(l+1)}, X),$$

where the matrix $\tilde{Z}_{(l)} = (U, \tilde{S}_{(l)})$ until convergence is reached. The converged values $(\tilde{C}, \tilde{S}, \tilde{R}, \tilde{\Sigma}, \tilde{\Phi})$ are joint posterior modal (maximum a posteriori) estimates of the model parameters. Conditional maximum a posteriori variance estimates can also be found. The conditional modal variance of the matrix containing the Regression and mixing coefficients is

$$var(C|\tilde{C}, \tilde{S}, \tilde{R}, \tilde{\Sigma}, \tilde{\Phi}, , X, U) = \tilde{\Sigma} \otimes (D^{-1} \otimes \tilde{Z}'\tilde{\Phi}^{-1}\tilde{Z})^{-1}$$

or equivalently

$$var(c|\tilde{c}, \tilde{S}, \tilde{R}, \tilde{\Sigma}, \tilde{\Phi}, X, U) = (D^{-1} \otimes \tilde{Z}'\tilde{\Phi}^{-1}\tilde{Z})^{-1} \otimes \tilde{\Sigma}$$
$$= \tilde{\Delta},$$

where $c = vec(C)$, $\tilde{S}$, $\tilde{R}$, and $\tilde{\Sigma}$ are the converged value from the ICM algorithm.

To determine statistical significance with the ICM approach, use the conditional distribution of the matrix containing the Regression and mixing coefficients which is

$$p(C|\tilde{C}, \tilde{S}, \tilde{R}, \tilde{\Sigma}, \tilde{\Phi}, X, U) \propto |D^{-1} + \tilde{Z}'\tilde{\Phi}^{-1}\tilde{Z}|^{\frac{1}{2}}|\tilde{\Sigma}|^{-\frac{1}{2}}$$
$$\times e^{-\frac{1}{2}tr\tilde{\Sigma}^{-1}(C-\tilde{C})(D^{-1}+\tilde{Z}'\tilde{\Phi}^{-1}\tilde{Z})(C-\tilde{C})'}.$$

$$(14.5.34)$$

That is,

$$C|\tilde{C}, \tilde{S}, \tilde{R}, \tilde{\Sigma}, \tilde{\Phi}, X, U \sim N\left(\tilde{C}, \tilde{\Sigma} \otimes (D^{-1} + \tilde{Z}'\tilde{\Phi}^{-1}\tilde{Z})^{-1}\right). \qquad (14.5.35)$$

General simultaneous hypotheses can be evaluated regarding the entire matrix containing the Regression and mixing coefficients, a submatrix, or the coefficients of a particular independent variable or source, or an element by computing marginal conditional distributions.

It can be shown [17, 41] that the marginal conditional distribution of any column of the matrix containing the Regression and mixing coefficients C, C_k is Multivariate Normal

$$p(C_k|\tilde{C}_k, \tilde{S}, \tilde{\Sigma}, \tilde{\Phi}, U, X) \propto |W_{kk}\tilde{\Sigma}|^{-\frac{1}{2}} e^{-\frac{1}{2}(C_k-\tilde{C}_k)'(W_{kk}\tilde{\Sigma})^{-1}(C_k-\tilde{C}_k)}, \quad (14.5.36)$$

where $W = (D^{-1} + \tilde{Z}'\tilde{\Phi}^{-1}\tilde{Z})^{-1}$ and W_{kk} is its kth diagonal element.

With the marginal distribution of a column of C, significance can be determined for a particular independent variable or source. Significance can be determined for a subset of coefficients by determining the marginal distribution of the subset within C_k which is also Multivariate Normal. With the subset being a singleton set, significance can be determined for a particular coefficient with the marginal distribution of the scalar coefficient which is

$$p(C_{kj}|\tilde{C}_{kj}, \tilde{S}, \tilde{\Sigma}_{jj}\tilde{\Phi}, U, X) \propto (W_{kk}\tilde{\Sigma}_{jj})^{-\frac{1}{2}} e^{-\frac{(C_{kj} - \tilde{C}_{kj})^2}{2W_{kk}\tilde{\Sigma}_{jj}}}, \qquad (14.5.37)$$

where $\tilde{\Sigma}_{jj}$ is the jth diagonal element of $\tilde{\Sigma}$. Note that $\tilde{C}_{kj} = \tilde{c}_{jk}$ and that

$$z = \frac{(C_{kj} - \tilde{C}_{kj})}{\sqrt{W_{kk}\tilde{\Sigma}_{jj}}}$$

$$(14.5.38)$$

follows a Normal distribution with a mean of zero and variance of one.

14.6   Generalized Priors and Posterior

Generalized Conjugate prior distributions are assessed in order to quantify available prior information regarding values of the model parameters. The joint prior distribution for the matrix of sources S, the within source vector covariance matrix R, the between source vector covariance matrix χ, the vector of Regression/mixing coefficients $c = vec(C)$, the within observation vector covariance matrix Σ, and the between observation vector Φ is given by

$$p(S, R, \chi, c, \Sigma, \Phi) = p(S|R, \chi)p(R)p(\chi)p(c)p(\Sigma)p(\Phi), \qquad (14.6.1)$$

where the prior distribution for the parameters from the generalized Conjugate procedure outlined in Chapter 4 are as follows

$$p(S|R, \chi) \propto |R|^{-\frac{n}{2}}|\chi|^{-\frac{m}{2}}e^{-\frac{1}{2}tr\chi^{-1}(S-S_0)R^{-1}(S-S_0)'}, \qquad (14.6.2)$$

$$p(R) \propto |R|^{-\frac{\eta}{2}}e^{-\frac{1}{2}trR^{-1}V}, \qquad (14.6.3)$$

$$p(\Sigma) \propto |\Sigma|^{-\frac{\nu}{2}}e^{-\frac{1}{2}tr\Sigma^{-1}Q}, \qquad (14.6.4)$$

$$p(c) \propto |\Delta|^{-\frac{1}{2}}e^{-\frac{1}{2}(c-c_0)'\Delta^{-1}(c-c_0)}, \qquad (14.6.5)$$

$$p(\Phi) \propto |\Phi|^{-\frac{\kappa}{2}}e^{-\frac{1}{2}tr\Phi^{-1}\Psi}, \qquad (14.6.6)$$

$$p(\chi) \propto |\chi|^{-\frac{\xi}{2}}e^{-\frac{1}{2}tr\chi^{-1}\Xi}, \qquad (14.6.7)$$

where χ, Ξ, Σ, R, V, Q, Δ, Φ, and Ψ are positive definite matrices. The hyperparameters S_0, η, V, ν, Q, c_0, Δ, ξ, and Ξ are to be assessed. Upon assessing the hyperparameters, the joint prior distribution is completely determined.

The prior distribution for the matrix of sources S is Matrix Normally distributed, the prior distribution for the within source vector covariance matrix R is Inverted Wishart distributed, the prior distribution for the between source vector covariance matrix χ is Inverted Wishart distributed, the prior distribution for the vector of Regression/mixing coefficients $c = vec(C)$, $C = (B, \Lambda)$ is Multivariate Normally distributed, the prior distribution for the error covariance matrix Σ is Inverted Wishart distributed, and the prior distribution for the between observation vector covariance matrix Φ is Inverted Wishart distributed.

Note that R, χ, Σ, and Φ, are full covariance matrices allowing within and between correlation for the observed mixed signals vectors (microphones) and also for the unobserved source vectors (speakers). The mean of the sources is often taken to be constant for all observations and thus without loss of generality taken to be zero. An observation (time) varying source mean is adopted here.

Upon using Bayes' rule the joint posterior distribution for the unknown parameters with generalized Conjugate prior distributions for the model parameters is given by

$$p(S, R, \chi, c, \Sigma, \Phi) = p(S|R, \chi)p(R)p(\chi)p(c)p(\Sigma)p(\Phi), \qquad (14.6.8)$$

which is

$$
\begin{aligned}
p(S, R, \chi, c, \Sigma, \Phi | U, X) \propto\ & |\Sigma|^{-\frac{(n+\nu)}{2}} e^{-\frac{1}{2} tr \Sigma^{-1}[(X-ZC')'\Phi^{-1}(X-ZC')+Q]} \\
& \times |R|^{-\frac{(n+\eta)}{2}} e^{-\frac{1}{2} tr R^{-1}[(S-S_0)'\chi^{-1}(S-S_0)+V]} \\
& \times e^{-\frac{1}{2}(c-c_0)'\Delta^{-1}(c-c_0)} |\Phi|^{-\frac{(p+\kappa)}{2}} e^{-\frac{1}{2} tr \Phi^{-1}\Psi} \\
& \times |\chi|^{-\frac{\xi}{2}} e^{-\frac{1}{2} tr \chi^{-1} \Xi} \qquad (14.6.9)
\end{aligned}
$$

after inserting the joint prior distribution and likelihood.

This joint posterior distribution is now to be evaluated in order to obtain parameter estimates of the matrix of sources S, the vector of Regression/mixing coefficients c, the within source vector covariance matrix R, the between source vector covariance matrix χ, the within observation covariance matrix Σ, and the between observation covariance matrix Φ.

14.7   Generalized Estimation and Inference

With the above posterior distribution, it is not possible to obtain marginal distributions and thus marginal estimates for all or any of the parameters in an analytic closed form or explicit maximum a posteriori estimates from differentiation. It is possible to use both Gibbs sampling to obtain marginal parameter estimates and the ICM algorithm for maximum a posteriori estimates. For both estimation procedures, the posterior conditional distributions are required.

14.7.1   Posterior Conditionals

Both the Gibbs sampling and ICM algorithms require the posterior conditional distributions. Gibbs sampling requires the conditionals for the generation of random variates while ICM requires them for maximization by cycling through their modes or maxima.

The conditional posterior distribution of the matrix of sources S is found by considering only the terms in the joint posterior distribution which involve S and is given by

$$p(S|B,R,\chi,\Lambda,\Sigma,\Phi,U,X) \propto p(S|R,\chi)p(X|B,S,\Lambda,\Sigma,\Phi,U)$$
$$\propto e^{-\frac{1}{2}tr\chi^{-1}(S-S_0)R^{-1}(S-S_0)'}$$
$$\times e^{-\frac{1}{2}tr\Phi^{-1}(X-UB'-S\Lambda')\Sigma^{-1}(X-UB'-S\Lambda')'},$$

$$(14.7.1)$$

which after performing some algebra in the exponent can be written as

$$p(s|B,R,\chi,\Lambda,\Sigma,\Phi,u,x) \propto e^{-\frac{1}{2}(s-\tilde{s})'(\chi^{-1}\otimes R^{-1}+\Phi^{-1}\otimes\Lambda'\Sigma^{-1}\Lambda)(s-\tilde{s})}, \quad (14.7.2)$$

where the vector $\tilde{s}$ has been defined to be

$$\tilde{s} = [\chi^{-1}\otimes R^{-1}+\Phi^{-1}\otimes\Lambda'\Sigma^{-1}\Lambda]^{-1}$$
$$\times[(\chi^{-1}\otimes R^{-1})s_0+(\Phi^{-1}\otimes\Lambda'\Sigma^{-1}\Lambda)\hat{s}], \quad (14.7.3)$$

and the matrix $\hat{S}$ has been defined to be

$$\hat{S} = (X-UB')\Sigma^{-1}\Lambda(\Lambda'\Sigma^{-1}\Lambda)^{-1}, \quad (14.7.4)$$

with the vector $\hat{s}$ has been defined to be

$$\hat{s} = vec(\hat{S}'). \quad (14.7.5)$$

That is, the matrix of sources S given the matrix of Regression coefficients B, the within source vector covariance matrix R, the between source vector covariance matrix χ, the matrix of mixing coefficients Λ, the within observation vector covariance matrix Σ, the between observation vector covariance matrix Φ, the matrix of observable sources U, and the matrix of data is Matrix Normally distributed.

The conditional posterior distribution of the within source vector covariance matrix R is found by considering only the terms in the joint posterior distribution which involve R and is given by

$$p(R|B,S,\chi,\Lambda,\Sigma,\Phi,U,X) \propto p(R)p(S|R,\chi)$$
$$\propto |R|^{-\frac{\nu}{2}}e^{-\frac{1}{2}trR^{-1}V}|R|^{-\frac{n}{2}}e^{-\frac{1}{2}trR^{-1}(S-S_0)'\chi^{-1}(S-S_0)}$$
$$\propto |R|^{-\frac{(n+\nu)}{2}}e^{-\frac{1}{2}trR^{-1}[(S-S_0)'\chi^{-1}(S-S_0)+V]}. \quad (14.7.6)$$

That is, the conditional posterior distribution of the within source vector covariance matrix R given the the matrix of Regression coefficients B, the matrix of sources S, the between source vector covariance matrix χ, the matrix of mixing coefficients Λ, the within observation vector covariance matrix Σ, the between observation vector covariance matrix Φ, the matrix of observable sources U, and the matrix of data X has an Inverted Wishart distribution.

The conditional posterior distribution of the vector of Regression/mixing coefficients c matrix is found by considering only the terms in the joint posterior distribution which involve c or C and is given by

$$p(c|S,R,\chi,\Sigma,\Phi,U,X) \propto p(c)p(X|Z,C,\Sigma,U)$$
$$\propto |\Delta|^{-\frac{1}{2}}e^{-\frac{1}{2}(c-c_0)'\Delta^{-1}(c-c_0)}$$
$$\times|\Sigma|^{-\frac{n}{2}}e^{-\frac{1}{2}tr\Sigma^{-1}(X-ZC')'\Phi^{-1}(X-ZC')}, \quad (14.7.7)$$

which after performing some algebra in the exponent becomes

$$p(c|S,R,\chi,\Sigma,\Phi,U,X) \propto e^{-\frac{1}{2}(c-\tilde{c})'[\Delta^{-1}+Z'\Phi^{-1}Z\otimes\Sigma^{-1}](c-\tilde{c})}, \quad (14.7.8)$$

where the vector $\tilde{c}$ has been defined to be

$$\tilde{c} = [\Delta^{-1} + Z'\Phi^{-1}Z\otimes\Sigma^{-1}]^{-1}[\Delta^{-1}c_0 + (Z'\Phi^{-1}Z\otimes\Sigma^{-1})\hat{c}], \quad (14.7.9)$$

and the vector $\hat{c}$ has been defined to be

$$\hat{c} = vec[X'Z(Z'\Phi^{-1}Z)^{-1}]. \quad (14.7.10)$$

The conditional posterior distribution of the vector of Regression/mixing coefficients given the matrix of sources S, the within source vector covariance matrix R, the between source vector covariance matrix χ, the within observation vector covariance matrix Σ, the between observation vector covariance matrix Φ, the matrix of observable sources U, and the matrix of data X is Multivariate Normally distributed.

The conditional posterior distribution of the within observation vector covariance matrix Σ is found by considering only the terms in the joint posterior distribution which involve Σ and is given by

$$p(\Sigma|C,Z,R,\chi,\Phi,U,X) \propto p(\Sigma)p(X|S,C,Z,\Sigma)$$
$$\propto |\Sigma|^{-\frac{(n+\nu)}{2}}e^{-\frac{1}{2}tr\Sigma^{-1}[(X-ZC')'\Phi^{-1}(X-ZC')+Q]}.$$
$$(14.7.11)$$

That is, the conditional distribution of the within observation vector covariance matrix Σ given matrix of Regression/mixing coefficients C, the matrix of sources S, the within source vector covariance matrix R, the between source covariance matrix χ, the between observation covariance matrix Φ, the matrix of observable sources U, and the matrix of data X has an Inverted Wishart distribution.

The conditional posterior distribution for the between observation vector covariance matrix Φ is found by considering only the terms in the joint posterior distribution which involve Φ and is given by

$$p(\Phi|S,C,R,\chi,\Sigma,U,X) \propto p(\Phi)p(X|S,C,\Sigma,\Phi)$$
$$\propto |\Phi|^{-\frac{\kappa}{2}}e^{-\frac{1}{2}tr\Phi^{-1}\Psi}$$
$$\times |\Phi|^{-\frac{p}{2}}e^{-\frac{1}{2}tr\Phi^{-1}(X-ZC')\Sigma^{-1}(X-ZC')'}$$
$$\propto |\Phi|^{-\frac{p+\kappa}{2}}e^{-\frac{1}{2}tr\Phi^{-1}[(X-ZC')\Sigma^{-1}(X-ZC')'+\Psi]}.$$
$$(14.7.12)$$

The conditional posterior distribution for the between observation vector covariance matrix Φ given the matrix of sources S, the matrix of Regression/mixing coefficients C, the within source vector covariance matrix R, the between source vector covariance matrix χ, the within observation vector covariance matrix Σ, the matrix of observable sources U, and the matrix of data X is Inverted Wishart distributed.

The conditional posterior distribution for the between observation vector covariance matrix χ is found by considering only the terms in the joint posterior distribution which involve χ and is given by

$$p(\chi|S,C,R,\Sigma,\Phi,U,X) \propto p(\chi)p(S|R,\chi)$$
$$\propto |\chi|^{-\frac{\xi}{2}}e^{-\frac{1}{2}tr\chi^{-1}\Xi}$$

$$\times |\chi|^{-\frac{m}{2}} e^{-\frac{1}{2} tr \chi^{-1}(S-S_0)R^{-1}(S-S_0)'}$$
$$\propto |\chi|^{-\frac{m+\xi}{2}} e^{-\frac{1}{2} tr \chi^{-1}[(S-S_0)R^{-1}(S-S_0)'+\Xi]}.$$

$$(14.7.13)$$

The conditional posterior distribution for the between source vector covariance matrix χ given the matrix of sources S, the matrix of Regression/mixing coefficients C, the within source vector covariance matrix R, the within observation vector covariance matrix Σ, the between observation vector covariance matrix Φ, the matrix of observable sources U, and the matrix of data X is Inverted Wishart distributed.

The modes of these conditional distributions are $\tilde{S}$, $\tilde{c}$, (both as defined above)

$$\tilde{R} = \frac{(S-S_0)'\chi^{-1}(S-S_0)+V}{n+\eta}, \qquad (14.7.14)$$

$$\tilde{\Sigma} = \frac{(X-ZC')'\Phi^{-1}(X-ZC')+Q}{n+\nu}, \qquad (14.7.15)$$

$$\tilde{\Phi} = \frac{(X-ZC')\Sigma^{-1}(X-ZC')'+\Psi}{p+\kappa}, \qquad (14.7.16)$$

and

$$\tilde{\chi} = \frac{(S-S_0)R^{-1}(S-S_0)'+\Xi}{m+\xi} \qquad (14.7.17)$$

respectively.

14.7.2   Gibbs Sampling

To find marginal mean estimates of the parameters from the joint posterior distribution using the Gibbs sampling algorithm, start with initial values for the vector of sources s, the within source vector covariance matrix R, and the within observation vector covariance matrix Σ, say $\bar{s}_{(0)}$, $\bar{R}_{(0)}$ and $\bar{\Sigma}_{(0)}$, and then cycle through

$$\bar{R}_{(l+1)} = \text{a random variate from } p(R|\bar{S}_{(l)},\bar{C}_{(l)},\bar{\Sigma}_{(l)},\bar{\Phi}_{(l)},\bar{\chi}_{(l)},U,X)$$
$$= A_R(Y_R'Y_R)^{-1}A_R', \qquad (14.7.18)$$
$$\bar{c}_{(l+1)} = \text{a random variate from } p(c|\bar{S}_{(l)},\bar{R}_{(l+1)},\bar{\Sigma}_{(l)},\bar{\Phi}_{(l)},\bar{\chi}_{(l)},U,X)$$
$$= A_c Y_c + M_c, \qquad (14.7.19)$$

$\bar{\Sigma}_{(l+1)}$ = a random variate from $p(\Sigma|\bar{S}_{(l)},\bar{R}_{(l+1)},\bar{C}_{(l+1)},\bar{\Phi}_{(l)},\bar{\chi}_{(l)},U,X)$

$$= A_\Sigma (Y'_\Sigma Y_\Sigma)^{-1} A'_\Sigma, \qquad (14.7.20)$$

$\bar{s}_{(l+1)}$ = a random variate from $p(s|\bar{R}_{(l+1)},\bar{C}_{(l+1)},\bar{\Sigma}_{(l+1)},\bar{\Phi}_{(l)},\bar{\chi}_{(l)},u,x)$

$$= A_s Y_s + M_s, \qquad (14.7.21)$$

$\bar{\Phi}_{(l+1)}$ = a random variate from $p(\Phi|\bar{S}_{(l+1)},\bar{R}_{(l+1)},\bar{C}_{(l+1)},\bar{\Sigma}_{(l+1)},\bar{\chi}_{(l)},U,X)$

$$= A_\Phi (Y'_\Phi Y_\Phi)^{-1} A'_\Phi, \qquad (14.7.22)$$

$\bar{\chi}_{(l+1)}$ = a random variate from $p(\chi|\bar{S}_{(l+1)},\bar{R}_{(l+1)},\bar{C}_{(l+1)},\bar{\Sigma}_{(l+1)},\bar{\Phi}_{(l+1)},U,X)$

$$= A_\chi (Y'_\chi Y_\chi)^{-1} A'_\chi, \qquad (14.7.23)$$

where

$$\bar{Z}_{(l)} = (U,\bar{S}_{(l)}),$$

$$\hat{c}_{(l)} = vec[X'\bar{Z}_{(l)}(\bar{Z}'_{(l)}\bar{\Phi}^{-1}_{(l+1)}\bar{Z}_{(l)})^{-1}],$$

$$\bar{c}_{(l+1)} = [\Delta^{-1} + \bar{Z}'_{(l)}\bar{\Phi}^{-1}_{(l+1)}\bar{Z}_{(l)} \otimes \bar{\Sigma}^{-1}_{(l)}]^{-1}[\Delta^{-1}c_0 + (\bar{Z}'_{(l)}\bar{\Phi}^{-1}_{(l+1)}\bar{Z}_{(l)} \otimes \bar{\Sigma}^{-1}_{(l)})\hat{c}_{(l)}],$$

$$A_c A'_c = (\Delta^{-1} + \bar{Z}'_{(l)}\bar{\Phi}^{-1}_{(l+1)}\bar{Z}_{(l)} \otimes \bar{\Sigma}^{-1}_{(l)})^{-1},$$

$$M_c = [\Delta^{-1} + \bar{Z}'_{(l)}\bar{\Phi}^{-1}_{(l+1)}\bar{Z}_{(l)} \otimes \bar{\Sigma}^{-1}_{(l)}]^{-1}[\Delta^{-1}c_0 + (\bar{Z}'_{(l)}\bar{\Phi}^{-1}_{(l+1)}\bar{Z}_{(l)} \otimes \bar{\Sigma}^{-1}_{(l)})\hat{c}],$$

$$A_\Sigma A'_\Sigma = (X - \bar{Z}_{(l)}\bar{C}'_{(l+1)})'\bar{\Phi}^{-1}_{(l)}(X - \bar{Z}_{(l)}\bar{C}'_{(l+1)})$$
$$\qquad + (\bar{C}_{(l+1)} - C_0)D^{-1}(\bar{C}_{(l+1)} - C_0)' + Q,$$

$$A_R A'_R = (\bar{S}_{(l)} - S_0)'\bar{\Phi}^{-1}_{(l)}(\bar{S}_{(l)} - S_0) + V,$$

$$A_s A'_s = (\bar{\chi}^{-1}_{(l)} \otimes \bar{R}^{-1}_{(l+1)} + \bar{\Phi}^{-1}_{(l)} \otimes \bar{\Lambda}'_{(l+1)}\bar{\Sigma}^{-1}_{(l+1)}\bar{\Lambda}_{(l+1)})^{-1},$$

$$\hat{s}_{(l)} = vec[(\bar{\Lambda}'_{(l+1)}\bar{\Sigma}^{-1}_{(l+1)}\bar{\Lambda}_{(l+1)})^{-1}\bar{\Lambda}'_{(l+1)}\bar{\Sigma}^{-1}_{(l+1)}(X - U\bar{B}'_{(l+1)})'],$$

$$M_s = [\bar{\chi}^{-1}_{(l)} \otimes \bar{R}^{-1}_{(l+1)} + \bar{\Phi}^{-1}_{(l)} \otimes \bar{\Lambda}'_{(l+1)}\bar{\Sigma}^{-1}_{(l+1)}\bar{\Lambda}_{(l+1)}]^{-1},$$
$$\qquad \times [(\bar{\chi}^{-1}_{(l)} \otimes \bar{R}^{-1}_{(l+1)})s_0 + (\bar{\Phi}^{-1}_{(l)} \otimes \bar{\Lambda}'_{(l+1)}\bar{\Sigma}^{-1}_{(l+1)}\bar{\Lambda}_{(l+1)})\hat{s}_{(l)}]$$

$$A_\Phi A'_\Phi = (X - \bar{Z}_{(l+1)}\bar{C}'_{(l+1)})\bar{\Sigma}^{-1}_{(l+1)}(X - \bar{Z}_{(l+1)}\bar{C}'_{(l+1)})' + \Psi,$$

$$A_\chi A'_\chi = [(\bar{S}_{(l+1)} - S_0)\bar{R}^{-1}_{(l+1)}(\bar{S}_{(l+1)} - S_0)' + \Xi,$$

while Y_C, Y_Σ, Y_R, Y_S, Y_Φ, and Y_χ are $p \times (m+q+1)$, $(n+\nu+m+q+1+p+1) \times p$, $(n+\eta+m+1) \times m$, $n \times m$, $(p+m+\kappa+n+1) \times n$, and $(m+\xi+n+1) \times n$ dimensional matrices respectively, whose elements are random variates from the standard Scalar Normal distribution. The formulas for the generation of random variates from the conditional posterior distributions are easily found from the methods in Chapter 6.

The first random variates called the "burn in" are discarded and after doing so, compute from the next L variates means of the parameters

$$\bar{s} = \frac{1}{L}\sum_{l=1}^{L}\bar{s}_{(l)} \qquad \bar{R} = \frac{1}{L}\sum_{l=1}^{L}\bar{R}_{(l)} \qquad \bar{c} = \frac{1}{L}\sum_{l=1}^{L}\bar{c}_{(l)}$$

$$\bar{\Sigma} = \frac{1}{L} \sum_{l=1}^{L} \bar{\Sigma}_{(l)} \qquad \bar{\Phi} = \frac{1}{L} \sum_{l=1}^{L} \bar{\Phi}_{(l)} \qquad \bar{\chi} = \frac{1}{L} \sum_{l=1}^{L} \bar{\chi}_{(l)}$$

which are the exact sampling-based marginal posterior mean estimates of the parameters.

Exact sampling-based estimates of other quantities can also be found. Similar to Bayesian Regression, Bayesian Factor Analysis, and Bayesian Source Separation, there is interest in the estimate of the marginal posterior variance of the matrix containing the Regression and mixing coefficients

$$\overline{var}(c|X,U) = \frac{1}{L} \sum_{l=1}^{L} \bar{c}_{(l)} \bar{c}'_{(l)} - \bar{c}\bar{c}'$$

$$= \bar{\Delta}$$

where $c = vec(C)$ and $\bar{c} = vec(\bar{C})$.

The covariance matrices of the other parameters follow similarly.

14.7.3   Maximum a Posteriori

The joint posterior distribution can also be maximized with respect to the vector of coefficients c, the vector of sources s, the within source vector covariance matrix R, the between source vector covariance matrix χ, the within observation vector covariance matrix Σ, and the between observation vector covariance matrix Φ, using the ICM algorithm. To maximize the joint posterior distribution using the ICM algorithm, start with initial values for S, Σ, c, and R, say $\tilde{S}_{(0)}$, $\tilde{\Sigma}_{(0)}$, $\tilde{c}_{(0)}$, $\tilde{R}_{(0)}$, and then cycle through

$$\tilde{\chi}_{(l+1)} = \overset{\text{Arg Max}}{\chi} \, p(\chi|\tilde{C}_{(l)}, \tilde{Z}_{(l)}, \tilde{R}_{(l)}, \tilde{\Sigma}_{(l)}, \tilde{\Phi}_{(l)}, X)$$

$$= \frac{[(\tilde{S}_{(l)} - S_0)\tilde{R}_{(l)}^{-1}(\tilde{S}_{(l)} - S_0)' + \Xi}{m + \xi},$$

$$\tilde{\Phi}_{(l+1)} = \overset{\text{Arg Max}}{\Phi} \, p(\Phi|\tilde{C}_{(l)}, \tilde{Z}_{(l)}, \tilde{R}_{(l)}, \tilde{\Sigma}_{(l)}, \tilde{\chi}_{(l+1)}, X)$$

$$= \frac{(X - \tilde{Z}_{(l)}\tilde{C}'_{(l)})\tilde{\Sigma}_{(l)}^{-1}(X - \tilde{Z}_{(l)}\tilde{C}'_{(l)})' + \Psi}{p + \kappa},$$

$$\tilde{S}_{(l+1)} = \overset{\text{Arg Max}}{S} \, p(S|\tilde{B}_{(l)}, \tilde{R}_{(l)}, \tilde{\Lambda}_{(l)}, \tilde{\Sigma}_{(l)}, \tilde{\Phi}_{(l+1)}, \tilde{\chi}_{(l+1)}, U, X),$$

$$\hat{s}_{(l)} = vec[(\tilde{\Lambda}'_{(l)}\tilde{\Sigma}_{(l)}^{-1}\tilde{\Lambda}_{(l)})^{-1}\tilde{\Lambda}'_{(l)}\tilde{\Sigma}_{(l)}^{-1}(X - U\tilde{B}'_{(l)})'],$$

$$\tilde{s}_{(l+1)} = [\tilde{\chi}_{(l+1)}^{-1} \otimes \tilde{R}_{(l)}^{-1} + \tilde{\Phi}_{(l+1)}^{-1} \otimes \tilde{\Lambda}'_{(l)}\tilde{\Sigma}_{(l)}^{-1}\tilde{\Lambda}_{(l)}]^{-1}$$

$$\times [(\tilde{\chi}_{(l+1)}^{-1} \otimes \tilde{R}_{(l)}^{-1})s_0 + (\tilde{\Phi}_{(l+1)}^{-1} \otimes \tilde{\Lambda}'_{(l)}\tilde{\Sigma}_{(l)}^{-1}\tilde{\Lambda}_{(l)})\hat{s}_{(l)}],$$

$$\tilde{\Sigma}_{(l+1)} = \overset{\text{Arg Max}}{\Sigma} p(\Sigma|\tilde{C}_{(l)},\tilde{S}_{(l+1)},\tilde{R}_{(l)},\tilde{\Phi}_{(l+1)},\tilde{\chi}_{(l+1)},U,X)$$

$$= \frac{(X - \tilde{Z}_{(l+1)}\tilde{C}'_{(l)})'\tilde{\Phi}_{(l+1)}^{-1}(X - \tilde{Z}_{(l+1)}\tilde{C}'_{(l)}) + Q}{n+\nu},$$

$$\tilde{c}_{(l+1)} = \overset{\text{Arg Max}}{c} p(c|\tilde{S}_{(l+1)},\tilde{R}_{(l)},\tilde{\Sigma}_{(l+1)},\tilde{\Phi}_{(l+1)},\tilde{\chi}_{(l+1)}U,X),$$

$$\hat{c}_{(l)} = vec[X'\tilde{Z}_{(l+1)}(\tilde{Z}'_{(l+1)}\tilde{\Phi}_{(l+1)}^{-1}\tilde{Z}_{(l+1)})^{-1}],$$

$$\tilde{c}_{(l+1)} = [\Delta^{-1} + \tilde{Z}'_{(l+1)}\tilde{\Phi}_{(l+1)}^{-1}\tilde{Z}_{(l+1)} \otimes \tilde{\Sigma}_{(l+1)}^{-1}]^{-1}$$

$$\times \{\Delta^{-1}c_0 + (\tilde{Z}'_{(l+1)}\tilde{\Phi}_{(l+1)}^{-1}\tilde{Z}_{(l+1)} \otimes \tilde{\Sigma}_{(l+1)}^{-1})\hat{c}_{(l)}\},$$

$$\tilde{R}_{(l+1)} = \overset{\text{Arg Max}}{R} p(R|\tilde{C}_{(l+1)},\tilde{Z}_{(l+1)},\tilde{\Sigma}_{(l+1)},\tilde{\Phi}_{(l+1)},\tilde{\chi}_{(l+1)},X)$$

$$= \frac{(\tilde{S}_{(l+1)} - S_0)'\tilde{\chi}_{(l+1)}^{-1}(\tilde{S}_{(l+1)} - S_0) + V}{n+\eta},$$

until convergence is reached with the joint modal estimator for the unobservable parameters $(\tilde{c},\tilde{S},\tilde{R},\tilde{\chi},\tilde{\Sigma},\tilde{\Phi})$.

14.8 Interpretation

Although the main focus after having performed a Bayesian Source Separation is on the separated sources, there are others. One focus as in Bayesian Regression is on the estimate of the Regression coefficient matrix B which defines a "fitted" line. Coefficients are evaluated to determine whether they are statistically "large" meaning that the associated independent variable contributes to the dependent variable or statistically "small" meaning that the associated independent variable does not contribute to the dependent variable. The coefficient matrix also has the interpretation that if all of the independent variables were held fixed except for one u_{ij} which if increased to u_{ij}^*, the dependent variable x_{ij} increases to an amount x_{ij}^* given by

$$x_{ij}^* = \beta_{i0} + \cdots + \beta_{ij}u_{ij}^* + \cdots + \beta_{iq}u_{iq}. \tag{14.8.1}$$

Another focus after performing a Bayesian Source Separation is on the estimated mixing coefficients. The mixing coefficients are the amplitudes which determine the relative contribution of the sources. A particular mixing coefficient which is relatively "small" indicates that the corresponding source does not significantly contribute to the associated observed mixed signal. If a particular mixing coefficient is relatively "large," this indicates that the corresponding source does significantly contribute to the associated observed mixed signal.

14.9   Discussion

Note that particular structures for Φ have been specified as has been done in the context of Bayesian Factor Analysis [50] in order to capture more detailed covariance structures and reduce computational complexity. This could have also been done for Σ and χ.

Exercises

1. For the Conjugate model, specify that Φ is a first order Markov correlation matrix with ii'^{th} element given by $\rho^{|i-i'|}$. Assess a generalized Beta prior distribution for ρ. Derive Gibbs sampling and ICM algorithms [50].

2. For the Conjugate model, specify that Φ is an intraclass correlation matrix with off diagonal element given by ρ. Assess a generalized Beta prior distribution for ρ. Derive Gibbs sampling and ICM algorithms [50].

3. Specify that Φ and χ have degenerate distributions,

$$p(\Phi) = \begin{cases} 1, \text{ if } \Phi = \Phi_0 \\ 0, \text{ if } \Phi \neq \Phi_0, \end{cases} \tag{14.9.1}$$

and

$$p(\chi) = \begin{cases} 1, \text{ if } \chi = \chi_0 \\ 0, \text{ if } \chi \neq \chi_0, \end{cases} \tag{14.9.2}$$

which means that Φ and χ are known. Derive Gibbs sampling and ICM algorithms.

15

Conclusion

There is a lot of material that needed to be covered in this text. The first part on fundamential material was necessary in order to properly understand the Bayesian Source Separation model. I have tried to provide a coherent description of the Bayesian Source Separation model and how it can be understood by starting with the Regression model.

Throughout the text, Normal likelihoods with Conjugate and generalized Conjugate prior distributions have been used. The coherent Bayesian Source Separation model presented in this text provides the foundation for generalizations to other distributions.

As stated in the Preface, the Bayesian Source Separation model incorporates available prior knowledge regarding parameter values and incorporates it into the inferences. This incorporation of knowledge avoids model and likelihood constraints which are necessary without it.

Appendix A

FMRI Activation Determination

A particular source reference function is deemed to significantly contribute to the observed signal if its (mixing) coefficient is "large" in the statistical sense. Statistically significant activation is determined from the coefficients for source reference functions. If the coefficient is "large," then the associated source reference function is significant; if it is "small," then the associated source reference function is not significant. The linear Regression model is presented in its Multivariate Regression format as in Chapter 7 where voxels are assumed to be spatially dependent.

Statistically significant activation associated with a particular source reference function is found by considering its corresponding coefficient. Significance of the coefficients for the linear model is discussed in which spatially dependent voxels are assumed (with independent voxels being a specific case) when the source reference functions are assumed to be observable (known). The joint distribution of the coefficients for all source reference function for all voxels is determined. From the joint distribution, the marginal distribution for the coefficients of a particular source reference function for all voxels is presented so that significance of a particular source reference function can be determined for all voxels. The marginal distribution of a subset of voxels for a particular source reference functions coefficients can be derived so that significant activation in a set of voxels can be determined for a given source reference function. With the above mentioned subset consisting of a single voxel, the marginal distribution is that of a particular source reference function coefficient in a given voxel. From this significance corresponding to a particular source reference function in each voxel can be determined.

A.1   Regression

Consider the linear multiple Regression model

$$x_{tj} = c_{j0} + c_{j1} z_{t1} + c_{j2} z_{t2} + \cdots + c_{j\tau} z_{t\tau} + \epsilon_{tj} \tag{A.1.1}$$

in which the observed signal in voxel j at time t is made up of a linear combination of the τ observed (known) source reference functions $z_{t1}, \ldots, z_{t\tau}$ plus an intercept term. In terms of vectors the model is

$$x_{tj} = c_j' z_t + \epsilon_{tj}, \tag{A.1.2}$$

where for the Source Separation model $c_j' = (\beta_j', \lambda_j')$ and $z_t' = (u_t', s_t')$.

The linear Regression model for a given voxel j at all n time points is written in vector form as

$$X_j = Z c_j + E_j, \tag{A.1.3}$$

where $X_j = (x_{1j}, \ldots, x_{nj})'$ is a $n \times 1$ vector of observed values for voxel j, $Z = (z_1, \ldots, z_n)'$ is an $n \times (\tau + 1)$ design matrix, $c_j = (c_{j0}, c_{j1}, \ldots, c_{j\tau})'$ is a $(\tau + 1) \times 1$ matrix of Regression coefficients, and E_j is an $n \times 1$ vector of errors. The model for all p voxels at all n time points is written in its matrix form as

$$X = Z C' + E, \tag{A.1.4}$$

where $X = (X_1, \ldots, X_p) = (x_1, \ldots, x_n)'$ is an $n \times p$ matrix of the observed values, $C = (c_1, \ldots, c_p)'$ is a $p \times (\tau + 1)$ matrix of Regression coefficients, and $E = (E_1, \ldots, E_p) = (\epsilon_1, \ldots, \epsilon_n)'$ is an $n \times p$ matrix of errors.

With the distributional specification that $\epsilon_t \sim N(0, \Sigma)$ as in the aforementioned Source Separation model, the likelihood of the observations is

$$p(X|Z, C, \Sigma) \propto |\Sigma|^{-\frac{n}{2}} e^{-\frac{1}{2} tr \Sigma^{-1} (X - Z C')' (X - Z C')}. \tag{A.1.5}$$

It is readily seen by performing some algebra in the exponent of the aforementioned likelihood that it can be written as

$$p(X|Z, C, \Sigma) \propto |\Sigma|^{-\frac{n}{2}} e^{-\frac{1}{2} tr \Sigma^{-1} [(C - \hat{C}) Z' Z (C - \hat{C})' + (X - Z\hat{C}')' (X - Z\hat{C}')]}, \tag{A.1.6}$$

where $\hat{C} = X' Z (Z' Z)^{-1}$. By inspection or by differentiation with respect to C it is seen that $\hat{C}$ is the value of C which yields the maximum of the likelihood and thus is the maximum likelihood estimator of the Regression coefficients C. It can further be seen by differentiation of Equation A.1.5 with respect to Σ that

$$\hat{\Sigma} = \frac{(X - Z\hat{C}')' (X - Z\hat{C}')}{n} \tag{A.1.7}$$

is the maximum likelihood estimate of Σ.

It is readily seen that the matrix of coefficients follows a Matrix Normal distribution given by

$$p(\hat{C}|X,Z,C,\Sigma) \propto |Z'Z|^{\frac{p}{2}} |\Sigma|^{-\frac{(\tau+1)}{2}} e^{-\frac{1}{2}tr\Sigma^{-1}(\hat{C}-C)Z'Z(\hat{C}-C)'} \qquad (A.1.8)$$

and that $G = n\hat{\Sigma} = (X - Z\hat{C}')'(X - Z\hat{C}')$ follows a Wishart distribution given by

$$p(G|X,Z,\Sigma) \propto |\Sigma|^{-\frac{(n-\tau-1)}{2}} |G|^{\frac{(n-\tau-1-p-1)}{2}} e^{-\frac{1}{2}tr\Sigma^{-1}G}. \qquad (A.1.9)$$

It can also be shown as in Chapter 7 that $\hat{C}|\Sigma$ and $G|\Sigma$ are independent.

The distribution of $\hat{C}$ unconditional of Σ as in Chapter 7 is the Matrix Student T-distribution given by

$$p(\hat{C}|X,Z,C) \propto |G + (\hat{C} - C)(Z'Z)(\hat{C} - C)'|^{-\frac{(n-\tau+1)+(\tau+1))}{2}} \qquad (A.1.10)$$

which can be written in the more familiar form

$$p(\hat{C}|X,Z,C) \propto \frac{1}{|W + (\hat{C} - C)'G^{-1}(\hat{C} - C)|^{\frac{(n-\tau+1)+(\tau+1)}{2}}}, \qquad (A.1.11)$$

where $W = (Z'Z)^{-1}$.

General simultaneous hypotheses (which do not assume spatial independence) can be performed regarding the coefficient for a particular source reference function in all voxels (or a subset of voxels) by computing marginal distributions. It can be shown [17, 41] that the marginal distribution of any column of the matrix of $\hat{C}$, $\hat{C}_k$ is Multivariate Student t-distributed

$$p(\hat{C}_k|C_k,X,Z) \propto \frac{1}{|W_{kk} + (\hat{C}_k - C_k)'G^{-1}(\hat{C}_k - C_k)|^{\frac{(n-\tau-p)+p}{2}}} \qquad (A.1.12)$$

where W_{kk} is the kth diagonal element of W. With the marginal distribution of a column of $\hat{C}$, significance can be determined for the coefficient of a particular source reference function for all voxels. Significance can be determined for a subset of voxels for a particular source reference function by determining the marginal distribution of the subset within $\hat{C}_k$ which is also Multivariate Student t-distributed. With the subset of voxels being a singleton set, significance can be determined for a particular source reference function for a given voxel with the marginal distribution of the scalar coefficient which is

$$p(\hat{C}_{kj}|C_{kj},X,Z) \propto \frac{1}{|W_{kk} + (\hat{C}_{kj} - C_{kj})G_{jj}^{-1}(\hat{C}_{kj} - C_{kj})|^{\frac{(n-\tau-p)+(\tau+1)}{2}}}, \qquad (A.1.13)$$

where $G_{jj} = (X_j - Z\hat{c}_j)'(X_j - Z\hat{c}_j)$ is the jth diagonal element of G. The above can be rewritten in the more familiar form

$$p(\hat{C}_{kj}|C_{kj}, X, Z) \propto \frac{1}{\left[(n - \tau - p) + \frac{(\hat{C}_{kj} - C_{kj})^2}{(n-\tau-p)^{-1}W_{kk}G_{jj}}\right]^{\frac{n-(\tau-1)+1}{2}}} \qquad (A.1.14)$$

which is readily recognizable as a Scalar Student t-distribution. Note that $\hat{C}_{kj} = c_{jk}$ and that

$$t = \frac{(\hat{C}_{kj} - C_{kj})}{\sqrt{W_{kk}G_{jj}(n - \tau - p)^{-1}}} \qquad (A.1.15)$$

follows a Scalar Student t-distribution with $n - \tau - p$ degrees of freedom and t^2 follows an F distribution with 1 and $n - \tau - p$ numerator and denominator degrees of freedom which is commonly used in Regression [39, 64, 68] derived from a likelihood ratio test of reduced and full models when testing a single coefficient, thus allowing a t statistic instead of an F statistic.

To determine statistically significant activation in voxels with the Source Separation model using the standard Regression approach, join the Regression coefficient and source reference function matrices so that $C = (B, \Lambda)$ are the coefficients and $Z = (U, S)$ are the (observable or known) source reference functions and $\tau = m + q$. The model and likelihood are now in the standard Regression formats given above.

A.2   Gibbs Sampling

To determine statistically significant activation with the Gibbs sampling approach, use the marginal distribution of the mixing coefficients given in Equation 12.4.2. General simultaneous hypotheses (which do not assume spatial independence) can be performed regarding the coefficient for a particular source reference function in all voxels by computing marginal distributions. It can be shown [17, 41] that the marginal distribution of the kth column of the mixing matrix $\bar{\Lambda}$, $\bar{\Lambda}_k$ is Multivariate Normal

$$p(\Lambda_k|\bar{\Lambda}_k, X, U) \propto |\bar{\Delta}_k|^{-\frac{1}{2}} e^{-\frac{1}{2}(\Lambda_k - \bar{\Lambda}_k)'\bar{\Delta}_k^{-1}(\Lambda_k - \bar{\Lambda}_k)}, \qquad (A.2.1)$$

where $\bar{\Delta}_k$ is the covariance matrix of $\bar{\Lambda}_k$ found by taking the kth $p \times p$ submatrix along the diagonal of $\bar{\Delta}$.

With the marginal distribution of a column of $\bar{\Lambda}$, significance can be determined for the coefficient of a particular source reference function for all voxels. Significance can be determined for a subset of voxels for a particular source reference function by determining the marginal distribution of the subset within $\bar{\Lambda}_k$ which is also Multivariate Normal. With the subset of voxels being a singleton set, significance can be determined for a particular source reference function for a given voxel with the marginal distribution of the scalar coefficient which is

$$p(\bar{\Lambda}_{kj}|\Lambda_{kj}, X, U) \propto (\bar{\Delta}_{kj})^{-\frac{1}{2}} e^{-\frac{(\Lambda_{kj}-\bar{\Lambda}_{kj})^2}{2\bar{\Delta}_{kj}}}, \qquad (A.2.2)$$

where $\bar{\Delta}_{kj}$ is the jth diagonal element of $\bar{\Delta}_k$. Note that $\hat{\Lambda}_{kj} = \hat{\lambda}_{jk}$ and that

$$z = \frac{(\bar{\Lambda}_{kj} - \Lambda_{kj})}{\sqrt{\bar{\Delta}_{kj}}} \qquad (A.2.3)$$

follows a Normal distribution with a mean of zero and variance of one.

A.3 ICM

To determine statistically significant activation with the iterated conditional modes (ICM) approach, use the conditional posterior distribution of the mixing coefficients given in Equation 12.4.6.

General simultaneous significance (which does not assume spatial independence) can be determined regarding the coefficient for a particular source reference function in all voxels by computing marginal distributions. It can be shown [17, 41] that the marginal distribution of any column of the matrix of $\tilde{\Lambda}$, $\tilde{\Lambda}_k$ is Multivariate Normal

$$p(\Lambda_k|\tilde{\Lambda}_k, \tilde{B}, \tilde{S}, \tilde{R}, \tilde{\Sigma}, U, X) \propto |W_{kk}\tilde{\Sigma}|^{-\frac{1}{2}} e^{-\frac{1}{2}(\Lambda_k-\tilde{\Lambda}_k)'(W_{kk}\tilde{\Sigma})^{-1}(\Lambda_j-\tilde{\Lambda}_j)}, \quad (A.3.1)$$

where $W = (D^{-1} + \tilde{Z}'\tilde{Z})^{-1}$ and W_{kk} is its kth diagonal element.

With the marginal distribution of a column of $\tilde{\Lambda}$, significance can be determined for the coefficient of a particular source reference function for all voxels. Significance can be determined for a subset of voxels for a particular source reference function by determining the marginal distribution of the subset within $\tilde{\Lambda}_k$ which is also Multivariate Normal. With the subset of voxels being a singleton set, significance can be determined for a particular source reference function for a given voxel with the marginal distribution of the scalar coefficient which is

$$p(\Lambda_{kj}|\tilde{\Lambda}_{kj}, \tilde{B}, \tilde{S}, \tilde{R}, \tilde{\Sigma}_{jj}, U, X) \propto (W_{kk}\tilde{\Sigma}_{jj})^{-\frac{1}{2}} e^{-\frac{(\tilde{\Lambda}_{kj}-\Lambda_{kj})^2}{2W_{kk}\tilde{\Sigma}_{jj}}}, \qquad (A.3.2)$$

where $\tilde{\Sigma}_{jj}$ is the jth diagonal element of $\tilde{\Sigma}$. Note that $\tilde{\Lambda}_{kj} = \tilde{\lambda}_{jk}$ and that

$$z = \frac{(\tilde{\Lambda}_{kj} - \Lambda_{kj})}{\sqrt{W_{kk}\tilde{\Sigma}_{jj}}} \qquad (A.3.3)$$

follows a Normal distribution with a mean of zero and variance of one.

After determining the test Statistics, a threshold or significance level is set and a one to one color mapping is performed. The image of the colored voxels is superimposed onto an anatomical image.

Appendix B

FMRI Hyperparameter Assessment

The hyperparameters of the prior distributions could be subjectively assessed from a substantive field expert, or by use of a previous similar set of data from which the hyperparameters could be assessed as follows. Denote the previous data by the $n_0 \times p$ matrix X_0 which has the same experimental design as the current data. The source reference functions corresponding to the experimental stimuli are chosen to mimic the experiment with peaks during the experimental stimuli and valleys during the control stimulus, typically a square, sine, or triangle wave function with unit amplitude and the same timing as the experiment. Other source reference functions could be assessed from a substantive field expert or possibly from a cardiac or respiration monitor.

Reparameterizing the prior source reference matrix in terms of columns instead of rows as $S_0 = (S_{01}, \ldots, S_{0m})$, each of these column vectors is the time course associated with a source reference function.

Using these a priori values for the source reference functions, the model for the previous data is

$$X_0 = UB' + S_0\Lambda' + E \tag{B.1}$$
$$= Z_0C' + E \tag{B.2}$$

with $Z_0 = (U, S_0)$, $C = (B, \Lambda)$, and other variables as previously defined. The likelihood of the previous data is

$$p(X_0|Z_0, C, \Sigma) \propto |\Sigma|^{-\frac{n_0}{2}} e^{-\frac{1}{2}tr\Sigma^{-1}(X-Z_0C')'(X-Z_0C')} \tag{B.3}$$

which after rearranging the terms in the exponent becomes

$$p(X_0|Z_0, C, \Sigma) \propto |(Z_0'Z_0)^{-1}|^{-\frac{(\tau+1)}{2}} |\Sigma|^{-\frac{n_0}{2}} e^{-\frac{1}{2}tr\Sigma^{-1}(C-\hat{C})(Z_0'Z_0)(C-\hat{C})'}, \tag{B.4}$$

where $\tau = m + q$ and $\hat{C} = X_0'Z_0(Z_0'Z_0)^{-1}$. The above can be viewed as the distribution of C (which is Normal) given the data and the other parameters with

$$E(C|X_0, Z_0, \Sigma) = \hat{C} \tag{B.5}$$
$$var(c|X_0, Z_0, \Sigma) = (Z_0'Z_0)^{-1} \otimes \Sigma, \tag{B.6}$$

where $c = vec(C)$. The likelihood can also be viewed as the distribution of Σ (which is Inverted Wishart) given the data and the other parameters

$$p(X_0|Z_0,C,\Sigma) \propto |\Sigma|^{-\frac{n_0}{2}} e^{-\frac{1}{2} tr \Sigma^{-1} G}, \tag{B.7}$$

where $G = (X - Z_0 C')'(X - Z_0 C')$ with

$$E(\Sigma|X_0,Z_0,C) = \frac{G}{n_0 - 2p - 2}, \tag{B.8}$$

$$var(\Sigma_{jj}|X_0,Z_0,C) = \frac{G_{jj}}{(n_0 - 2p - 2)^2(n_0 - 2p - 4)}. \tag{B.9}$$

Other second order moments are possible and slightly more complicated [41] but are not be needed.

Hyperparameters are assessed from the previous data using the means and variances given above, namely,

$$
\begin{align}
C_0 &= (B_0, \Lambda_0) \tag{B.10} \\
&= X_0' Z_0 (Z_0' Z_0)^{-1} = \hat{C}, \tag{B.11} \\
D &= (Z_0' Z_0)^{-1}, \tag{B.12} \\
Q &= (X - Z_0 \hat{C}')'(X - Z_0 \hat{C}') = G, \tag{B.13} \\
\nu &= n_0. \tag{B.14}
\end{align}
$$

Under the assumption of spatially independent voxels,

$$
\begin{align}
c_{j0} &= (\beta_{j0}, \lambda_{j0}) \tag{B.15} \\
&= (Z_0' Z_0)^{-1} Z_0' X_j = \hat{c}_j, \tag{B.16} \\
Q_{jj} &= (X_j - Z_0 \hat{c}_j)'(X_j - Z_0 \hat{c}_j) = G_{jj}, \tag{B.17}
\end{align}
$$

while D, and ν are as defined above.

The hyperparameters η and V which quantify the variability of the source reference functions around their mean S_0 must be assessed subjectively. The mean and variance of the Inverted Wishart prior distribution as listed in Chapter 2 for the covariance matrix of the source reference functions are

$$E(R) = \frac{V}{\eta - 2m - 2}, \quad var(r_{kk}) = \frac{2v_{kk}^2}{(\eta - 2m - 2)^2(\eta - 2m - 4)}. \tag{B.18}$$

For ease of assessing these parameters V is taken to be diagonal as $V = v_0 I_m$ thus $r_{kk} = r$ and $v_{kk} = v_0$.

The above means and variances becomes

$$E(r) = \frac{v_0}{\eta - 2m - 2}, \quad var(r) = \frac{2v_0^2}{(\eta-2m-2)^2(\eta-2m-4)}, \tag{B.19}$$

a system of two equations with two unknowns. Solving for η and v_0 yields

$$\eta = \frac{[E(r)]^2}{2var(r)} + 6, \quad v_0 = E(r)(\eta - 4) \tag{B.20}$$

and the hyperparameter assessment has been transformed to assessing a prior mean and variance for the variance of the source reference functions.

Bibliography

[1] Lee J. Bain and Max Engelhardt. Introduction to Probability and Mathematical Statistics. PWS-Kent Publishing Company, Boston, Massachusetts, USA, second edition, 1992.

[2] Peter Bandettini, Andrzej Jesmanowicz, Eric Wong, and James Hyde. Processing strategies for time-course data sets in functional mri of the human brain. Magnetic Resonance in Medicine, 30, 1993.

[3] Thomas Bayes. An essay towards solving a problem in the doctrine of chance. Philosophical Transactions of the Royal Society of London, 53:370–418, 1763.

[4] G. M. Becker, M.H. De Groot, and M. Marshak. Measuring utility by a single response sequential method. Behavioral Science, 9:226–232, 1964.

[5] George E. Box and M. A. Muller. A note on the generation of random normal deviates. Annals of Mathematical Statistics, 29(610), 1958.

[6] Pierre Common. Independent component analysis: a new concept? Signal Processing, 36:11–20, 1994.

[7] Robert W. Cox. AFNI: Software for analysis and visualization of functional magnetic resonance neuroimages. Computers and Biomedical Research, 29, 1996.

[8] Robert W. Cox, Andrzej Jesmanowicz, and James S. Hyde. Real-time functional magnetic resonance imaging. Magnetic Resonance in Medicine, 33, 1995.

[9] Ward Edwards, Harold Lindman, and Leonard J. Savage. A note on choosing the number of factors. Psychological Review, 70(3), 1963.

[10] Leonard J. Savage et al. The Foundations of Statistical Inference. John Wiley and Sons (Methuen & Co. London), New York, New York, USA, 1962.

[11] Ross L. Finney and George B. Thomas. Calculus. Addison-Wesley Publishing Company, Reading, Massachusetts, USA, 1990.

[12] Seymour Geisser. Bayesian estimation in multivariate analysis. Annals of Mathematical Statistics, 36:150–159, 1965.

[13] Alan E. Gelfand and Adrian F. M. Smith. Sampling based approaches to calculating marginal densities. Journal of the American Statistical Association, 85:398–409, 1990.

[14] Stuart Geman and Donald Geman. Stochastic relaxation, Gibbs distributions and the Bayesian restoration of images. IEEE Transactions on Pattern Analysis and Machine Intelligence, 6:721–741, 1984.

[15] Christopher R. Genovese. A Bayesian time-course model for functional magnetic resonance imaging. Journal of the American Statistical Association, 95(451):691–719, 2000.

[16] Wally R. Gilks and Pascal Wild. Adaptive rejection sampling for Gibbs sampling. Journal of the Royal Statistical Society C, 41:337–348, 1992.

[17] Arjun K. Gupta and D. K. Nagar. Matrix Variate Distributions. Chapman & Hall/CRC Press, Boca Raton, Florida, USA, 2000.

[18] David A. Harville. Matrix Algebra from a Statisticians Perspective. Springer-Verlag Inc., New York, New York, USA, 1997.

[19] W. Keith Hastings. Monte Carlo sampling methods using Markov chains and their applications. Biometrika, 87:97–109, 1970.

[20] Kentaro Hayashi. The Press-Shigemasu Bayesian Factor Analysis Model with Estimated Hyperparameters. PhD thesis, Department of Psychology, University of North Carolina, Chapel Hill, North Carolina, USA, 1997.

[21] Kentaro Hayashi and Pranab K. Sen. Bias-corrected estimator of factor loadings in Bayesian factor analysis model. Educational and Psychological Measurement, 2001.

[22] Robert V. Hogg and Allen T. Craig. Introduction to Mathematical Statistics. Macmillan, New York, New York, USA, fourth edition, 1978.

[23] Harold Hotelling. Analysis of a complex of statistical variables into principal components. Journal of Educational Psychology, 24:417–441, 1933.

[24] John Imbrie and Nilva Kipp. A new micropaleontological method for quantitative paleoclimatology: Application to a late pleistocene caribbean core. In The Late Cenozoic Glacial Ages, chapter 5. Yale University Press, New Haven, Connecticut, USA, 1971.

[25] Harold Jeffreys. Theory of Probability. Clarendon Press, Oxford, third edition, 1961.

[26] Maurice Kendall. Multivariate Analysis. Charles Griffin & Company, London, second edition, 1980.

[27] Kevin Knuth. Bayesian source separation and localization. In A. Mohammad-Djafari, editor, SPIE'98 Proceedings: Bayesian Inference for Inverse Problems, San Diego, California, pages 147–158, July 1998.

[28] Kevin Knuth. A Bayesian approach to source separation. In J. F. Cardoso, C. Jutten, and P. Loubaton, editors, Proceedings of the First International Workshop on Independent Component Analysis and Signal Separation: ICA'99, Aussios, France, pages 283–288, 1999.

[29] Kevin Knuth and Herbert G. Vaughan. Convergent Bayesian formulations of blind source separation and electromagnetic source estimation. In W. von der Linden, V. Dose, R. Fischer, and R. Preuss, editors, Maximum Entropy and Bayesian Methods, Munich 1998. Kluwer Dordrecht, 1999, pages 217–226.

[30] Subrahmaniam Kocherlakota and Kathleen Kocherlakota. Bivariate Discrete Distributions. Marcel Dekker, Inc., New York, New York, USA, 1992.

[31] Samuel Kotz and Norman Johnson, editors. Encyclopedia of Statistical Science, volume 5. John Wiley and Sons, Inc., New York, New York, USA, 1985, pages 326–333.

[32] Erwin Kreyszig. Advanced Engineering Mathematics. John Wiley & Sons, Inc., New York, New York, USA, sixth edition, 1988.

[33] A. S. Krishnamoorthy. Multivariate Binomial and Poisson distributions. Sankhya: The Indian Journal of Statistics, 11(3), 1951.

[34] Sang Eun Lee. Robustness of Bayesian Factor Analysis Estimates. PhD thesis, Department of Statistics, University of California, Riverside, California, USA, December 1994.

[35] Sang Eun Lee and S. James Press. Robustness of Bayesian factor analysis estimates. Communications in Statistics – Theory and Methods, 27(8), 1998.

[36] Dennis V. Lindley and Adrian F. M. Smith. Bayes estimates for the linear model. Journal of the Royal Statistical Society B, 34(1), 1972.

[37] Ali Mohammad-Djafari. A Bayesian estimation method for detection, localisation and estimation of superposed sources in remote sensing. In SPIE 97 annual meeting, (San Diego, California, USA, July 27-August 1, 1997), 1997.

[38] Ali Mohammad-Djafari. A Bayesian approach to source separation. In Proceedings of the Nineteenth International Conference on Maximum Entropy and Bayesian Methods, (August 2-6, 1999, Boise, Idaho, USA), 1999.

[39] Raymond H. Myers. Classical and Modern Regression with Applications. Duxbury, Pacific Grove, California, USA, 1990.

[40] Anthony O'Hagen. Kendalls' Advanced Theory of Statistics, Volume 2B Bayesian Inference. John Wiley and Sons, Inc., New York, New York, USA, second edition, 1994.

[41] S. James Press. Applied Multivariate Analysis: Using Bayesian and Frequentist Methods of Inference. Robert E. Krieger Publishing Company, Malabar, Florida, USA, second edition, 1982.

[42] S. James Press. Bayesian Statistics: Principles, Models, and Applications. John Wiley and Sons, Inc., New York, New York, USA, 1989.

[43] S. James Press and K. Shigemasu. Bayesian inference in factor analysis. In Leon J. Gleser and Michael D. Perlman, editors, Contributions to Probability and Statistics: Essays in Honor of Ingram Olkin, chapter 15. Springer-Verlag, New York, New York, USA, 1989.

[44] S. James Press and K. Shigemasu. Bayesian inference in factor analysis-Revised. Technical Report No. 243, Department of Statistics, University of California, Riverside, California, USA, May 1997.

[45] S. James Press and K. Shigemasu. A note on choosing the number of factors. Communications in Statistics – Theory and Methods, 29(7), 1999.

[46] Gerald S. Rogers. Matrix Derivatives. Marcel Dekker, New York, New York, USA, 1980.

[47] Vijay K. Rohatgi. An Introduction to Probability Theory and Mathematical Statistics. John Wiley and Sons, Inc., New York, New York, USA, 1976.

[48] Sheldon M. Ross. Introduction to Probability Models. Academic Press, Inc., San Diego, California, USA, fourth edition, 1989.

[49] T. J. Rothenberg. A Bayesian analysis of simultaneous equation systems. Report 6315, Netherlands School of Economics, Econometric Institute, 1963.

[50] Daniel B. Rowe. Correlated Bayesian Factor Analysis. PhD thesis, Department of Statistics, University of California, Riverside, California, USA, December 1998.

[51] Daniel B. Rowe. A Bayesian factor analysis model with generalized prior information. Social Science Working Paper 1099, Division of Humanities and Social Sciences, Caltech, Pasadena, California, USA, August 2000.

[52] Daniel B. Rowe. Bayesian source separation of fmri signals. In A. Mohammad-Djafari, editor, Maximum Entropy and Bayesian Methods. American Institute of Physics, College Park, Maryland, USA, 2000.

[53] Daniel B. Rowe. Factorization of separable and patterned covariance matrices for Gibbs sampling. Monte Carlo Methods and Applications, 6(3), 2000.

[54] Daniel B. Rowe. Incorporating prior knowledge regarding the mean in Bayesian factor analysis. Social Science Working Paper 1097, Division of Humanities and Social Sciences, Caltech, Pasadena, California, USA, July 2000.

[55] Daniel B. Rowe. On estimating the mean in Bayesian factor analysis. Social Science Working Paper 1096, Division of Humanities and Social Sciences, Caltech, Pasadena, California, USA, July 2000.

[56] Daniel B. Rowe. A Bayesian model to incorporate jointly distributed generalized prior information on means and loadings in factor analysis. Social Science Working Paper 1110, Division of Humanities and Social Sciences, Caltech, Pasadena, California, USA, February 2001.

[57] Daniel B. Rowe. A Bayesian observable/unobservable source separarion model and activation determination in FMRI. Social Science Working Paper 1120R, Division of Humanities and Social Sciences, Caltech, Pasadena, California, USA, June 2001.

[58] Daniel B. Rowe. Bayesian source separarion with jointly distributed mean and mixing coefficients via MCMC and ICM. Social Science Working Paper 1118, Division of Humanities and Social Sciences, Caltech, Pasadena, California, USA, April 2001.

[59] Daniel B. Rowe. Bayesian source separation for reference function determination in fmri. Magnetic Resonance in Medicine, 45(5), 2001.

[60] Daniel B. Rowe. A model for Bayesian factor analysis with jointly distributed means and loadings. Social Science Working Paper 1108, Division of Humanities and Social Sciences, Caltech, Pasadena, California, USA, January 2001.

[61] Daniel B. Rowe. A model for Bayesian source separarion with the overall mean. Social Science Working Paper 1118, Division of Humanities and Social Sciences, Caltech, Pasadena, California, USA, April 2001.

[62] Daniel B. Rowe. A Bayesian approach to blind source separation. Journal of Interdisciplinary Mathematics, 5(1):49–76, 2002.

[63] Daniel B. Rowe. Bayesian source separation of functional sources. Journal of Interdisciplinary Mathematics, 5(2), 2002.

[64] Daniel B. Rowe and Steven W. Morgan. Computing fmri activations: Coefficients and t-statistics by detrending and multiple regression. Technical Report 39, Division of Biostatistics, Medical College of Wisconsin, Milwaukee, WI, USA, June 2002.

[65] Daniel B. Rowe and S. James Press. Gibbs sampling and hill climb-
     ing in Bayesian factor analysis. Technical Report No. 255, Department
     of Statistics, University of California, Riverside, California, USA, May
     1998.

[66] George W. Snedecor. Statistical Methods. Iowa University Press, Ames,
     Iowa, USA, 1959.

[67] Aage Volund. Multivariate bioassay. Biometrics, 36:225–236, 1980.

[68] B. Douglas Ward. Deconvolution analysis of FMRI time series data.
     AFNI 3dDeconvolve Documentation, Medical College of Wisconsin, Mil-
     waukee, Wisconsin, USA, July 2000.

Index